Kulturwissenschaftliche
Japanstudien

Herausgegeben von
Stephan Köhn, Martina Schönbein
und Chantal Weber

Band 13

2024

Harrassowitz Verlag · Wiesbaden

Articulations of the Nuclear

Postwar Japan under the Spell
of the Atomic Age

Edited by
Stephan Köhn and Katharina Hülsmann

2024

Harrassowitz Verlag · Wiesbaden

Published with kind support of the Deutsche Forschungsgemeinschaft (DFG).

Cover illustration: © Donated by Akito Kawagoe, Hiroshima Peace Memorial Museum.

Bibliografische Information der Deutschen Nationalbibliothek
Die Deutsche Nationalbibliothek verzeichnet diese Publikation in der Deutschen
Nationalbibliografie; detaillierte bibliografische Daten sind im Internet
über https://dnb.de abrufbar.

Bibliographic information published by the Deutsche Nationalbibliothek
The Deutsche Nationalbibliothek lists this publication in the Deutsche
Nationalbibliografie; detailed bibliographic data are available in the internet
at https://dnb.de.

For further information about our publishing program consult our
website https://www.harrassowitz-verlag.de
© Otto Harrassowitz GmbH & Co. KG, Wiesbaden 2024
This work, including all of its parts, is protected by copyright.
Any use beyond the limits of copyright law without the permission
of the publisher is forbidden and subject to penalty. This applies
particularly to reproductions, translations, microfilms and storage
and processing in electronic systems.
Printed on permanent/durable paper.
Printing and binding: docupoint, Magdeburg
Printed in Germany
ISSN 1860-2320
ISBN 978-3-447-12199-6

Contents

Foreword VII

Articulations of the Nuclear: Postwar Japan under the Spell of the Atomic Age—An Introduction to this Volume—
Stephan Köhn and Katharina Hülsmann 1

From John Hersey's Hiroshima to Fukushima: General Leslie R. Groves' Long Shadow over Nuclear Narratives
Atsuko Shigesawa 31

From Cities in Ruins to Ambassadors of World Peace: The Daily Press in Hiroshima and Nagasaki from 1945 to 1949
Marie-Christine Dreßen 55

Depictions of the Nuclear in Children's Manga from the Occupation Period
Katharina Hülsmann 81

Depicting the Atomic Bombings of Hiroshima and Nagasaki: History and Continued Significance
Christopher P. Hood 105

Looking for Possibilities: Urban Planning in Manchuria and Tange Kenzō's Plan for Hiroshima
Chantal Weber 127

The Atomic Bomb Victims in Nagasaki and Hiroshima on Display: Differing Exhibition Strategies and International Trends in Musealization
André Hertrich 151

Preserving *hibakusha* memories: Ōishi Matashichi and the Daigo Fukuryū Maru Exhibition Hall
Lauren Constance 175

At the Crossroads to Oblivion—Ōta Yōko's Literature as a
Counter-narrative to National History Writing in Japan
Stephan Köhn   195

Narrating Nuclear Geographies and Victims: The Late Fiction and
Essays of Tsushima Yūko
Rachel DiNitto   215

How Shifting U. S. Perceptions of the Atomic Bombings of Hiroshima
and Nagasaki have Influenced Americans' Attitudes toward the
Nuclear Arms Race, 1945–2022
Peter Kuznick   233

Splitting the Atom: Nishioka Yuka's Manga Rebuke of the
Nuclear-Weapon / Power Divide
Michele M. Mason   253

Rhetorical Perspectives on Post-nuclear Japan: Embracing Hiroshima
and Nagasaki Within Fukushima
Hiroko Okuda   271

Responsibility for the Fukushima Nuclear Disaster in the Japanese Media:
TEPCO and the "Nuclear Village" or Prime Minister Kan and the DPJ?
Tobias Weiß   297

Living with the Nuclear: Life Experiences of Farmers in Fukushima
Prefecture
Anna Wiemann   329

Transcending the Nuclear Fallacy: Japan's Atomic Legacy as Thematized
in Hayashi Kyōko's Late Work—In Lieu of an Epilogue—
Stephan Köhn   351

Contributors   365

Index   371

Foreword

This publication is composed of contributions presented at the conference "Hiroshima—Nagasaki—Fukushima: Articulations of the Nuclear" which took place in May 2022 in Cologne, Germany, as well as papers that have grown out of the discussions and exchanges we had during the joint research project "Japan's Split Society: Discursive Constitution between the Atomic Bombs (*genbaku*) and Nuclear Power Plants (*genpatsu*)" of the University of Cologne and Leipzig University, funded by the Deutsche Forschungsgemeinschaft (DFG).

The research project was of course affected by the global Covid-19 pandemic. From one day to the next, global travel became impossible and research institutions and archives inaccessible. Meetings were quickly moved to the virtual realm of Zoom and other conferencing systems. Planning for future field research and in-person-exchanges was very difficult. The Department of Japanese Studies of Leipzig University under Steffi Richter and project research associate Felix Jawinski carried the torch by organizing an online lecture series with the title "Japan's Split Society Between Genbaku and Genpatsu: Media, Propaganda and Science" in the summer semester of 2021. Through this lecture series we could begin sharing our research with other scholars on an international level and, in turn, also receive their insights.

In autumn 2021, the Department of Japanese Studies at the University of Cologne under Stephan Köhn and project research associate Katharina Hülsmann, drawing on extensive preparations by research assistant Marie-Christine Dreßen, began the ambitious planning process for the first hybrid conference that would be organized by the department after pandemic travel bans had been lifted. It was a confusing time, with very few definite answers and guarantees by the administration on how and with what scope a conference could be held in May 2022. Yet, with a lot of optimism and commitment to ensuring the attendance of the conference would be as safe as possible for the in-person attendees—as well as accessible to a wide audience online—we went ahead with the planning. In February 2022, while planning for the conference was in full swing, news about Russia's invasion of Ukraine put atomic issues—such as the safety of nuclear power plants in war zones and, above all, the threat of nuclear weapons—very clearly to the fore. The research project seemed more topical than ever.

The conference did not simply consist of the papers compiled in this volume. It included an excursion to a site that serves as a symbol for recollecting the Cold War anxieties of the late 1970s: The Civilian Shelter at the Documentation

Center DOKK, which is preserved and located in the heart of Cologne, under the subway station Kalk-Post. This civilian shelter was supposed to house 2000 people for two weeks in the event of a nuclear attack on Cologne. As became evident from the guided tour through the shelter, the German government at the time tried its best to plan withstanding a possible nuclear attack—yet the chances of survival remain rather slim and there was simply no plan for what would happen once the two weeks had passed. Seeing this historical site in person and experiencing the atmosphere of the shelter illuminated the Cold War way of thinking to the conference attendees and pointed us to unresolved questions that have, once again, become pressing.

We are proud to present our research on the articulations of the nuclear in this edited volume. We would first like to thank all the contributors to the conference for sharing their research and their insights and criticisms with us: Lauren Constance, Rachel DiNitto, Marie-Christine Dreßen, André Hertrich, Christopher P. Hood, Robert Jacobs, Felix Jawinski, Franziska König, Peter Kuznick, Michele M. Mason, Hiroko Okuda, Steffi Richter, Atsuko Shigesawa, Tobias Weiss and Anna Wiemann. We would also like to thank those who helped us make the conference a success: Chantal Weber, Martin Thomas, Paul Schoppe, Imân Nur Bodenstein, Bianca M. Hanzic, Marvin Kamphausen, Eda Tanriverdi, Alison Tonnet and Sylvia Schütz. Furthermore, we would like to thank Michael Dorrity for the English language proofreading. We would also like to express our gratitude to Ursula Flache and Christian Dunkel for their guidance in researching the relevant literature—especially during the trying lockdown periods—and for making it possible to access microfilms from the Gordon W. Prange Collection and from the 20th Century Media Information Database through the Staatsbibliothek zu Berlin—Preußischer Kulturbesitz. Similarly, we would like to thank the curator of the Gordon W. Prange Collection, Kana Jenkins, for providing her guidance and aiding our research at the Hornbake Library of the University of Maryland. Last but not least, we would like to extend our thanks to Luna Sartor for her extensive help with administrative matters and her diligent work in proofreading the galley proofs. Any remaining flaws in the manuscript are the editors' responsibility.

Finally, we would also like to thank Steffi Richter and Felix Jawinski for their insights and support in organizing the conference, their ongoing cooperation in this research project, and their sharing of funding for travel costs.

Cologne, March 2024

The editors

Articulations of the Nuclear:
Postwar Japan under the Spell of the Atomic Age
—An Introduction to this Volume—

Stephan Köhn and Katharina Hülsmann

> Our government behaved extremely cold-heartedly showing no charity towards us as victims. We were totally abandoned and forgotten. Of course, people were not aware of this fact at that time. For them, only the war was to be blamed. Nothing would change anyway, so they stopped complaining about the situation. Of course, they felt grief and resentment deep in their hearts but they were unable to find words for this feeling. Every time I think about this situation, my heart beats faster and my whole body begins to tremble. Even today, Japanese people are generally quite indifferent to the damage actually caused by the atomic bombings of Hiroshima and Nagasaki. But that's not everything, even the residents of Hiroshima are indifferent. They do not want to hear any scientific explanation for the very thing that provoked their innermost fears and horrors.
> The damage caused by an atomic bomb is peculiar [compared to conventional bombs] insofar as it is not the day of the explosion that will become the most horrendous, but the days that follow. Because even if you survive the explosion itself, you will have no guarantee of surviving [nuclear power] overall. In some respect, it is like having cancer. You are thrown into the deepest misery, into a world of unfathomable and unprecedented confusion and horror. (ŌTA 1949: 48)

In August 1949, ŌTA Yōko's 大田洋子 (1903–63) short report "August 6, 8:15" (*Hachigatsu muika, hachiji jūgofun* 8月6日8時１５分) was published in the general interest magazine "Reconstruction" (*Kaizō* 改造). At the time of publication, she was undoubtedly one of the most prominent—and perhaps influential— voices from Hiroshima. ŌTA Yōko had already been an accomplished writer before the atomic bomb was dropped on Hiroshima on August 6, 1945. On August 30, 1945, she published the first written report on the atomic bombing and the aftermath in the daily newspaper *Asahi shinbun* 朝日新聞 (Tokyo edition), titled: "An Unfathomable Deep Light" (*Kaitei no yō na hikari* 海底のやうな光). It was made abundantly clear to ŌTA Yōko from the very beginning that writing about the atomic bomb would be anything but easy as her articulations of the nuclear immediately became a target of suspicion and suppression. Her short story "Riverbank" (*Kawara* 河原) illustrates this most clearly. The short novel ought to have been published as early as November 1946, in the first issue of the literary magazine "Novel" (*Shōsetsu* 小説). The manuscript of the entire issue was confiscated by the censors, however, and the publication inevitably delayed, such that "Riverbank" was not ultimately published until February 1948, in the second issue of "Novel". It is striking that both Ōta Yōko's short story as well as the short novel "Back home" (*Kokoku e* 故国へ) by Tamura Taijirō 田村泰次郎 (1911–83) fell afoul of the censors. The latter was deemed to constitute "militaristic propaganda".

Ōta's "Riverbank" is quite an unusual account of August 6, 1945, as it tells the rather unpretentious story of three survivors of the atomic bombing of Hiroshima who found refuge at a riverbank. Ōta painstakingly describes their daily struggle for survival, their fears of the past, and their dreams for the future.[1] Ōta's story succeeded in the feat of speaking about a taboo without explicitly mentioning it. That is, although neither the time, the city, the atomic bomb itself, nor the aftermath are mentioned, the reader is keenly aware of the fact that the story is about the abovementioned "unknown and unexperienced confusion and terror" with which all survivors of August 6 were confronted. It is very likely that Ōta Yōko deliberately erased every written trace of the "nuclear" (i. e. atomic bomb, exposure to radiation et cetera) to avoid invoking any of the common censorship practices of the time. Needless to say, this form of "self-restraint" (*jishuku* 自粛) amounted to a form of internalized censorship practiced by many survivors, and writers, of the bombing of Hiroshima on August 6 and Nagasaki on August 9, 1945.

Two years later, in November 1948, Ōta Yōko published what is probably her best-known account on the atomic bombing of Hiroshima, "City of Corpses" (*Shikabane no machi* 屍の街), printed by the renowned publisher Chūō kōronsha. However, the author herself did not practice self-restraint on this occasion, with the publisher instead removing every expression or passage that may have posed a problem in terms of obtaining the censors' approval (Nagaoka 1982: 241–67). The greatest such intervention was the removal of the entire chapter, "Apathetic faces" (*Muyoku ganbō* 無欲顔貌), which had originally contained a great deal of information—including direct citations—given by renowned scientists and well-known newspaper articles on the subject of the bombings. It would take a further two years before "City of Corpses" would finally be published unabridged by the publisher Tōga shobō in 1950.

Remarkably, Ōta decided to include a paragraph containing, of all things, large parts of the deleted chapter "Apathetic faces" in her short account of the bombing titled "August 6, 8:15". Of course, the August issue of "Reconstruction" also had to be filed for approval in July 1949. However, as can be seen from the censors' notes, Ōta Yōko's account was not classified as a violation. It was, rather, pieces such as the short stories of Satomi Ton 里見弴 (1888–1983) and Mikasanomiya Takahito 三笠宮崇仁 (1915–2016) that were "censored and deleted in part" before publication. By all appearances, the rather oppressive climate for authors and media outlets seeking to provide accounts of the atomic bombings eased somewhat in 1949. Of course, this does not mean that censorship ceased to exist but it was no longer so rigid as to force many authors and publishers into employing such drastic forms of self-restraint when addressing the events of August 6.

The Allied Occupation of Japan officially began on September 2, 1945, and postwar censorship was institutionalized shortly thereafter with the creation of the Civil Censorship Detachment (CCD) on September 10. The CCD would serve to censor undesirable views supposedly contradicting or undermining Japan's transformation into a democratic state committed to freedom of speech. Censorship was—as was pointed out by Monica BRAW in her seminal study—also institutionalized in the American occupation zones of defeated Germany at a very early stage of the occupation. However, in marked contrast to the German case, the scope of censorship in Japan—and thus the number of personnel involved—diverged significantly, given the intention of the Allied Powers. BRAW formulates the Allied injunction in the following terms: "The whole of the Japanese people must be reformed. It would not be enough just to refuse certain publishers' licenses. All Japanese, including publicists, must be reeducated" (BRAW 1986: 60).

It goes without saying that censorship was not exactly new for the Japanese population given that newspapers, magazines, film, and radio had been strictly regulated by the government during the Fifteen Years War (1931–45). Just how opaque the categories for censorship had become, however, was unprecedented. On September 15, 1945, the Liberal Democratic MP—and subsequent Prime Minister, HATOYAMA Ichirō (1945: 2)—publicly rebuked the U. S. government, writing in the *Asashi shinbun* that "the Americans can no longer deny that the use of the atomic bombs and the killing of innocent civilians were even a greater violation of international law and a greater war crime than the bombings of hospital ships or the use of poison gas". Immediately thereafter, on September 19, 1945, the Allied Powers released a ten-point Press Code declaring, rather vaguely, that news would now "adhere strictly to the truth" (1) and do nothing to "disturb public tranquility" (2).[2] Needless to say, more often than not it was left to the judgement of the individual censor whether or not a certain text or picture would to be treated as a violation of the press code. Ambiguity and the unpredictability of the censors created a climate fostering self-restraint—the "educational effect" for the Japanese as BRAW puts it—as an anticipatory obedience towards the occupation policy of the Allied Powers.

The CCD, which at its peak employed well over 10,000 censors controlling media and telecommunications of every ilk, was ultimately dissolved on October 31, 1949.[3] This did not mean, however, that censorship came to an end. On the contrary, censorship under the Civil Information and Educational Section (CIE) as well as the Civil Intelligence Section (CIS) continued until the official end of the Allied Occupation in April 1952 (ETŌ 2015: 20). The effects of censorship on the historical awareness of an entire nation cannot be overstated. In his eponymous work, "Sealed Space for Verbal Utterances" (*Tozasareta gengo kūkan* 閉された言語空間), ETŌ Jun speaks to the climate created by Allied censorship in Japan (ETŌ 1994: 347–66). The impossibility of "articulating" supposedly contestable facts or memories paved the way for a nescience towards, and a distortion of, recent war history for broad swathes of society. The effects of censorship for the atomic bombings, in particular, were disastrous. As SHIGESAWA Atsuko (2010: 136–74) has pointed out, the majority of Japanese society became "ignorant" (*muchi* 無知) and "indifferent" (*mukanshin* 無関心) to the bombs, the aftermath and, above all, the victims (*hibakusha* 被爆者). The renowned author, INOUE Hisashi 井上ひさし (1934–2010), serves as an illustrative example in this context. In a round-table talk on atomic bomb literature with the famous writer, HAYASHI Kyōko 林京子 (1930–2017), INOUE had to admit that he had not learned about the real extent of the devastation unleashed in Hiroshima on August 6 and Nagasaki on August 9 in 1951 until a full six years after the bombings (INOUE et al.2004: 25).

The entire "nuclear issue" was an extremely sensitive matter from the very beginning. After the atomic bombing of Hiroshima on August 6, the Japanese government did its best to conceal the true nature of the "enemy's new type of bomb" (*teki shingata bakudan* 敵新型爆弾)[4] given that acknowledging the use of an atomic bomb on Hiroshima would have been tantamount to admitting the enemy's technical and strategical superiority. This, in turn, would have inevitably caused a general decline in morale on the side of the civilian population (SODEI 1975: 269). On August 11, 1945, the former "new type of bomb" was finally acknowledged for the first time as an "atomic bomb" (*genshi bakudan* 原子爆弾) in Japanese newspapers, following Harry S. TRUMAN's famous radio address to the American people on August 9. Nonetheless, coverage in the Japanese papers on the extent of the devastation continued to downplay the extent of the damage, comparing it to that of a conventional bomb. The real effects of radiation in the aftermath of the explosions—or in ŌTA Yōko's words, the main reason for the "unfathomable and unprecedented confusion and horror" among survivors—were more or less hushed up, such that knowledge about the true nature of the bomb remained restricted to the disaster areas of Hiroshima and Nagasaki. Interestingly, the Cabinet Intelligence Bureau (Naikaku jōhōkyoku 内閣情報局) that was responsible for this kind of media control and censorship in Japan continued its work until its ultimate dissolution on October 1, 1945. As TAKAKUWA Kōkichi (1984: 35–49) has indicated, the bureau, for whatever reason, thus continued in its unwavering control of the mass media even after the Allied Occupation of Japan officially began on September 2.⁵

It goes without saying that after the CCD was established, covering the atomic bombs became even more difficult for the media than before. Reports on the radioactive aftereffects of the bomb were strictly suppressed—in both the United States and Japan—since these kinds of reports would undermine the officially propagated image of a new and clean weapon that could be used in a just war primarily for the destruction of military targets. The famous case of the Australian journalist Wilfred BURCHETT illustrates just how sensitively the General Headquarters of the Supreme Commander of the Allied Powers could react. BURCHETT had visited the devastated city of Hiroshima in August 1945 and subsequently published a report in London's *The Daily Express* on September 5, 1945, titled "The Atomic Plague", in which he wrote:

> In these hospitals I found people who, after the bomb fell, suffered absolutely no injuries, but now are dying from the uncanny after-effects. For no apparent reason their health began to fail. They lost appetite. Their hair fell out. Bluish spots appeared on their bodies. And the bleeding began from the ears, nose, and mouth. [...] All these phenomena, they told me, were due to the radioactivity released by the atomic bomb's explosion of the uranium atom. (BURCHETT 2008 [1945]: 3; 5)

Only one week later, on September 12, 1945, a counterstatement was published by William L. LAWRENCE (by order of Thomas F. FARRELL from the Manhattan Project) in *The New York Times* that rejected any aftereffects of radioactivity on the human body and discredited BURCHETT as an unreliable source (SASAMOTO 1999: 456–62).

The CCD filtered out virtually every kind of "articulation of the nuclear" that had the potential to become a powerful counter-narrative to the official version of the atomic bombings. Film materials, photographs, reports, novels, and even pieces of poetry were painstakingly scrutinized before being given approval for publication or release. Confiscated material, on the other hand, was archived in the CCD and, ultimately, brought to the United States where it became part of the Gordon W. Prange Collection (University of Maryland) after Allied Occupation reached its conclusion in 1952.

However, as we have seen in the case of ŌTA Yōko's "August 6, 8:15", the CCD censorship of the "nuclear issue" grew less rigid over time. Deep-rooted distrust of former U. S. ally Joseph STALIN—undoubtedly another central motivation for concealing information about the atomic bombings and their aftereffects—as well as the fear of communism rapidly spreading throughout East Asia (including Japan), becoming the driving forces for a gradual shift in censorship practices. As early as 1959, the German-Austrian philosopher Günther ANDERS (1902–92)—who had visited Hiroshima and Nagasaki in 1958—exposed the hypocrisy behind the American justification that the second atomic bombing on Nagasaki was a necessary means to quickly end the war as deceptive, stating:

> It is common practice [...] to consider Hiroshima as the ultimate symbol, and therefore mention it first, as in the saying "Hiroshima and Nagasaki". However, this common practice is not justified. Because for a moralist, Nagasaki should be mentioned first. In other words, one should say instead: "Nagasaki and Hiroshima". The simple reason is that the second atomic bomb was an even greater evil than the first. [...] The 70,000 people that were murdered [...] were not murdered because they had been enemies, let alone dangerous enemies. On the contrary, they were murdered because this mass murder would make an example of them. It would assign 70,000 corpses an entirely new function. These corpses could now be used, used as a threat. (ANDERS 1995 [1959]: 110; 111–12)

For ANDERS, the atomic bombing of Nagasaki was nothing more than a threat to Joseph STALIN who had ordered his troops to invade Manchuria on August 8, 1945. It was a strategic move to keep the Soviet Union at bay. At the same time, it ushered the whole world into a new age: the Cold War.

It is, thus, not surprising that, from the very beginning, employees working for the CCD were frequently hypersensitive to any association with communism. HIRANO Kyōko's study on film censorship during the occupation period takes a look at the interesting case of the 1946 documentary film "The Tragedy of Japan" (*Nippon no higeki* 日本の悲劇), directed by KAMEI Fumio 亀井文夫 (1908–87). The film production was enthusiastically encouraged by the head of the motion and theater branch of the CIE at the time, David W. CONDE (1906–81). The script for the film had no difficulty in passing the CIE's screening process. Problems started, however, when the CCD began to review the script again in June 1946, demanding major changes to "problematic scenes". Yet even after the changes were made, the problems did not cease. In August 1946, the film was reviewed once again by CCD censors, which ultimately resulted in a complete ban, with all copies confiscated (HIRANO 1998: 210–24). The film's serious claims that many war criminals—including Emperor Hirohito—had held onto their positions, even after the war had ended, was the main reason for the extreme reaction of the CCD. The International Military Tribunal for the Far East began its activities in May 1946 and the question of war guilt, as such, was an extremely sensitive issue, both for Japanese politicians such as Prime Minister YOSHIDA Shigeru 吉田茂 (1878–1967), and for the supreme commander of the Allied Powers in Japan, Douglas MACARTHUR (1880–1964). Indeed, the issue was more sensitive still, given that the Soviet Union, for example, was calling for a conviction of the emperor (and an abolishment of the emperor system). The assertations made in "The Tragedy of Japan", thus, seemed suspiciously leftist. For the sake of keeping communism at bay, the Allied Powers—or, more aptly, the American Forces in Japan—persecuted individuals that, due to their personal history in the war, were not necessarily compatible with the U. S. vision of a new and democratic Japan.⁶ Incidentally, David W. CONDE himself was forced to resign from his job in July 1946—and return to the United States in spring 1947—on suspicion of being a communist (HIRANO 1992: 142).

According to MATSUURA Sōzō, an obvious change in censorship policies occurred in 1947 (MATSUURA 1974: 299–308). In 1947, the Japanese Communist Party and the Japanese Labor Union planned a national strike for February 1 that aimed to overthrow the YOSHIDA cabinet. The incident became a wake-up call for the Allied Powers in terms of an urgent need to more closely examine leftist movements in the country. In the event, MACARTHUR was able to intervene at the very last second to avert the strike and, thus, the planned takeover of the Japanese government. The incident is often considered pivotal for the well-cited "reverse course" (*gyaku kōsu* 逆コース) in occupation policy, which led to a withdrawal of the central reforms that had been implemented at the beginning of the occupation period, and to an inevitable strengthening of right-wing, conservative forces in Japan. The "red scare" that had already come to constitute a central issue in domestic U. S. politics in the first postwar period soon took on a protagonistic role in occupation politics and censorship measures in Japan. Anti-leftist resentment during the occupation period undoubtedly reached its climax with the onset of the Korean War in June 1950 when tens of thousands of alleged communists were removed from their jobs as part of the so-called "red purge". The equation of pacifism—and later antinuclear activism—with communism became a fatal nexus that silenced many voices, including survivors of the atomic bombings.⁷ In sum, the occupation period in all its hypocrisy fostered, nolens volens, a reconceptualization of history insofar as that the atomic bombings were relativized in their historical singularity due to a lack of information. Ultimately, very few military leaders were sentenced for war crimes in the Tokyo Trial and the whole nation, as such, felt exculpated from any war guilt. Indeed, the Fifteen Years War was generally misconceived as a war without perpetrators, as a series of incidents that occurred simply by chance.⁸ As has been quite astutely indicated by ARIYAMA Teruo (2003: 9–12), the occupation period became the "incubation phase" (*senpukuki* 潜伏期) for Japan's postwar re-nationalism and self-victimization, and this despite any initial intention thereof on the part the Allied Powers. In other words, Japan had chosen a way "of remembering Japanese suffering while forgetting the suffering that the Japanese caused others" (DOWER 1993: 27). In his eponymous work, KATŌ Norihiro speaks of the dire consequences of the "American Shadow" (*Amerika no kage* アメリカの影) that was cast over Japan in the first postwar years. According to KATŌ (1995: 294–300), the "democracy" that was imposed on Japan after its defeat was nothing more than an "atomic bomb-based democracy" (*genbaku minshu shugi* 原爆民主主義), which was unable to change deep-rooted fascist structures among

Japan's political elite. The result was, thus, a political system that was neither willing to learn its lessons from the past nor was it particularly inclined to turn a new leaf for the future. However, the U.S. influence KATŌ depicts in his study did not diminish with the end of the occupation period in 1952 as one could be forgiven for thinking. On the contrary, its influence on domestic policy in Japan continued unabated.

For the "articulations of the nuclear", which depicted the "evil" side of nuclear power, the end of occupation marked a new era given that medical reports, eyewitness accounts, novels, and poems could be published without running the risk of censorship or confiscation. Yet, the new era brought challenges of its own for such expressions, as a new kind of "articulation of the nuclear" emphasizing the "good" side of nuclear power was now enthusiastically encouraged by the political elite. Dwight D. EISENHOWER's famous speech, "Atoms for Peace", held at the Plenary Meeting of the United Nations General Assembly on December 8, 1953, marked the beginning of a new nuclear age in Japan. Needless to say, the United States' commitment to providing their knowledge for the peaceful use of nuclear energy was, more or less, a reaction to the Soviet Union's commitment—declared on October 5, 1952—to supporting "the peaceful use of nuclear energy for the advancement of society". The first commercial nuclear power plant in Obninsk (near Moscow) began producing electricity on an industrial scale as early as in 1954 and it was incumbent on the United States, as such, to take action proactively in order to protect their sphere of influence, at least in the so-called Western world (ICHIKAWA 2016: 28–35). However, a spanner was thrown into the works for EISENHOWER by the Bikini Incident in which the crew of the Japanese fishing boat Lucky Dragon Number 5 (Daigo Fukuryūmaru 第五福竜丸) was contaminated by radioactive fallout caused by the U.S. hydrogen bomb test in the Bikini Atoll in March 1954.[9] The antinuclear-weapons movement—which started with a signature campaign launched by concerned housewives of Tokyo's Suginami Ward—temporarily created a climate in which promotion campaigns for the peaceful use of nuclear energy seemed to be anything but advisable. Hence, for Japan, the plans for a tour of the exhibition, "Atoms for Peace" (Genshiryoku heiwa riyō hakurankai 原子力平和利用博覧会), had to be postponed to 1955 (ZWIGENBERG 2012).[10] Having first been displayed in Japan's larger cities, the exhibition finally was shown in the Hiroshima Peace Memorial Museum (Hiroshima heiwa kinen shiryōkan 広島平和記念資料館)—which had just opened its doors one year prior—from May 27 to June 17, 1956. The question as to whether the special exhibition ought to be held at the Hiroshima Peace Memorial Museum at all was a controversial one, and much discussed. The initial exhibits had to be removed and stored externally to provide enough space, which simultaneously meant avoiding questions about the potential dangers of nuclear energy that could have arisen from viewing the exhibits from the initial collection. Yet, in the end, the influence of so many supporters outweighed the critical voices, and it would seem that the success of the exhibition proved them right. With well over one hundred thousand visitors in three weeks, "Atoms for Peace" was more successful in Hiroshima than in many other cities, such as Osaka, for instance, despite the fact that the latter was, and still is, the larger of the two (ZWIGENBERG 2014: 123).[11]

On a governmental level, the path towards the peaceful use of nuclear energy in Japan had already been paved in 1954 by politician, NAKASONE Yasuhiro 中曽根康弘 (1918–2019), and media mogul, SHŌRIKI Matsutarō 正力松太郎 (1885–1969), the latter of whom history would later reveal to have been a CIA agent. These two men would together constitute the driving force for the continued promotion of peaceful nuclear energy in Japan, which had first been set in motion by the United States. The signing of the atomic energy agreement between the United States and Japan and the ratification of the Atomic Energy Basic Act (Genshiryoku kihon hō 原子力基本法) in November and December of 1955 respectively, followed by the establishment of the government's Nuclear Energy Commission (Genshiryoku iinkai 原子力委員会) in January 1956, were undoubtedly crucial steps in leading Japan toward a bright and promising nuclear-based future (*akarui seikatsu* 明るい生活).[12] However, following the Bikini Incident in 1954, Hiroshima had hosted the first World Conference on the Prohibition of Nuclear and Hydrogen Bombs (Gensuibaku kinshi sekai taikai 原水爆禁止世界大会) in 1955, going on to become a hub for antinuclear movements. Indeed, the touring exhibition had become a litmus test for Japan's disposition to the promotion of nuclear power plant construction. As shown above, Hiroshima successfully passed the test. In an article co-authored with Yuki TANAKA, Peter KUZNICK summed up the issue in the following rather sober terms: "Wanting their country to be a modern scientific-industrial power and being aware of the fact that Japan lacked energy resources, the public allowed itself to be convinced that nuclear power was safe and clean. It had forgotten the lessons of Hiroshima and Nagasaki" (KUZNICK / TANAKA 2011). Bizarre though it sounds from a modern point of view, in 1955, Sidney R. YATES initially proposed the construction of Japan's first nuclear power plant, of all places, in Hiroshima. Following the failure of the initiative, it was not until an entire decade later that Japan's first nuclear power plant—constructed in Tōkaimura (Ibaraki Prefecture)—began to produce electricity on an industrial scale in 1965.[13]

As we have seen, the first postwar decade saw the development of Japan—under the permanent "guidance" of the United States—into one of the world's foremost producers of nuclear energy. Indeed, the atom was irreversibly split into, on the hand, an "evil" energy epitomized by the atomic bomb (*genbaku* 原

爆) and, on the other, a form of "good" energy, represented by the nuclear power plant (*genpatsu* 原発). The former came to symbolize a remote past that humanity had now overcome, while the latter symbolized a bright and promising future. It goes without saying that this discursive splitting of the atom inevitably led to a splitting of society with, in the one corner, those for whom the aftermath of this "evil" energy was still affecting their daily lives and, in the other, those for whom a praiseworthy "good" energy signified the ultimate key to a new, electrically powered, middle-class life (YOSHIMI 2012: 143–67).

The first postwar decade also saw a striking synchronicity in the development of events that seemed—at least at first glance—to have very little to do with one other. Upon closer inspection, however, it soon becomes clear that these developments were somewhat triggered by the processes depicted above. We shall now flesh out some of the most crucial of these developments with a view to gaining a better overall picture.

We begin with the transformation of Hiroshima and Nagasaki into sacred places for national mourning. The reinvention of these two cities as centers for peace accelerated a musealization of both the atomic bombings and their victims.[14] Especially in Hiroshima, the musealization of ground zero as a new center for world peace and peace tourism, respectively, had fatal consequences for those survivors that settled in various slums in the city center. The slums in question—a thorn in the side of city planners—were forcibly evacuated and dissolved in order to enable the construction of the planned peace areas (NISHII 2020: 143–49). And yet, it was primarily non-victims of the bombings (*hi-hibakusha* 非被爆者) that first availed themselves of the newly constructed commemoration spaces—including parks, museums, statues, and cenotaphs—to commemorate an emotionally shared memory.[15] The presence of the "real" victims of the bomb, as ŌTA Yōko (1982 [1954]: 276–77) dryly remarked in her novel "Half human" (*Han ningen* 半人間), was apparently only necessary at the annual nationwide memorial services held on August 6 (and August 9, respectively). The musealization process ended, more or less, with the opening of the Hiroshima Peace Memorial Museum and the Nagasaki Atomic Bomb Museum (Nagasaki genbaku shiryōkan 長崎原爆資料館) in 1955. With the inauguration of these two institutions, August 6 and August 9 were put into a time capsule, preserved for eternity. Yet, as FUKUMA Yoshiaki (2015: 84) has pointed out, this preservation—epitomized by the impressive compound of architectural entities in the two peace areas—conceals the fact that an irreversible erosion of memory had already taken place throughout Japanese society. The places of commemoration merely re-enacted a memory that had already faded away.

A second noteworthy development in this context is the social ostracism many intellectuals, artists, authors, and even survivors began to experience in the first postwar decade. The questions of "war guilt" and "war responsibility" were a delicate matter for most people in the first postwar years. The sentences issued at the Tokyo Trial fueled, as mentioned above, the widespread idea that only a small group of fanatic military leaders was ultimately responsible for the war.[16] As Tsurumi Shunsuke (1959: 80–81) has stated, it was for this reason that bureaucrats and officials who had supported the military system lived their lives in the absence of any real guilt about the war. Especially the exoneration of the emperor at the Tokyo Trial—thanks to the enormous influence exercised by MacArthur in this very special case—reinforced this feeling of collective innocence in Japan (Kojima 1991: 81–95). If the emperor himself—the very man at the head of the military who had signed off on the war through his holy decisions (*seidan* 聖断)—is not responsible, who is?[17] In contrast to the majority of those in postwar Japanese society, intellectual circles began fiercely discussing why and how an entire nation could be seduced by military leaders. The leading literary magazines of that time, "New Literature of Japan" (*Shin Nihon bungaku* 新日本文学) and "Modern Literature" (*Kindai bungaku* 近代文学), became major battlegrounds on which the issue of the "conversion" (*tenkō* 転向) among Japan's intelligentsia during the Fifteen Years War was relentlessly debated.[18] Mutual denunciation of intellectuals and artists was quite common practice in the journals. However, the debates did not only center on the question of intellectual backwardness—that is, not being "individual" in a modern sense—but also on the extent to which modern literature had failed to become the central means by way of which intellectuals could develop "individuality" (*shutaisei* 主体性). As a result, quite a few intellectuals, artists, and authors were ostracized either by their colleagues or by their readers. Perceived as dissenting voices from the past, they appear to have reminded this new "community of memory" (*kioku no kyōdōtai* 記憶の共同体) of an uncomfortable historical episode that they had otherwise relegated to oblivion. By the mid-1950s at the latest, Japan's sense of identity as a nation full of war victims no longer left any room for debates concerning the problem of "conversion", "war guilt", or "the atomic bombings".[19]

The third development requiring explication here is the radical change of course in the educational sector. From the very beginning of the occupation period, the Allied Powers placed a focus on peace and democracy in order to prevent further indoctrination of children in Japanese schools. The Basic Act on Education (Kyōiku kihon hō 教育基本法), which came into effect in March 1947, became the legal framework for this new educational concept while the new screening system for teaching materials (*kyōkasho kentei seido* 教科書検定制度), implemented in August 1948, served as the necessary instrument for controlling teaching contents (Köhn 2022: 174–76). However, here again, the abovementioned "reverse course" in occupation policy had severe consequences, this time for the teaching materials and the teaching staff. In line with the red purge in Japan, anti-communism became a new issue to be taught in school classes from 1950 onwards, and the supposed ideological orientation of the teachers quickly became an issue of the utmost importance (MATSUI 1969: 21–23). For teachers, teaching peace, as promulgated in the Basic Act, became a highly difficult task as teachers generally ran the risk of being discredited as a communist by colleagues or parents. Especially after the Japanese government had taken complete control over the screening system for teaching materials at the end of the occupation period, the content of the textbooks began to change dramatically insofar as the new "war without perpetrators" became the predominant narrative in all officially approved textbooks. Even supplementary teaching materials required an official approval before they could be used. Knowledge about the (aftermath of the) atomic bombings in Hiroshima and Nagasaki, as such, inevitably declined among students and August 6 and 9 were, thus, reduced to abstracts events of a remote past within the national curriculum. In June 1954, the Double Education Act (kyōiku nihō 教育二法)—as agreed in preliminary talks between Walter S. ROBERTSON (1893–1970), Assistant Secretary of State, and IKEDA Hayato 池田勇人 (1899–1965), representative of prime minister YOSHIDA Shigeru, in October 1953—was put into effect. From then on, committed teachers had to live with the fear of losing their jobs due to alleged bias in their teaching methods. In this sense, the IKEDA-ROBERTSON Talks, as ISHIDA Takeshi (2000: 260–63) has astutely pointed out, paved the way for a new kind of postwar nationalism that affected the entirety of public life. After the Guidelines for Teaching and Learning (Gakushū shidō yōryō 学習指導要領) had been accordingly "adjusted" to this new educational policy in 1958, articulations of the nuclear ultimately vanished from textbooks and classes for several years.[20]

As these three developments paradigmatically illustrate, the first postwar decade was a time in which the discursive splitting of the atom (and society) had an enormous impact on virtually all spheres of public life. With all its contradictory decisions and unpredictable changes of course, the occupation period fostered the emergence of power structures that pervaded the entirety of Japanese society for decades to come. These structures are the key mechanisms for rendering "texts" or "images" (as understood in the sense promulgated within critical culture studies) of the "good" and "evil" nuclear energy visible or invisible. In this regard, visibility should be considered just as untrustworthy as invisibility given that neither of these states are coincidental. On the contrary, they are the inevitable results of these underlying power structures. To put it differently, articulations become visible or invisible as the result of a deliberate choice for representation or suppression.

A case in point for one type of "untrustworthy" visibility is the sudden change to the way in which Hiroshima and Nagasaki were represented in written Japanese in the first postwar decade. The prewar cities were formerly written in *kanji* as 廣島 and 長崎 but were transformed through public discourse into the postwar *katakana* form—ヒロシマ and ナガサキ—shortly after the atomic bombing. Inconspicuous though this change of notation may seem at first, the new notation irreversibly transformed the historical cities of Hiroshima and Nagasaki. They became detached from time and space in Japanese collective memory to become places of national commemoration and mourning. It also symbolized a specific historical and national awareness. While the *kanji* notation referred to the historical pre- and interwar cities of Hiroshima and Nagasaki, the *katakana* notation referred to the reconstructed postwar cities. According to MATSUMOTO Hiroshi (1995: 14–15), the *katakana* notation of Hiroshima (and Nagasaki) was the result of a deliberate depoliticization of August 1945. This depoliticization was promoted in a bid to emphasize victimhood and, thus, conceal Japan's role as the aggressor during the war. Needless to say, this new visibility for Hiroshima and Nagasaki was not coincidental, but rather a deliberate strategy to render contested memories invisible and to shape a new postwar history and, as such, a new postwar identity.

On March 11, 2011, a 9.0 magnitude earthquake shook the Tōhoku region and, with it, the myth of the safety of Japanese nuclear power plants (*anzen shinwa* 安全神話) that had lulled postwar Japanese society into a false sense of security. The earthquake triggered a tsunami that led to a meltdown in three reactors of the Fukushima Daiichi Nuclear Power Plant. The scope in the ensuing leakage of radioactive materials was unprecedented. The (aftermath of the) "Triple Disaster"—as it came to be known—ought to have disillusioned the nation insofar as it testified to the ongoing successful operation of those same mechanisms outlined above, which continued to obfuscate the dark side of nuclear energy 66 years after the atomic bombings in August 1945. As the avalanche of publications surrounding the catastrophe have impressively shown, it did not take long for Fukushima to follow the example of Hiroshima and Nagasaki with a change to its *katakana* notation. Needless to say, the country's nuclear legacy quickly ensured that the new *katakana* rendition of Fukushima—フクシマ—made of the city a highly contested abstract place for Japanese society.

The structural similarities between August 1945 and March 2011 are striking. In 2011, mass media was once again controlled and censored, the real danger of radioactivity downplayed, and dissenting voices were silenced. In public discourse, vested interests in party politics did their best to conceal the human-made nature (*jinsai* 人災) of the Triple Disaster. They sought, thereby, to establish a public narrative of an event on par with a mere natural disaster (*tensai* 天

災). Indeed, the most commonly used name for the event—the Great East Japan Earthquake (Higashi Nihon daishinsai 東日本大震災)—reinforces the feeling that 3/11 was first and foremost a natural disaster. So formulated, the event is perfectly in line with other natural disasters such as the Great Kantō Earthquake of 1923 (Kantō daishinsai 関東大震災) or the 1995 Great Hanshin-Awaji Earthquake (Hanshin Awaji daishinsai 阪神淡路大震災).

It goes without saying that this narrow view of the Triple Disaster had severe consequences for many of the survivors, especially those being evacuated from the restricted areas around the power plant due to the massive radioactive fallout. Now, as then, victims of radiation exposure were targets of ostracization and discrimination. The hypersensitivity towards radioactivity in Japanese society became the driving force for this process. Even the undoubtedly well-intended debate about "harmful rumors" (*fūhyō higai* 風評被害)—that is, the public defamation of victims through the spread of "false" information—turned out to be a double-edged sword. Instead of protecting the reputation of the affected people in the disaster area, the government quickly began to instrumentalize the phrase to suppress any negative coverage capable of revealing the real impact of radioactive contamination in the disaster-stricken region.[21] Of course, the official "protection" of victims from "harmful rumors" was a perfect opportunity for the government to deflect any accountability for the nuclear disaster and failure in crisis management.

The Triple Disaster revealed, more than any other nuclear incident in Japan, the Janus-faced character of "the nuclear". It epitomized the risk of relying on nuclear energy, a risk that the majority of Japanese society had willingly accepted in exchange for a bright, nuclear-based future. As a classical "risk society" in the Beckian sense, Japan welcomed prosperity and growth as their reward for a risk that was fundamentally borne by the (remote) countryside communities where virtually all nuclear power plants were erected. Takahashi Tetsuya (2012: 27–28) coined the term "sacrificial system" (*gisei no shisutemu* 犠牲のシステム) to describe the structural mechanisms that produced human "sacrifices" for the benefit of the collective. According to Takahashi, these sacrifices can be either glorified by a society as noble, or rendered invisible as unpleasant truths. However, as the annual commemoration ceremonies show, the latter is the case. Inhabitants, workers, or liquidators that had been exposed to the radiation were rendered both invisible and speechless in public discourse.

As could be seen from the above, the "nuclear issue" pervaded (and continues to pervade) all spheres of public and private life in postwar Japan. It epitomizes power *per se* and exercises, at the same time, power on subjects and objects. In this regard, it is probably more appropriate to speak of the "nuclear dispositive" rather than the "nuclear issue", given the material and ideational infrastructure of central discourses and discourse formations in the postwar period. According to Michel FOUCAULT, a dispositive is a heterogenous ensemble, that is, a network of discourses / non-discourses, practices / non-practices, institutions, laws and / or other agencies that form and influence the making of, in this case, the postwar society. A dispositive, as FOUCAULT (1978: 120) would have it, always emerges out of a specific "urgency" in times of political, social, or ideological crisis. It has a primary strategic function and is embedded in power relations and power plays, respectively. A dispositive takes control of the actions, thoughts, or mindsets of the individuals who are surrounded by this network of discourses and non-discourses. In so doing, it shapes each and every individual's identity. The dispositive has the power to govern and change a society in accordance with the "urgency" that originally induced the emergence of the dispositive. Despite its invisibility as such, it is a powerful and primary mechanism of governance. As Andrea D. BÜHRMANN and Werner SCHNEIDER (2008: 53) fittingly put it, a dispositive is a governing strategy without a strategist.

Rethinking Japan's postwar period from the FOUCAULTian concept of the dispositive helps us to unveil the interrelated nature of very heterogeneous elements and developments under the spell of the nuclear. For Giorgio AGAMBEN (2008: 24), the primary "objective [of the dispositive is] to conduct, govern, and control people's behavior, gestures, and thoughts so that they can be steered into a supposedly beneficial direction". It is, thus, first and foremost a governing machine that produces knowledge / unknowledge and controls the process of subjectification / objectification. In order to gain a more comprehensive overview of Japan's "nuclear paradox", one would need to engage in an in-depth analysis—today more than ever—of this all-pervasive, and yet invisible net of power relations and knowledge production that can be classified as the "nuclear dispositive". However, this Herculean task would be far beyond the scope of a single book. In lieu thereof, this publication focuses on selected knots—or "ensembles"—in this net. The 15 contributions provide thorough analyses of: single discourses and non-discourses; practices and non-practices and; pivotal institutions in this context. Our contributors demonstrate that this net is, in point of fact, an impactful governing machine that generates and regulates "knowledges", "practices", and processes of "subjectification / objectification" in the atomic age.

Our volume begins with an examination of General Leslie R. GROVES' editorial comments of John HERSEY's well-known article *Hiroshima* (1946) and the influence this exerted on the public's perception of the nuclear. Initially published in *The New Yorker* in the wake of the atomic bombings, *Hiroshima* was the first piece to garner significant attention in the United States for the suffering of *hibakusha*. However, as Atsuko SHIGESAWA points out, General GROVES took issue with some of the information included in HERSEY's initial manuscript.

SHIGESAWA's analysis makes clear that by compelling HERSEY to change certain wordings and semantics, GROVES downplayed the health risks of the prolonged exposure to radio nucleoids inherent to internal radiation. For SHIGESAWA, the general misconceptions fueled by GROVES' intervention continue to play a role in public assessment of health hazards to this day, and can clearly be seen in the case of the Fukushima nuclear disaster. SHIGESAWA's paper thus delineates a tendency that will crop up again and again in the many papers in this volume. To wit, a process by way of which certain narratives on—and perceptions of—the nuclear bombings were favored over others was set into motion immediately after the event. Especially conceptions that allowed for a splitting of the atom into "good" and "evil" were placed front and center.

Marie-Christine DRESSEN's paper testifies to the ostensible seamlessness of the process by way of which Hiroshima and Nagasaki went from marketing themselves as cities in ruins to coveting an image as cities of peace and culture. By examining (front-page) newspaper articles from 1945 to 1949—particularly those that appeared on relevant dates such as the anniversaries of the bombings—DRESSEN articulates the light in which the bombings were generally shown in the daily press, and this at a time when CCD censorship was still in operation. DRESSEN's analysis points to certain key narratives that emerged during that period, such as the tendency to frame the bombings within an international context as events of primary importance to world history. The city of Hiroshima, in particular, took on the mantle of ambassador for world peace while at the same time cultivating a narrative of peace in a bid to attract international tourists. Destined to live somewhat in the shadow of Hiroshima's aggressive peace-branding, Nagasaki cultivated its image as a city of culture. As DRESSEN shows, while the construction of these images of peace and culture necessitated a nationalization of atomic bomb commemoration, it also silenced the perspectives of other groups of *hibakusha*—such as Korean and *burakumin* 部落民 voices—that did not fit with the mainstream narratives that were established.

For her part, Katharina HÜLSMANN looks to depictions of the nuclear in children's manga of the occupation period, delineating which narratives about the nuclear can be found in children's literature in the years immediately after the dropping of the atomic bombs. In particular, she examines two manga aimed at young boys that were thematically quite similar to the well-known wartime manga classic "Adventurous Dankichi" (1933–39). As a medium, manga came to the attention of public in the U. S. and Europe by way of NAKAZAWA Keiji's seminal work "Barefoot Gen" (1973–87) in which he provides a detailed depiction of the horrors of the atomic bombs. In the stories that HÜLSMANN examines, however, the nuclear does not appear as a bomb or explosive weapon, but instead as an almost mystical power that can animate everyday life objects and help the young protagonists triumph in their adventures, a narrative that was obviously approved by the censorship apparatus. As HÜLSMANN points out, very few children's publications at the time were censored. Indeed, the stories with which the authorities took issue were mainly samurai stories and, in particular, the sword fights contained therein, which were deemed a glorification of feudal Japan. Typical adventure story tropes, such as racist depictions of the inhabitants of island nations, and the glorification of colonialism, gave censors little pause for thought. Indeed, they were seen as entirely fit for young eyes.

Christopher P. HOOD then takes a closer look at depictions of the atomic bombings in a broad swathe of films, both live action and animated. He pays particular attention to the ways in which different elements of the bombings, such as the flash (*pika*), the bang (*don*), the shockwave, and especially the characteristic mushroom cloud are reproduced in the depictions. Over the years, films depicting the bombings have featured many different kinds of images, some taken from the actual bombings that were filmed for documentary purposes, some drawn and animated, some created by practical effects and computer-generated images, and some that employ a blend of these different kinds of images. HOOD points out the impact in terms of which point of view is employed when showing the bombing on screen. The only existing documentary footage of the bombings was shot from planes accompanying the actual bomber and the angle of the footage is accordingly acute. Certain subsequent renditions, such those which appear in the historical manga series, *Barefoot Gen*, however, depict the development of the mushroom cloud from a low angle. The manga equally portrays the chaos and destruction as it is experienced by those on the ground, that is, those who experienced the bombings firsthand. As HOOD shows, while documentary footage of the bombings exists, many renditions eschew authenticity as regards the shape of the mushroom cloud, favoring instead a depiction in line with the images—and correlating shape—that had come to be known from the extensive nuclear testing and other pop-cultural images that followed August 1945.

After examining some of the narratives that appeared in the daily papers, in the media, and in popular culture in the aftermath of the bombings, this volume turns to urban planning and the architecture of memorialization. Chantal WEBER takes a look at architect, TANGE Kenzō, who was profoundly influenced by the French architect and designer Le CORBUSIER, and who is perhaps best-known as the man behind the Hiroshima Peace Memorial Museum. WEBER focuses, in particular, on TANGE's early creative work in Manchuria during Japanese occupation, outlining the impact of this period on TANGE's plan for new buildings in Hiroshima. While Manchuria's lack of existing infrastructure provided TANGE, to a certain extent, with a seemingly blank canvas, the destruction wrought by the atomic bombs on Hiroshima would again give the architect an opportunity to build a city from the ground up. Of course, he was constrained by the practical and political demands that were prevalent in each context. Stylistically speaking, TANGE's concept for the Peace Memorial Museum incorporated ideas influenced by Western concepts of urban planning (such as the Charter of Athens) yet it also harkened back to examples of traditional Japanese architecture, such as the Ise Shrine. As WEBER indicates, the end of the war did not necessarily engender an ideological fissure. Indeed, the debate on tradition and modernity in Japanese architecture continued, before eventually taking a turn with the arrival of the Metabolist movement. WEBER's paper concludes by posing the question as to whether the wave of reconstruction efforts that ensued in Tōhoku—another "blank canvas" that appeared in the wake of the 2011 earthquake and tsunami—will inspire another round of debate on the nature of "Japan-ness".

For his part, André HERTRICH examines the contents of the aforementioned Hiroshima Peace Memorial Museum and the lesser known Nagasaki Atomic Bomb Museum. In particular, HERTRICH considers the different approaches to memorialization employed by each museum. Though both rely on the exhibition of victims' personal items and stories, the Hiroshima Peace Memorial Museum avails itself of the individualization approach that has become very common in memory culture, and which can be seen, for example, in the United States Holocaust Museum in Washington DC or Yad Vashem in Jerusalem. HERTRICH compares the use of victims' belongings in both cases to demonstrate that items on display in Nagasaki—such as a burned lunchbox with rice charred by the fire from the bomb—are first and foremost used to document the effects of the atomic bombs. At the Hiroshima museum, on the other hand, these items are framed by the stories of individual survivors, and are designed to evoke a dramatic emotional reaction.

Lauren CONSTANCE's paper on the Daigo Fukuryū Maru Exhibition Hall focuses on the preservation of *hibakusha* memories through eyewitness testimony. The hall's foremost display presents a fishing vessel that was contaminated with nuclear fallout from the Castle Bravo hydrogen bomb test at the Bikini Atoll in 1954, an event known as the Bikini incident. Having once purportedly inspired the Godzilla franchise, the Bikini incident has since largely fallen by the wayside when it comes to the memorialization of *hibakusha*. Eyewitness testimony of the incident is sparse, which stems, on the one hand, from the fact that only two of the original 23 survivors are alive today. There are also, as CONSTANCE points out, a number of significant barriers to the provision of testimony. Following the Bikini incident, the *hibakusha* were both subjected to discrimination and actively discouraged from sharing their testimony given the widespread belief that radiation poisoning was contagious. This misconception had already cropped up in the context of atomic bomb *hibakusha* and would later resurface with fears surrounding radiation in the wake of the Fukushima nuclear disaster. While other nuclear memorial museums heavily rely on eyewitness testimony, CONSTANCE shows that is not applicable to the Daigo Fukuryū Maru Exhibition Hall. Largely conceptualized without the involvement of *hibakusha*, the institution instead focuses on documenting the Bikini incident and framing it as one nuclear incident among many that have caused incredible suffering for thousands of people around the globe.

Stephan KÖHN's paper examines the writings of ŌTA Yōko, who personally experienced the bombing of Hiroshima and shared her experiences in her work. As KÖHN points out, even during Allied censorship and occupation, ŌTA's works continued to be published, which goes to show that certain depictions of the atomic bombs did, indeed, manage to avoid suppression. In particular, KÖHN examines the novel "City and People in the Evening Calm" (1955) as well as its reception and the context in which it was published, exposing thereby the strategies ŌTA used to represent Hiroshima and Japan. The paper emphasizes the importance of considering ŌTA's work against the backdrop of national history writing in late 1940s and early 1950s Japan, which was taking a new course toward the nationalization of memory. Following the bombings, a narrative was established according to which Hiroshima was a necessary sacrifice that allowed the emperor to achieve peace by surrendering to the Allies. As KÖHN indicates, the narrative marginalizes the voices of the actual victims, who were regarded as little more than unpleasant reminders of suppressed war guilt. According to to the author, ŌTA Yōko was also one of the few dissenting voices as regards the change in public opinion that saw those who had protested the Bikini incident in 1954 flocking to the "Exhibition on the peaceful use of nuclear energy" in 1955. She would continue to use her writing to point out the discursive "gaps" in the articulation of Japan's legacy of the nuclear.

Rachel DINITTO's examination of nuclear literature focusses on TSUSHIMA Yūko's successful efforts to unravel the "myth of safety" as well as the anti-nuclear activism of her later fiction and essays. The Fukushima nuclear disaster was an eye-opener for TSUSHIMA as it incited her to articulate connections between the various instances of nuclear harm that occurred during her lifetime, including Chernobyl and Three Mile Island. TSUSHIMA's writing was influenced by that of fellow writer HAYASHI Kyōko—especially HAYASHI's essay "To Rui, Once Again" (2013). In this piece, HAYASHI expresses her shock the government would discuss the dangers of internal radiation in the context of the Fukushima nuclear disaster given the lengths to which they went to downplay the effects that this radiation had on atomic bomb victims. DINITTO shows how HAYASHI and other literary voices, such as ŌE Kenzaburō, encouraged TSUSHIMA to develop her criticism of the widespread "myth of safety". Beyond drawing a line from Hiroshima to Nagasaki and, ultimately, Fukushima, DiNitto also articulates the harm suffered by various people throughout the nuclear fuel cycle. This includes those who are harmed by: uranium extraction; the production and testing of weapons; the construction of power plants; and, ultimately, the disposal of nuclear waste. As DiNitto points out, Tsushima also rejects, thereby, a sense of Japanese exceptionalism or national identity, focusing instead on the global history of the nuclear.

Peter Kuznick's paper interrogates changes in perceptions of the atomic bombings of Hiroshima and Nagasaki and what impact these changes have had on attitudes towards the Nuclear Arms Race in the United States. As is outlined in Dressen's and Köhn's paper, Japan quickly developed a narrative of peace and sacrifice around the atomic bombings. Though opinion polls may have shown consistent shifts in public opinion over the decades, the myth of the atomic bombings as instrumental to ending the war has long been proliferated in U. S. media, particularly in the case of Hiroshima. Kuznick's paper begins with an examination of the extreme tension between the U. S. and North Korea in 2017 at which time President Trump boasted that the U. S. had nuclear capacity enough to completely annihilate its enemies. As the author demonstrates, however, the subject of the atomic bombs and, in particular, the justification for dropping them was always relevant to the U. S.'s international image, as evidenced by Nobel Peace Prize laureate Obama's insistence not to apologize for dropping the bombs during his visit to Hiroshima in 2016. Kuznick pays special attention to the publication of John Hersey's 1946 article, *Hiroshima*, as a turning point for public opinion with regard to the atomic bombs. Indeed, the article prompted Henry Stimson to defend the decision to use the bombs in the February 1947 issue of *Harper's*. Kuznick goes on to show that the decision to drop the atomic bombs is the subject of increasingly critical opinions in the U. S., particularly among the younger generation. Moreover, the idea of officially apologizing to Japan enjoys broad support. Nonetheless, the myth of the bombs as a necessary evil to end the war persists to this day, despite continued efforts on the part of academics and the antinuclear movement to dismantle it.

Michele M. Mason's paper reflects on the work of mangaka, Nishioka Yuka. Born in Nagasaki in 1965, Nishioka followed in the footsteps of other critical mangaka, such as Katsumata Susumu, who used the medium of graphic narratives to convey an antinuclear message in the wake of the Fukushima nuclear disaster. Already engaged in peace activism and nuclear abolitionism prior to Fukushima, Nishioka created educational manga for young people, highlighting issues such as the memorialization of Nagasaki and discrimination against Korean *hibakusha*. As is demonstrated by Mason, Nishioka's post-3 / 11 work is an indictment of male technocratic fantasies and the delusional myth according to which the power of nuclear energy can be safely harnessed. Nishioka's "Goodbye, Atomic Dragon" (2012) uses the dragon—a powerful mythical beast—to symbolize nuclear power as a similarly devastating force awakened by humankind and invites readers to rethink the strict binary between life and matter, according to which only humans are afforded agency. Mason equally points to Nishioka's technique of depicting the nuclear as 'lively materiality', which the artist uses to show the hidden connection between *genbaku* and *genpatsu* in her work, disarming mainstream narratives such as that of the "Atoms for Peace" campaign spearheaded by the U. S.

Hiroko Okuda's paper on media discourse in the wake of the Fukushima nuclear disaster then opens the final section of papers in our volume, all of which focus on particular connections between Fukushima and the legacy of the atomic bombs. Okuda points to the importance of nuclear power for the imperatives of reconstruction and nation-building in the wake of Japan's defeat. As such, it is particularly important to examine reactions to the triple disaster and what it meant in terms of the image according to which nuclear energy is peaceful and safe. As Okuda argues, faith in nuclear energy in Japan was not completely destroyed by the disaster. Indeed, the image of the evil power of atomic mass destruction and the positive power of nuclear energy proliferated in the wake of the atomic bombings in 1945. Okuda examines how Japan's four biggest daily newspapers covered the events, focusing, in particular, on photo journalism. She shows that Japan is among the many nations that still peddle the myth of nuclear power as safe while depending, by the same token, on technocratic notions of progress. She points to a shift in photo coverage away from front-page depictions of the damaged reactor buildings in favor of images supporting narratives endorsed by the Tokyo Electric Power Company (TEPCO) as well as images aligned with a shift in attention toward technical issues. Finally, Okuda shows that more than a decade on from the 2011 nuclear crisis, Fukushima is now a symbol for the fact that nuclear disasters can supposedly be overcome. Indeed, a great deal of public discourse in Japan centered around the safety of local Japanese food for public consumption as well as the suitability of the country for the 2022 Olympics and for tourism more generally. As such, Japan remains largely unchanged in terms of its view of nuclear power.

While Okuda's paper obviates the fact that neither the Japanese government nor TEPCO can be said to have excelled in their handling of the Fukushima nuclear disaster, Tobias Weiss takes a look at the allocation of responsibility for the accident. Providing a close reading of developments between March 11 and March 15, 2011, Weiss identifies two primary camps—the office of Prime Minister Kan and TEPCO/the nuclear village—from which information about the unfolding nuclear disaster emerged. In particular, he looks at three questions that were raised in later reports: 1) Who was responsible for any eventual delay in venting at the Fukushima Daiichi reactor? 2) Who was responsible for the delay in the injection of sea water into the reactor? 3) Did the TEPCO leadership intend to withdraw their personnel from the plant at the height of the nuclear crisis? WEISS unravels the contrasting narratives by giving a detailed examination of the differing accounts presented in the more conservative / right wing papers, *Yomiuri shinbun* and *Sankei shinbun*, and the more progressive outlets, *Asahi shinbun* and *Mainichi shinbun*. WEISS delineates claims that KAN's visit to the Fukushima plant on 12 March—one day after the tsunami struck—was portrayed by certain parties as having distracted the plant's staff, leading to delays in appropriate management of the crisis. In turn, this version of events absolves TEPCO of its responsibility in terms of crucial decisions in crisis management, and with regards to irregularities in the chain of communication.

Anna WIEMANN provides a longitudinal narrative case study of the memory of the Fukushima nuclear disaster among farmers in the Fukushima Prefecture. She draws on the archive of the "Voices from Tohoku" (2012 and 2013, administered by David SLATER) and on her own interviews with such farmers that were willing to give a second interview ten years later. WIEMANN's analysis delineates the farmers' reactions to the disaster as it unfolded, their technical understanding thereof, and the changes to their views in relation to the various ways it has impacted their daily lives. She also investigates which social frames were used by the farmers, how they positioned themselves against the broader landscape of public opinion, and the broader shifts to the social framing of the disaster. Taking two regional farmers as her example, WIEMANN traces the struggle to grasp the invisible threat of the radioactive nuclides and portrays the economic disruption brought to farmers in the region by nuclear contamination. As WIEMANN shows, the events surrounding the disaster led to significant mistrust between farmers—as producers—and the consumer of their products. Fukushima was, as such, an essentially social phenomenon for many of the farmers concerned. For some, criticizing the government over Fukushima was an unbreakable taboo and the disaster was, thus, a cause for complete resignation. For others, the events of 2011 led to profound disillusionment with consumerism, with global politics, and with society itself.

In lieu of an epilogue, Stephan KÖHN traces the ways in which HAYASHI Kyōko transcended the nuclear fallacy in her later works. Starting with an examination of her 2002 short novel "Harvest" (*Shūkaku* 収穫), which examines the nuclear accident at the Tōkaimura Nuclear Fuel Processing Plant, KÖHN illustrates HAYASHI's trajectory from her beginnings as a *hibakusha* writer focused on August 9, 1945, to her role in unravelling the myth of nuclear safety as well as her work in demonstrating the fallacious nature of the discursive splitting of the atom into 'good' and 'evil'. HAYASHI's writing was strongly criticized as clumsy and unliterary at the time, despite the fact that she was awarded the 73rd Akutagawa Prize for her work "Festival Ground" (*Matsuri no ba* 祭りの場) in 1975. However, as Rachel DiNitto's paper shows, HAYASHI's writing and particularly her shift in thinking on nuclear issues profoundly influenced TSUSHIMA Yūko and many other writers, helping them to transcend the nuclear dispositive. In 1999, HAYASHI visited the Trinity Test site in New Mexico and came to perceive of herself as something quite different to the almost exclusively white visitors and staff, namely a Japanese *hibakusha*. HAYASHI's "From Trinity to Trinity" speaks to her identification not as a solitary *hibakusha* but as a *hibakusha* in a global context who had recognized July 16, 1945, as the first detonation of an atomic bomb.

HAYASHI continued to highlight instances where this distinction between the 'good' and 'evil' sides of the nuclear was at play. Her last literary piece, "To Rui, Once Again", for example, draws a link between August 9, 1945, and March 11, 2011, thereby unmasking the Fukushima nuclear disaster as man-made rather than as a unforeseeable "accident".

As the collected studies in this volume aim to show, unraveling the nuclear dispositive that has dominated postwar Japan is of the upmost importance. Doing so not only illustrates that the atomic bomb and nuclear energy are but two sides of the same coin, it also shows that this discursive construct transcends national boundaries. Global nuclear colonialism creates global *hibakusha*, and not just as a result of bombs and bomb tests. It does so by contaminating the workforce it deems disposable during both the extraction and processing of nuclear fuel. And it does so through the ill-executed decommissioning of power plants and clean-ups of "unforeseeable accidents". As a consequence of these actions, many future generations will find themselves not only exposed to nuclear contamination but faced with the terrible conundrum as to what precisely should be done with the stockpiles of nuclear waste left by their forebearers.

NOTES

1 Owing to the unusual publication process of this short story, it is not surprising that "Riverbank" is included neither in the four-volume, "Complete Work of Ōta Yōko" (*Ōta Yōko shū* 大田洋子集), published by San'ichi shobō in 1982, nor is it thematized in any other research work on ŌTA Yōko. For a translation of this work in German and a historical contextualization of its publication, see ŌTA (2022).
2 The Press Code is available at the website of the Gordon W. Prange Collection / University of Maryland; https://www.lib.umd.edu/sites/default/files/2022-03/censor-docs_presscode.jpg (last access 2023 / 08 / 15).

3 The figures of the employees largely differ in previous research. YAMAMOTO, who based his study on a thorough examination of the CCD's rosters, gives a total number of 13,946 employees for the CCD. See YAMAMOTO 2021: 10.
4 According to NISHINA Yoshio, nuclear scientist and leader of the Japanese atomic bomb program during war, the first time he heard of the atomic bombing of Hiroshima was on August 7. The next day, he arrived in Hiroshima, measured the levels of radiation in the city and confirmed that the bomb used in Hiroshima was indeed atomic. See NISHINA 1999 [1946]: 24–26.
5 A good example for this kind of censorship is ŌTA's "An Unfathomable Deep Light", which was "republished" the next day, on August 31, 1945, in the *Asahi shinbun* (Osaka edition) under the title "Suffered under the Atomic Bomb" (*Genshi bakudan o abite* 原子爆弾を浴びて). A comparison of the two texts reveals that passages of ŌTA's manuscript were obviously removed from the version that was published in the Tokyo edition. The removed passages contain, for example, depictions of melting iron bridges, burning freight trains, and mass panic among survivors—in other words, depictions that would have intensified the horrors caused by the enemy's atomic bomb. It seems to be very likely that the Tokyo headquarter of the *Asahi shinbun* was more strictly controlled by the Cabinet Intelligence Bureau (also based in Tokyo) than the Osaka branch of this newspaper.
6 Even in Japan, the question of the emperor's war guilt was controversially discussed in the leading newspapers of that time. In an article titled "Opinions regarding the emperor system from all over the world" (*Tennōsei ni kan suru sekai no koe* 天皇制に關する世界の聲), published in the *Asahi shinbun* on November 5, 1945, the Japanese Communist Party is cited as follows: "If the Americans do not put the emperor, as the worst of all war criminals, on public trial, surely the Japanese people will drag him to the courts" (p. 1).
7 For the effects of this policy change on the implementation of early "atomic bomb education" / "peace education", see KÖHN (2022).
8 This development stands in marked contrast to the initial impulse of the Allied forces to lay out the historical truth about the Fifteen Years War. For that very reason, the Allied Forces took the trouble to launch the series "History of the Pacific War" (*Taiheiyō sensō shi* 太平洋戦争史) in the *Asahi shinbun* at a very early stage of the occupation period, namely from December 8 to December 17, 1945.
9 Intellectuals in Japan were not really averse to the promises of nuclear power plants as the case of leftist author and literary critic, NOMA Hiroshi 野間宏 (1915–91), illustrates. For NOMA, the simultaneity of the American hydrogen bomb tests in the Bikini Atoll and the inauguration of the first nuclear power plant in the Soviet Union 1954 revealed the fundamental difference between the U.S. and Soviet Union in dealing with the nuclear contradiction (that is "good" vs "bad" energy) stating: "It has become quite obvious that the Americans were unable to resolve the contradiction of nuclear energy by finding a way that would lead humankind to unlimited happiness. The Soviets, however, understood this contradiction and showed the world that they finally found a way to resolve it" (NOMA 1983 [1954]: 152).

10 The touring exhibition, "Atoms for Peace", came to Japan in November 1955 via various European countries, India, Pakistan, and Brazil.

11 Interestingly enough, the exhibition had its greatest impact after its conclusion. Given that the Hiroshima Peace Memorial Museum was the final stop on its itinerary, some of the exhibits were subsequently kept at the museum instead of being shipped back to the United States. They became part of the permanent exhibition, which was reinstalled after "Atoms for Peace", for more than one decade, until they were finally removed from the museum in 1967. These exhibits "complemented" the collection and the information on nuclear science, showing now its use in war and its future economic potential. In doing so, the exhibits (and the museum) reinforced the dichotomy of the "good" and the "evil" atom.

12 For details on the agreement and its effects on domestic politics in Japan, see YAMAZAKI 2009: 15–18.

13 According to TANAKA Toshiyuki (2011: 28–39), the nuclear power plant proposal and the exhibition for peaceful use of nuclear energy were, in point of fact, parts of a two-step strategy launched by the American government to break the dam of resistance in Japanese society. What is striking in TANAKA's depiction is the fact that, more or less right from beginning, very few survivors of the atomic bombings were particularly supportive of the peaceful use of an energy that had only ten years before destroyed their hometowns and ruined their health.

14 Although the cities of Hiroshima and Nagasaki shared a similar fate, their relationship was characterized by competition rather than by cooperation. This became particularly apparent in the year 1949, when both cities tried to obtain funding from the government to reconstruct the devastated areas in their cities. Since the city of Hiroshima submitted single-handedly the Hiroshima Peace Memorial City Construction Act (Hiroshima heiwa kinen toshi kensetsu hō 広島平和記念都市建設法) to the national diet, the city of Nagasaki had no other choice than to submit a slightly different kind of bill: the Nagasaki International Culture City Construction Act (Nagasaki kokusai bunka toshi kensetsu hō 長崎国際文化都市建設法). Formally, Nagasaki is the city of "international culture" but, in reality, Nagasaki has equally reinvented itself as a city of peace. See FUKUMA 2011: 241–46; and DIEHL 2018: 29–35.

15 As a sidenote, it is worth mentioning that two people, who had supported the former war regime with their works, were involved in the construction of those sacred places in Hiroshima and Nagasaki. These were 1) architect, TANGE Kenzō 丹下健三 (1913–2005), who was responsible for the conception of the Hiroshima Peace Park and had come up with the design for the "Sacred place for the loyal souls that had helped to build the Greater East-Asia Co-Prosperity Sphere (Dai Tōa kensetsu chūrei shin'iki 大東亜建設忠霊神域) in 1942, and 2) sculptor, KITAMURA Seibō 北村西望 (1884–1987), who was responsible for the Nagasaki Peace Statue (Nagasaki heiwa kinen zō 長崎平和記念像) and who had created numerous militaristic, heroic sculptures during the war period. See ODA 2018:74–75; and NORO / NAKAZATO / YAMADA 1974: 98–99.

16 As early as 1948, literary critic HANADA Kiyoteru stated in his visionary essay "Guilt and punishment" (Tsumi to batsu 罪と罰) about the lacking feeling of guilt among his

contemporaries: "The relationship between guilt and punishment is extremely delicate. To put it bluntly, if guilt is not perceived as guilt, then there is no need to make the respective person realize that he is guilty by giving him a sort of punishment. Because the guilt is already their punishment. In the end, all people that are guilty think they will ascend to heaven, but will finally end up in hell. Even if they realize that this is not heaven but hell, they will never admit that their guilt is real. Instead, they will make excuses and try to burden someone else with this guilt. [...] They will lament their misfortune and deny that they must blame themselves for anything" (1976 [1948]: 286).

17 For poet KURIHARA Sadako (1978: 39–47), the emperor's transformation from a warmonger to a creator of peace by way of his famous "Humanity Declaration" (*ningen sengen* 人間宣言) on January 1, 1946, was the crucial building block for a "system of irresponsibility" (*musekinin taisei* 無責任体制) that became a pillar of postwar Japanese society's identity and historical awareness.

18 According to YOSHIMOTO Takaaki (1972: 385–87), the term "conversion" signified not only the process of radically changing one's political conviction (for example, Marxism or Communism) for fear of political persecution, but also a sudden kind of sympathy and even curiosity towards the oppressive power structures during wartime.

19 This is one of the reasons why, for example, eyewitnesses (*shōgensha* 証言者) of the atomic bombings long hesitated to give personal accounts of what happened on August 6 and 9, respectively. They feared becoming a scapegoat for a postwar society that had chosen to forget rather than remembering the historical events that had, ultimately, led to Japan's surrender.

20 According to HOSHINO Yasusaburō (1969: 92–93) and ŌSUKA Akira (1969: 98–100), for the Japanese government the screening system was the perfect instrument to conceal Japan's war guilt and war responsibility. It distorted the ideals of peace as represented by the Japanese constitution. For an analysis as to what extent articulations of the nuclear were removed from textbooks from the mid-1950s onwards, see the analyses in KÖHN (2021a; 2022).

21 The so-called "nose bleeding problem" (*hanaji mondai* 鼻血問題) of 2013 might illustrate this best. The two-volume special edition "The Truth about Fukushima" (*Fukushima no shinjitsu* 福島の真実) of the bestselling manga "The Foodie" (*Oishinbo* 美味しんぼ; 1983–2014)—written by KARIYA Tetsu 雁屋哲 and HANASAKI Akira 花咲アキラ—was grounds for serious contestation. In particular, this was to the protagonist Yamaoka Shirō stating, after his on-site inspection of the disaster area, that radioactivity remains the biggest problem for people. Under the pretext of "harmful rumors", Prime Minister ABE Shinzō expressed his "deepest regret" concerning this "inappropriate expression", which, he claimed, damaged the reputation of the people of Fukushima. As a result, the publisher bowed to political pressure and stopped the serialization of this long-running manga. For details, see KÖHN 2021b: 192–94.

References

AGAMBEN, Giorgio (2008): *Was ist ein Dispositiv*. Zürich / Berlin: diaphanes.
ANDERS, Günther (1995 [1959]): *Hiroshima ist überall* (Beck'sche Reihe 1112). München: Verlag C. H. Beck.
ANONYMOUS (1945): "Tennōsei ni kan suru sekai no koe 天皇制に關する世界の聲". In: *Asahi shinbun* 朝日新聞, 11 / 05: 1.
ARIYAMA, Teruo 有山輝雄 (2003): "Sengo Nihon ni okeru rekishi, kioku, media 戦後日本における歴史・記憶・メディア". In: *Mediashi kenkyū* メディア史研究 14: 1–26.
BRAW, Monica (1986): *The Atomic Bomb Suppressed. American Censorship in Japan, 1945–1949* (Lund Studies in International History 23). Malmö: Liber Förlag.
BÜHRMANN, Andrea D. / SCHNEIDER, Werner (2008): *Vom Diskurs zum Dispositiv. Eine Einführung in die Dispositivanalyse*. Bielefeld: transcript.
BURCHETT, Wilfred (2008 [1945]): "The Atomic Plague". In: BURCHETT, George (ed.): *Rebel Journalism. The Writings of Wilfred Burchett*. Cambridge: Cambridge University Press, 1–5 (originally published in: *The Daily Press*, 1945 / 09 / 05).
DIEHL, Chad R. (2018): *Resurrecting Nagasaki. Reconstruction and the Formation of Atomic Narratives*. Ithaca, London: Cornell University Press.
DOWER, John (1993): "Peace and Democracy in Two Systems". In: DOWER, John (ed.): *Post-war Japan as History*. Berkeley: University of California Press, 3–33.
ETŌ, Jun 江藤淳 (1994): *Tozasareta gengo kūkan: Senryōgun no ken'etsu to sengo Nihon* 閉された言語空間：占領軍の検閲と戦後日本 (Bunshun bunko 文春文庫). Tōkyō: Bungei shunjū.
ETŌ, Jun 江藤淳 (2015): *1946-nen kenpō: Sono kōsoku* 一九四六年憲法：その拘束 (Bunshun gakugei raiburarī shisō 文春学芸ライブラリー思想 13). Tōkyō: Bungei shunjū.
FOUCAULT, Michel (1978): *Dispositive der Macht. Über Sexualität, Wissen und Wahrheit*. Berlin: Merve Verlag.
FUKUMA, Yoshiaki 福間良明 (2011): *Shōdo no kioku: Okinawa, Hiroshima, Nagasaki ni utsuru sengo* 焦土の記憶：沖縄・広島・長崎に映る戦後. Tōkyō: Shin'yōsha 2011.
FUKUMA, Yoshiaki 福間良明 (2015): *"Senseki" no sengoshi: Semegiau ikō to monyumento* 「戦跡」の戦後史：せめぎあう遺構とモニュメント (Iwanami gendai zensho 岩波現代全書 072). Tōkyō: Iwanami shoten.

HANADA, Kiyoteru 花田清輝 (1976 [1948]): *Hanada Kiyoteru chosaku shū* 花田清輝著作集, vol. II. 5th ed. Tōkyō: Miraisha, 286–95 (originally published in: *Shakai* 社会, 1948 / 12).

HATOYAMA, Ichirō 鳩山一郎 (1945): "Shintō kessei no kōsō 新黨結成の構想 1". In: *Asahi shinbun* 朝日新聞 09 / 15: 2.

HIRANO, Kyōko (1992): *Mr. Smith goes to Tokyo. Japanese Cinema under the American Occupation, 1945–1952*. Washington / London: Smithsonian Institution Press.

HIRANO, Kyōko 平野共余子 (1998): *Tennō to seppun: Amerika senryōka no Nihon eiga ken'etsu* 天皇と接吻：アメリカ占領下の日本映画検閲. Tōkyō: Sōshisha.

HOSHINO, Yasusaburō 星野安三郎 (1969): "Heiwa kyōiku to kentei no kinō 平和教育と検定の機能". In: *Hōritsu jihō* 法律時報 41 (10): 89–94.

ICHIKAWA, Hiroshi (2016): "Obninks, 1955: The World's First Nuclear Power Plant and 'The Atomic Diplomacy' by Soviet Scientists". In: *Historia Scientiarum* 26 (1): 25–41.

INOUE, Hisashi 井上ひさし et al. (2004): "Genbaku bungaku to Okinawa bungaku 原爆文学と沖縄文学". In: INOUE, Hisashi 井上ひさし / KOMORI, Yōichi 小森陽一 (eds.): *Zadankai: Shōwa bungaku shi* 座談会：昭和文学史. Tōkyō: Shūeisha, 9–105.

ISHIDA, Takeshi 石田雄 (2000): *Kioku to bōkyaku no seijigaku* 記憶と忘却の政治学 (Akashi raiburarī 明石ライブラリー 23). Tōkyō: Akashi shoten.

KATŌ, Norihiro 加藤典洋 (1995): *Amerika no kage: Sengo saiken* アメリカの影：戦後再見 (Kōdansha gakujutsu bunko 講談社学術文庫). Tōkyō: Kōdansha.

KÖHN, Stephan (2021a): "How to Teach Peace? On the Difficulties of Implementing Peace Education (*heiwa kyōiku*) in Early Post-war Japan". In: *Bochumer Jahrbuch zur Ostasienforschung* 44: 107–32.

KÖHN, Stephan (2021b): "Questioning the politics of popular culture: Tatsuta Kazuto's manga 1F and the national discourse on 3 / 11". In: ROSENBAUM, Roman (ed.): *The Representation of Japanese Politics in Manga. The Visual Literacy of Statecraft*. New York: Routledge, 183–202.

KÖHN, Stephan (2022): "How to Fill the Void in National History: Japanese Peace Education at the Crossroads in the 1970s". In: *Japonica Humboldtiana* 24: 167–91.

KOJIMA, Noboru 小島襄 (1991): *Sensō sekinin to tennō* 戦争責任と天皇 (Bunshun bunko 文春文庫). Tōkyō: Bungei shunjū.

KURIHARA, Sadako 栗原貞子 (1978): *Kaku, tennō, hibakusha* 核・天皇・被爆者. Tōkyō: San'ichi shobō.

KUZNICK, Peter / TANAKA, Yuki (2011): "Japan, the Atomic Bomb, and the 'Peaceful Uses of Nuclear Power'". In: *The Asia-Pacific Journal: Japan*

Focus 9 (18.1); https://apjjf.org/2011/9/18/Yuki-Tanaka/3521/article.html (last access 2023 / 08 / 20).

MATSUI, Eiichi 松井栄一 (1969): "Rekishi kyōiku to kyōkasho kentei 歴史教育と教科書検定". In: *Hōritsu jihō* 法律時報 41 (10): 17–23.

MATSUMOTO, Hiroshi 松元寛 (1995): *Hiroshima to iu shisō* ヒロシマという思想. Tōkyō: Tōkyō sōgensha.

MATSUURA, Sōzō 松浦総三 (1974): *Zōho ketteiban: Senryōka no genron dan'atsu* 増補決定版：占領下の言論弾圧. Tōkyō: Gendai jānarizumu shuppankai.

NAGAOKA, Hiroyoshi 長岡弘芳 (1982): *Genbaku bunken o yomu* 原爆文献を読む. Tōkyō: San'ichi shobō.

NISHII, Marina 西井麻里奈 (2020): *Hiroshima: Fukkō no sengoshi* 広島：復興の戦後史. Kyōto: Jinbun shoin.

NISHINA, Yoshio 仁科芳雄 (1999 [1946]): "Genshi bakudan 原子爆弾". In: *Nihon genbakuron taikei* 日本原爆論大系, vol. 1. Tōkyō: Nihon tosho sentā 1999, 23–38 (originally published in: *Sekai* 世界, 1946 / 03).

NOMA, Hiroshi 野間宏 (1983 [1954]): "Suibaku to ningen 水爆と人間". In: *Nihon no genbaku bungaku* 日本の原爆文学, vol. 15. Tōkyō: Horupu shuppan, 149–53.

NORO, Kuninobu 野呂邦暢 / NAKAZATO, Kishō 中里喜昭 / YAMADA, Kan 山田かん (1974): "Zadankai: Bungaku to genbaku taiken 座談会：文学と原爆体験". In: *Minshu bungaku* 民主文学 107: 84–105.

ODA, Tomoharu 小田智敏 (2018): "Gunto = gakuto to shite no Hiroshima 軍都＝学都としての広島". In: HIGASHI, Takuma 東琢磨 / KAWAMOTO, Takashi 川本隆史 / SENBA, Nozomu 仙波希望 (eds.): *Bōkyaku no kioku: Hiroshima* 忘却の記憶：広島. Tōkyō: Getsuyōsha, 74–96.

ŌSUKA, Akira 大須賀明 (1969): "Kentei kijun to shite no gakushū shidō yōryō 検定基準としての学習指導要領". In: *Hōritsu jihō* 法律時報 41 (10): 95–100.

ŌTA, Yōko 大田洋子 (1945a): "Kaitei no yō na hikari 海底のやうな光". In: *Asahi shinbun* 朝日新聞 (Tokyo edition), 08 / 30: 2.

ŌTA, Yōko 大田洋子 (1945b): "Genshi bakudan wo abite 原子爆弾を浴びて". In: *Asahi shinbun* 朝日新聞 (Osaka edition), 08 / 31: 2.

ŌTA, Yōko 大田洋子 (1949): "Hachigatsu muika hachiji jūgofun 8月6日8時１５分". In: *Kaizō* 改造 30 (8): 42–49.

ŌTA, Yōko 大田洋子 (1982 [1954]): "Han ningen 半人間". In: ŌTA, Yōko 大田洋子: *Ōta Yōko shū* 大田洋子集, vol. 1. Tōkyō: San'ichi shobō, 261–334.

ŌTA, Yōko 大田洋子 (2022 [1948]): "Das Flussufer" [*Kawara* 河原, 1948]. Translated by Stephan KÖHN et al. In: *Hefte für Ostasiatische Literatur* 73: 64–92.

SASAMOTO, Yukuo 笹本征男 (1999): "Genbaku hōdō to puresu kōdo 原爆報道とプレス・コード". In: *Nihon genbakuron taikei* 日本原爆論大系, vol. 1. Tōkyō: Nihon tosho sentā, 455–500 (originally published in: *Tsūshi: Nihon no kagaku gijutsu* 通史：日本の科学技術 1).

SHIGESAWA, Atsuko 繁沢敦子 (2010): *Genbaku to ken'etsu: Amerikajin kishatachi ga mita Hiroshima, Nagasaki* 原爆と検閲：アメリカ人記者たちが見た広島・長崎 (Chūkō shinsho 中公新書). Tōkyō: Chūō kōron shinsha.

SODEI, Rinjirō 袖井林次郎 (1975): "Genbaku wa ika ni hōdō sareta ka 原爆はいかに報道されたか". In: *Genbaku taiken o tsutaeru kai* 原爆体験を伝える会 (ed.): *Genbaku kara genpatsu made: Kaku-seminā no kiroku* 原爆から原発まで：核セミナーの記録, vol. 1. Tōkyō: Agune, 266–76.

TAKAHASHI, Tetsuya 高橋哲哉 (2012): *Gisei no shisutemu: Fukushima, Okinawa* 犠牲のシステム：福島・沖縄 (Shūeisha shinsho 集英社新書). Tōkyō: Shūeisha.

TAKAKUWA, Kōkichi 高桑幸吉 (1984): *Makkāsā no shinbun ken'etsu* マッカーサーの新聞検閲. Tōkyō: Yomiuri shinbunsha.

TANAKA, Toshiyuki 田中利幸 (2011): "'Genshiryoku heiwa riyō' to Hiroshima 「原子力平和利用」とヒロシマ". In: TANAKA, Toshiyuki 田中利幸 / KUZNICK, Peter ピーター・カズニック: *Genpatsu to Hiroshima* 原発とヒロシマ (Iwanami bukkuretto 岩波ブックレット 819). Tōkyō: Iwanami shoten, 23–58.

TSURUMI, Shunsuke 鶴見俊輔 (1959): "Sensō sekinin no mondai 戦争責任の問題". In: *Shisō no kagaku* 思想の科学 1 (1): 79–87.

YAMAMOTO, Taketoshi 山本武利 (2021): *Ken'etsukan: Hakken sareta GHQ meibo* 検閲官：発見されたGHQ名簿 (Shinchō shinsho 新潮新書). Tōkyō: Shinchōsha.

YAMAZAKI, Masakatsu 山崎正勝 (2009): "Nihon ni okeru 'Heiwa no tame no genshi' seisaku no tenkai 日本における「平和のための原子」政策の展開". In: *Kagakushi kenkyū* 科学史研究 48: 11–20.

YOSHIMI, Shun'ya 吉見俊哉 (2012): *Yume no genshiryoku* 夢の原子力 (Chikuma shinsho ちくま新書 971). Tōkyō: Chikuma shobō.

YOSHIMOTO, Takaaki 吉本隆明 (1972): "Tenkōron 転向論". In: *Gendai no bungaku* 現代の文学, vol. 25 (Yoshimoto Takaaki 吉本隆明). Tōkyō: Kōdansha, 384–99.

ZWIGENBERG, Ran (2012): "'The Coming of a Second Sun': The 1956 Atoms for Peace Exhibit in Hiroshima and Japan's Embrace of Nuclear Power". In: *The Asia-Pacific Journal: Japan Focus* 10 (6.1); https://apjjf.org/2012/10/6/Ran-Zwigenberg/3685/article.html (last access 2023 / 08 / 20).

ZWIGENBERG, Ran (2014): *Hiroshima: The Origins of Global Memory Culture*. Cambridge: Cambridge University Press.

From John Hersey's Hiroshima to Fukushima: General Leslie R. Groves' Long Shadow over Nuclear Narratives

Atsuko Shigesawa

1 Introduction

From the outset of the accident at the Tokyo Electric Power Company (TEPCO) Fukushima Daiichi Nuclear Power Station in March 2011, the Japanese government has made light of the effects of radiation, saying that "there are no immediate health effects" and it "does not concern us" (Prime Minister of Japan and His Cabinet 2011; Edano 2011). Those statements do not appear to be entirely true, however. At least two workers at the power plant have died of cancer. Six others have been adversely affected by leukemia and cancer (Murakami 2018; Hashimoto 2021). They were exposed to radiation after the accident and their claims for workers' compensation have been approved by the Ministry of Health, Labor and Welfare. On the other hand, six individuals with thyroid cancer, all between the ages of six and 16 at the time of the accident, filed a lawsuit against TEPCO in January 2022. They argue that the cancer resulted from radioactive substances from the nuclear disaster.[1] Possible effects of radiation from the plant on nearby residents remain highly controversial.

The same is true for the 1986 Chernobyl accident. According to Kate Brown (2019: 308–09), the Soviet government and international organizations have underestimated the accident's human toll. The UN Scientific Committee on the Effects of Atomic Radiation (UNSCEAR) acknowledged in 2008 (2011: 64–65) that 28 plant staff and emergency workers had died of acute radiation sickness because of the accident, and that contaminated milk led to more than 6,000 cases of thyroid cancer among children, of which 15 were fatal. However, "there has been no persuasive evidence of any other health effects in the general population that can be attributed to radiation exposure" (UNSCEAR 2011: 65). Brown (2019: 310) asserts that 35,000 to 150,000 people have died from Chernobyl radiation in Ukraine alone.

Brown (2019: 33–35) attributes this disparity to the Life Span Study of atomic bomb survivors in Japan that has been used as "the gold standard" for international radiation dose assessment. The Life Span Study is based almost solely on direct exposure to a large amount of initial radiation, but not to residual radiation. Studies conducted by UN agencies have not considered the effects of long-term low-dose exposure, such as internal exposures through the food chain, a major exposure factor at Chernobyl.

In Japan, too, the Life Span Study has been used for years as a dosimetry standard to recognize survivors of the atomic attacks who are eligible for state health care benefits. Therefore, people who were exposed to radioactive black rain outside a state-designated area, or people who have had health problems attributable to internal radiation exposure, or people who entered the cities immediately after the bombings and were exposed thereafter to residual radiation, had not qualified for the benefits.

Over the last two decades, however, the system has shown faults. Many plaintiffs in class action lawsuits have won their cases. The Japanese government chose not to appeal a 2021 high court ruling, which found that all plaintiffs—who had been exposed to the black rain outside a state-designated area—qualified for the benefits, supporting a landmark 2020 lower court ruling (SAITŌ 2018: 113–33; ŌTAKI 2020: 2–7; *Asahi shinbun*, 2021 / 07 / 27).

Earlier studies have indicated that the root of this problem lies in suppressing related information and underestimating damage caused by the 1945 atomic bombings (see HOOK 1991, TAKAHASHI 2008, and BRODIE 2015). After the war, the Manhattan Project leader, Brigadier General Leslie R. GROVES, took charge of protecting national security with regard to the atomic bomb and actively sought to control information related to the bomb's radiation. GROVES did, however, review and subsequently permit the publication of a now world-famous report written by the Pulitzer Prize-winning journalist John HERSEY, titled *Hiroshima*. Why did GROVES, so motivated to cover up the most hideous results of the bombings, not attempt to suppress an account of nuclear horror?

Hiroshima is most certainly a masterpiece of atomic bomb literature. The story of the survival of six *hibakusha* gives us a glimpse of nuclear war, and what people are (in)capable of in the face of such a crisis. Contrary to the popular belief that it "described the full range of the bomb's effects" (LIFTON / MITCHELL 1994: 89–90), however, it barely contradicted the official line at all. Those effects of the bomb it described had already been laid out in *The Effects of Atomic Bombs on Hiroshima and Nagasaki* by the United States Strategic Bombing Survey (USSBS), which was published as an official U.S. report in June 1946.[2] *Hiroshima* gave very little attention to what was considered taboo at that time, namely: residual radiation. This aspect has been little examined in earlier studies, which often reproduced what I call the *Hiroshima* "myth". But would HERSEY have tried to describe the full range of the bomb's effects if it was not for GROVES' censorship? Did GROVES have something to do with this negligence of the most troubling features of the new weapon?

This paper examines how GROVES' censorship[3] of HERSEY's *Hiroshima* changed the article. The results demonstrate that it did not change much—it was already in line with the official report when the galley proofs were submitted for censorship clearance. Yet evidence displays traces of efforts on the part of HERSEY and *The New Yorker* editors to tread a narrow path whereby they could relate as many facts as possible without crossing the official line. But this "self-censorship" would not have been necessary were it not for GROVES' tight grip. Examination of the galley proofs also implies that GROVES had his own agenda to bring the manuscript more into line with his version of the atomic bomb narrative.

Both Chernobyl, which occurred 40 years after its publication, and Fukushima, 65 years later, as well as the responses of governments and international organizations to these accidents, all took place in the long shadow cast by the atomic general. HERSEY achieved the incomparable accomplishment of bringing home the story of the nuclear war, but what the account did and did not include must be elucidated to characterize the nuclear threat to our world today.

2 HERSEY AND THE BIRTH OF THE *HIROSHIMA* "MYTH"

Hiroshima is one of the best-known documentations of the atomic bombings. As YAVENDITTI (1974: 33) points out, HERSEY "recreated the entire experience of atomic bombing from the victims' point of view", which had not been done before. When it appeared exclusively in the August 31, 1946, issue of *The New Yorker*, it provoked a huge sensation in the United States and abroad (1974: 31–32). I do not discuss the repercussions here, arguing instead against the "myth" according to which HERSEY "described the full range of the bomb's effects" or exposed the effects of radiation that the U.S. government was set to conceal (YAVENDITTI 1974: 38; SHARP 2007: 139; FORDE 2011: 565; BLUME 2020: 2, 102–04).

The widely held belief that the article was published without being subjected to censorship has also supported the "myth" (YAVENDITTI 1974: 36; SHARP, 2007: 139; FORDE 2011: 566–67; SHIBATA 2012: 123). Reports that espouse such a view often argue that *Hiroshima* cast ethical doubts about the atomic attacks (LIFTON / MITCHELL 1994: 88–92, 113; SHARP 2007: 149; FORD 2011: 568; BLUME 2020: 114). Patrick B. SHARP, for example, notes that HERSEY "criticized the widely held view that the atomic bomb was a justified, science-fiction-style attack against an evil and militaristic Yellow Peril" (2000: 434) and "challenged the official representation of the atomic attacks" (2007: 139).

The probable origin of the "myth" is an article by YAVENDITTI who interviewed HERSEY and wrote that he "did not submit his material to the U. S. government for censorship clearance" (1974: 36). *The New Yorker* papers at the New York Public Library include a document that supports this statement. In response to an inquiry from Harold ROSS, the magazine's founder and editor, as to whether it should be submitted for "censorship clearance", the magazine's legal counsel reviewed the provisions of the newly enacted Atomic Energy Act and subsequently wrote: "I do not think there is any 'restricted data' in the Hersey articles [...]. We are not publishing anything in the articles 'with intent to injure the United States [...] or to secure an advantage to any foreign nation'" (Letter to Milton GREENSTEIN from H. W. ROSS, 1946/08/01; Letter to ROSS from GREENSTEIN, 1946/08/12).

The papers include a cable from HERSEY while he was in China, telling William SHAWN, the magazine's editor, that it would be "more advantageous" to write the article in the United States rather than in Japan (Cabled message from HERSEY to SHAWN, 1946/03/22). HERSEY was aware of the notorious censorship enacted by the occupation forces in Japan. He was, after all, seeking to write a story the likes of which had not previously been published.

All these facts apparently contributed to the establishment of the "myth" that HERSEY *exposed* the bomb's effects in *full*. More recent studies have tended to emphasize the fact that a myth is generally reinforced over time. In particular, BLUME's *Fallout*, which was published in 2020, lionizes HERSEY by equating the scale of the repercussions of *Hiroshima* with the assumed extent of the exposure of information. This amounts to a simplistic confrontation between a journalist delivering truth versus a government authority suppressing it. It does not, however, explain the prevailing circumstances in the country.

During World War II, the Office of Censorship (OC), a wartime agency, imposed censorship of the American press to safeguard information that could benefit enemy countries. The soft-spoken approach that the office took to enlist cooperation from the press created an ideal relationship between the two parties, rather than an antagonistic one.[4] Around the time of the first nuclear testing, the Army's Manhattan Engineer District (MED) took charge of its publicity. The American media in general accepted this new phase of information control (WELLERSTEIN 2021: 51–131) and *The New Yorker* was no exception. For example, GROVES reviewed and subsequently cleared the manuscripts for a series of articles on nuclear facilities written by Daniel LANG, another *New Yorker* writer.[5]

In a paper of mine that was published in 2012, I had, in fact, followed suit and concluded that *Hiroshima* had escaped censorship (SHIGESAWA 2012: 23). However, subsequent research led me to *Hiroshima*'s galley proofs with GROVES' written notes.[6]

The publication of *Hiroshima* was top secret, even within *The New Yorker* (KUNKEL 2000: 294). However, the Army knew that HERSEY had visited Hiroshima to write a story: any individual movements in occupied Japan required the permission of the military (Invitational Travel Orders, 1946/05/21 & 24). As shall be explained below, GROVES was interested in having a story on the devastated cities published in the United States. When *Hiroshima* was in the final stage of editing in early August 1946, *The New Yorker* editors submitted to GROVES the proofs of an article written by LANG (Letter to GROVES from SHAWN, 1946/08/06). No record of GROVES' request for galley proofs of HERSEY's *Hiroshima* has been found, but such a request was probably made during communications about LANG's story. Considering LANG's access to the sources of his stories, it is not difficult to imagine that the editors had no choice in handing the proofs over to GROVE. The galley proofs of HERSEY's *Hiroshima* were submitted to GROVES' office on August 15, 1946 (HERSEY 1946b).

HERSEY himself knew that his story had been subjected to the Army's censorship clearance process (Letter to Jay CASSINO, 1947/01/10). That HERSEY would have grown increasingly troubled by this would hardly be surprising however. YAVENDITTI's interview took place on September 19, 1967, when the anti-Vietnam war movement was at its height. The media were serving as a watchdog over the government (KNIGHTLEY 1975: 390–426; SWEENEY 2006: 121–49). It would be understandable if HERSEY thought the fact would undermine the work's reputation and did not want it come out. Ironically, the writer himself would thereby take part in creating the "myth" at hand.

3 CENSORSHIP OF *HIROSHIMA*

The galley proofs of *Hiroshima*—then titled "Some Events at Hiroshima"—consist of 62 single-column pages. GROVES' comments are marked on nine paragraphs. In the following, I examine GROVES' comments in each paragraph and compare the texts with those in the published article and in the book edition.

> Account 1: They were actually drops of *condensation* falling from the turbulent tower of dust, heat, and fission fragments that had already risen miles into the sky above Hiroshima.[7] (my emphasis)

In the right margin, GROVES wrote "I doubt truth of this". The word "condensation" was then changed to "condensed moisture" in the published article (1946c: 20; 1989: 18). The original word could imply black rain that contains radioactive fallout, but it was changed to a phrase that more or less implies normal "rain".

> Account 2: But the drops were palpably water, and as they fell, the wind grew stronger and stronger, and suddenly, probably because of the tremendous convection *currents* set up by the blazing city—a *tornado* ripped through the park. (my emphasis)

GROVES' comment is "no basis for this". The word "currents" was deleted and "tornado" was changed to "whirlwind"[8] (1946c: 25; 1989: 39). These changes downplay the effects of the atomic bomb. In the aftermath of the atomic bombing, firestorms occurred in Hiroshima (GLASSTONE / DOLAN 1977: 304). According to Lynn EDEN (2004: 37–42), the U. S. armed forces generally tend to treat nuclear weapons as blast rather than fire weapons—GROVES' comment demonstrates that he was no exception.

> Account 3: The asphalt of the streets was still so soft and *hot* that walking was uncomfortable. [...] In the garden, on the way to the shelter, he [Father Kleinsorge] noticed a pumpkin roasted on the vine. He and Father Cieslik tasted it and it was good. They were surprised at their hunger; they ate quite a bit. They got out their rice and gathered up several cooked pumpkins and some potatoes which were nicely baked under the ground [...]. (my emphasis)

In the right margin with three pencil marks over these texts, GROVES wrote, "This was due to the subsequent fires, not the bomb itself". In the published article (1946c: 26; 1989: 40), "from the fires" was added after the underlined "hot". The latter part remained intact. Again, GROVES seems to want to avoid attention on the fire-related effects of the bomb.

> Account 4: That bomb had more power than *two* thousand tons of TNT. (my emphasis)

This is a simple factual error either on HERSEY's part or that of the editors. On GROVES' suggestion, the "two" was changed to "twenty" (1946c: 32; 1989: 49). In his August 6, 1945, statement, President Harry S. TRUMAN (1945: 621) stated that the atomic bomb dropped on Hiroshima had more power than 20,000 tons of TNT.

> Account 5: [... E]ven if they had known the truth, most of them were too busy or too badly hurt to care that they were the objects of the *first great experiment in the use of atomic power*, which (as the *bragging* voices on the short wave *said*) no country except the United States, with its industrial know-how, its willingness to throw two billion gold dollars into an important wartime gamble, could possibly have developed. (my emphasis)

GROVES suggested cutting "bragging" as well as the text in parenthesis through to the end, writing: "as far as I know not true. Certainly in poor facts. Recommend omission [sic]". Thereby, "bragging" was deleted (1946c: 32; 1989: 49–50), but the section in parenthesis and the text thereafter remained, with only the word "said" replaced by "shouted". In the months after the war, GROVES assumed it would take five to 20 years for the Soviet Union to develop the atomic bomb (1945a: 18). His comment might mean that he had thought otherwise and expected other countries to develop the weapons in a not-too-distant future. On the other hand, GROVES left the *highlighted* phrase—which called the bombing a "great experiment"—intact. SHARPE (2000: 447) argues that HERSEY implicitly condemned the United States for the Hiroshima attack with this passage. It seems GROVES was not at all bothered by this denunciation.

> Account 6: The doctors realized in retrospect that even though most of these dead had also suffered from burns and blast effects, they had absorbed enough radiation to kill them *anyhow*. (my emphasis)

GROVES' comment reads: "This sounds like S. B. S. propaganda. Can't it be revised, just like saying everybody dies of old age [*sic*]". By "S. B. S.", GROVES meant the USSBS, which published the above-mentioned official report. Informing this comment is GROVES' deep-seated dislike of the presidential commission, partly because it placed more importance on the radiological effects of the atomic bomb.[9] In its report, USSBS (1946a: 15) said radiation probably accounted for no less than 15% to 20% of the deaths, a figure that was higher than any other estimations made up to that point.[10] An estimate made earlier by Colonel Stafford L. WARREN, Chief of the MED's Medical Section, who led one of MED's survey teams in Japan described radiation as responsible for five to seven percent of the total deaths in the two cities (WARREN 1946: 510).[11] In the report, USSBS (1946a: 15) referred to the estimate as "far too low".

HERSEY drew much information from the USSBS report (SHIGESAWA 2012: 23–27). The account obviously came from the USSBS report, which said: "Indeed, many of these people undoubtedly died several times over, theoretically, since each was subjected to several injuries, any one of which would have been fatal" (1946a: 15). GROVES apparently caught a scent of the USSBS narrative in HERSEY's account.

The passage, however, remained almost intact, without the "anyhow" at the end (1946c: 56; 1989: 76). There are no indications as to why the rest of the text remained as it was but there would have been no point in changing it as the information had already appeared in the official report. Moreover, GROVES' comment rather suggests his dislike of the USSBS than any real desire that the text read "everybody dies of old age".

> Account 7: [... A]s if nature were protecting man against his own ingenuity, the reproductive processes were affected for a time; men became sterile, women had miscarriages, menstruation stopped.

In the blank space, GROVES wrote, "temporarily? should be inserted for truth, I think [*sic*]". The account was left in (1946c: 58; 1989: 78). No information is available as to why the editors of *The New Yorker* did not comply with his suggestion, but this fact was equally to be found already in the official USSBS report. The report (1946a: 19) reads: "Sperm counts done in Hiroshima [...] revealed low sperm counts or complete aspermia for as long as 3 months afterward in males who were within 5,000 feet of the center of the explosion"; "Of women in various stages of pregnancy who were within 3,000 feet of ground zero, all known cases have had miscarriages"; and "there are signs of an increase [in sterility] in the Hiroshima and Nagasaki areas to be attributed to the radiation" (1946a: 19).[12]

> Account 8: The American public still does not know, as they do, the exact height at which the bomb was dropped and the approximate weight of the uranium used. The Japanese scientists have had these and other facts which are still subject to security in the United States printed and mimeographed and bound into little books; tracing them and seeing that they did not fall into the wrong hands would have obliged the Americans to set up, for this one purpose alone, an enormous police system in Japan. Indeed, trying to keep security on atomic fission appears to be fruitless. *The United States Strategic Bombing Survey, which recently issued a report on some of its findings on the atomic bombing, and which tiptoed around Japan in an atmosphere of absolute huggermugger, derived most of its information—even the secret, unpublished parts—from our recent enemies*". (my emphasis)

Concerning the entire passage, GROVES wrote: "This is so far out. It should be eliminated". For the italicized part, he stated: "Out. That's why it is". In the published article, the account was changed to the formulation below. The height of detonation was classified information (Memorandum by the U. S. Chief of Staff to Combined Chiefs of Staff, 1946/06/19). Presumably, the fissionable material weights were also classified. GROVES' suggestion implies that he wanted to keep from the public not only specific information but also the fact that it had been classified.

> They also knew that theoretically one ten times as powerful-or twenty-could be developed. The Japanese scientists thought they knew the exact height at which the bomb at Hiroshima was exploded and the approximate weight of the uranium used. They estimated that, even with the *primitive bomb used at Hiroshima, it would require a shelter of concrete fifty inches thick to protect a human being entirely from radiation sickness*. The scientists had these and other details which remained subject to security in the United States printed and mimeographed and bound into little books. Americans knew of the existence of these, but tracking them and seeing that they did not fall into the wrong hands would have obliged the occupying authorities to set up, for this one purpose alone, an enormous police system in Japan. Altogether, the Japanese scientists were somewhat amused at the efforts of their conquerors to keep security on atomic fission (1946c: 62; 1989: 82–83). (my emphasis)

It is interesting to find that the italicized passage includes classified information—the thickness of a concrete wall needed for a shelter (Memorandum by the U. S. Chief of Staff to Combined Chiefs of Staff, 1946/06/19). GROVES would have demanded its removal were it to have been included in the galley proof.

> Account 9: Many citizens of Hiroshima, however, feel a hatred for Americans which nothing could possibly erase. *Dr. Sasaki is one of the few who are not too polite to tell an American the truth. "I hear," he said recently*, "that they are holding a trial for war criminals in Tokyo just now. I think they ought to try the men who decided to use the bomb and they should hang them all. (my emphasis)

GROVES wrote in the right margin, "I personally don't like this idea". It is not clear what GROVES specifically meant by "this idea". But it presumably referred to the idea of having the men who decided to use the bomb tried and hanged. In the published article, however, the passage was left almost untouched, with the word "feel" changed to "continued to feel" and the italicized clause shortened to: "'I see,' Dr. Sasaki once said" (1946c: 67–68; 1989: 89).

4 *HIROSHIMA* NEGLECTS RESIDUAL RADIATION

As described above, many of GROVES' suggestions were related to radiation and the effects of the atomic bomb. Those changes certainly weakened the description of the effects. However, they were so few and minor that they did not influence the work's nature, style, and power. Even without GROVES' censorship, the content of *Hiroshima* did not diverge from the precedents set by the USSBS report. HERSEY did not reveal any secret information (other than the thickness of concrete walls for a shelter, which is rather a matter of national security).

Yet the question here is why GROVES had no qualms with its overall publication. As we saw above, GROVES' censorship was never oppressive in nature: he offered suggestions, rather than engaging in overt suppression. Moreover, *The New Yorker* editors did not always accept GROVES' suggestions, often leaving the text as it had been written, and GROVES to have had no complaints with this. He did not intend to suppress it entirely and the editors seem to have understood this intention.

In fact, GROVES' attitude is also surprising in this instance given that it differs from his attitude toward an article that was written by one of the MED scientists, Dr. Harold JACOBSON, and which was "almost" suppressed. The article claimed, "Hiroshima will be a devastated area not unlike our conception of the moon for nearly three-quarters of a century" (JACOBSON 1945: 1). JACOBSON was the first person to openly suggest lingering radiation from the bomb when he argued: "The terrific force of the explosion irradiates every piece of matter in the area. Investigators [who enter the city to study damage caused by the bomb] will [...] die in the same way victims of leukemia die" (1945: 1). JACOBSON's assertion that the radiation would make the city uninhabitable for 70 years was ultimately inaccurate. However, what he referred to as "secondary radiation" (1945: 1, 3), a phenomenon that would break up the red corpuscles in the blood, was true (GLASSTONE / DOLAN 1977: 587, 615–16).

GROVES' office was "greatly agitated" by the story (KOOP 1946: 282). As soon as it was distributed by International News Service (INS) on August 7, 1945, FBI and Army Military Intelligence personnel held JACOBSON for violating the Espionage Act. On the following day, JACOBSON explained that it merely reflected his personal view (KOOP 1946: 282; WASHBURN 1988: 32–33). My research using online databases shows that of more than 100 newspapers in the United States that mentioned JACOBSON in the following days, only a handful carried his statement alone. Most others focused on a statement by the War Department, which contradicted JACOBSON by quoting Robert J. OPPENHEIMER, head of Los Alamos Laboratory, who had said that "there is every reason to believe that there was no appreciable radioactivity on the ground at Hiroshima and what little there was decayed very rapidly" (*Washington Post*, 1945: 1).

The Office of Censorship (OC) approved the story because the radioactive properties of uranium were already known. They assumed that JACOBSON had been informed by the Army of what can be openly talked about (WASHBURN 1988: 32).[13] However, the OC was totally helpless when the INS asked them to move in between JACOBSON and the FBI (Press Memorandum by Theodore F. KOOP, 1945/08/08). As wartime censorship came to an end with the end of OC's operation on August 15, 1945 (OFFICE OF CENSORSHIP 1945: 18), the Army took over the role of protecting atomic secrecy (War Department Bureau of Public Relations, "Note to Editors", 1945/09/14). From then on, GROVES would become a crusader against the bomb's lingering effects, which could prove a problematic reminder to the public of chemical warfare.[14] The country's press also learned certain lessons from JACOBSON's case in terms of enforced silence around certain topics as well as the consequences of not adhering to it (KOOP 1946: 282).

It was, thus, no surprise that HERSEY's *Hiroshima* did not discuss any troubling features of the bomb. For example, HERSEY did not seek to explore the issue of residual radiation, accepting instead the U. S. government's view that the initial radiation emissions were the only cause of damages. The following accounts were included in *Hiroshima* before GROVES' censorship.

> Since radiation of at least a thousand times the natural 'leak' would be required to cause serious effects on the human body, the scientists announced that people could enter Hiroshima without any peril at all. (1946b; 1946c: 52; 1989: 73)

> [... I]t was the direct reaction to the bombardment of the body, at the moment when the bomb went off, by neutrons, beta particles, and gamma rays. (1946b; 1946c: 56; 1989: 76)

Both accounts appear to be based on Japanese studies that HERSEY obtained in Japan (Preliminary report by the Research Commissions of the Imperial University of Kyoto). The announcement by Japanese scientists in early September 1945 was a good sign for the residents that marked the beginning of their city's reconstruction. But the citation is misleading given that the announcement did not signal the safe level immediately after the bombings, or during the month before the announcement. Even if there was no malicious intent on the part of

HERSEY or the Japanese scientists, this account could imply that allegations such as JACOBSON's were groundless.

GROVES and his staff had persistently denied the existence and effects of residual radiation in the Japanese target cities. They argued that the atomic bombs were detonated at such high altitude that radiation did not reach the ground and the bomb's effects were limited to those exposed at detonation (GROVES 1945b: 33–42; a statement by Brig. Gen. T. F. FARRELL, 1945/09/12). The second of the above accounts would also appear to support this assertion.

SHARPE argues that the following scene—in which Mrs. Nakamura and her son vomit after drinking water—challenges the U. S. government's denial of the effects of residual radiation (2007: 145). However, the phrase in the parenthesis works to contravene the Japanese speculation on the use of gas and gives the impression that it was a mere "smell".

> They all felt terribly thirsty, and they drank from the river. At once they were nauseated and began vomiting, and they retched the whole day. Others were also nauseated; they all thought (probably because of the strong odor of ionization, an "electric smell" given off by the bomb's fission) that they were sick from a gas the Americans had dropped. (1946b; 1946c: 24; 1989: 35)

The phrase was apparently included as an addition to his first manuscript when ROSS suggested further clearance on the reason why they retched.[15] It was thus already there when the galley proofs were submitted to GROVES, who made no notes on the passage.

5 HIROSHIMA AND THE USSBS

In fact, every American and British postwar survey mission that worked in Hiroshima and Nagasaki denied the effects of residual radiation.[16] For example, the USSBS report (1946a: 18) acknowledged the presence of residual or induced radiation in Takasu, Hiroshima, and Nishiyama, Nagasaki, even weeks after the bombing, but stated that "the degree of activity in these areas was insufficient to produce casualties" and "caused no authenticated fatalities". It also denied the effects on people who were not in the cities during the explosions, saying that such stories "have been disproved by investigation" (1946a: 28).

Even the official line of the USSBS report did not come easy (SHIGESAWA 2019: 289–91), however. While it denied the effects of residual radiation, the USSBS report described radiation sickness from direct exposure in greater detail than any other study. This description included the relative importance of various causes of death and symptoms of delayed deaths. As discussed earlier, the bomb affected reproduction (1946a: 15, 18–20).[17] When reviewing the manuscripts of the USSBS report, Groves held that USSBS placed "undue emphasis on the effects of gamma radiation from the bombs, particularly with respect to impairment of the sex functions" (Memorandum to the Secretary of War from L. R. Groves, 1946/06/19). The memo continued: "Too much emphasis on the sex phase will supply the more lurid news publications with openings for sensational stories".

But what upset Groves more was the tone of the report with regard to the bomb's role in the war. The USSBS report (1946a: 22) reads: "It cannot be said [...] that the atomic bomb convinced the leaders who effected the peace of the necessity of surrender." Two other reports from the USSBS Chairman's Office that followed concluded that "Japan would have surrendered even without the atomic bombs" (1946b: 26; 1946c: 13). This narrative alarmed Groves. It would invite criticism of the atomic bombings of Japan and jeopardize the country's postwar nuclear development program (Shigesawa 2019: 289–91, 302–08). It was also a time when the first postwar nuclear tests at Bikini Atoll came as a disappointment, reinforcing the view that it was "just another weapon" (Laurence 1946: 3).

For Groves, Hersey's *Hiroshima* was a godsend. At a time when the USSBS report was being published as the government white paper, Groves was seeking to occupy the contested terrain of the official narrative of the atomic bomb—with his version (Shigesawa 2019: 306–08). When Groves learned that an award-winning author was writing about *Hiroshima*, he is very likely to have seen it as an opportunity to remind the public of the bomb's war-time role. With all the destruction and suffering described in the article, readers would inevitably come away with the impression that the weapon ended the war.

In December 1946, a professor at the Army War College asked Groves for copies of *Hiroshima* for distribution to students (Letter from T. D. Stamps to Groves, 1946/12/31). Groves agreed, saying that "the book is an excellent piece of reporting". He asked the professor to distribute them with two strong admonitions: firstly "[t]hat an equally sympathetic piece of work could have been done on the American victims of Bataan or the British who were trapped in Hong Kong [...] and others"; and, secondly, "that the atomic bomb ended the war" (Letter from Groves to T. D. Stamps, 1947/01/07).

Groves also achieved another goal: any reference to the USSBS completely disappeared from *Hiroshima*. They were now described only as "statisticians" or "statistical workers" (1946c: 61–62; 1989: 80–81).

Groves' grip on *Hiroshima* did not loosen with his comments on the nine paragraphs. He wanted to make sure that the would-be bestseller included more of his narratives. Although there is no record, Groves likely called Shawn to request "changing the article a little", just as he did for LANG's. The following three amendments were newly added to the article *after* GROVES' censorship.

> Account A: Doctors could not be certain whether some of these symptoms were the result of radiation or nervous shock. (1946c: 56; 1989: 76–77)

> Account B: The ruined city had flourished—and had been an inviting target—mainly because it had been one of the most important military command and communications centers in Japan, and would have become the Imperial headquarters had the islands been invaded and Tokyo been captured. Now there would be no huge military establishments to help revive the city. (1946c: 61; 1989: 80)

> Account C: Since many people died of a combination of causes, it was impossible to figure exactly how many were killed by each cause, but the statisticians calculated that about twenty-five per cent had died of direct burns from the bomb, about fifty per cent from other injuries, and about twenty per cent as a result of radiation effects. (1946c: 62; 1989: 81)

These three accounts—one that casts doubt on the effect of radiation, one that justifies the decision to use the bomb, and one that places more importance on radiation than other studies—attest to a last-minute tug of war that is likely to have taken place between *The New Yorker* and GROVES. It is my belief that accounts A and B were added at the request of GROVES, while C was added by HERSEY.

Account A is the most bizarre statement in *Hiroshima* and stands in stark contradiction with much of its writing, especially with the sentences immediately preceding it: "The rays simply destroyed body cells—caused their nuclei to degenerate and broke their walls. Many people who did not die right away came down with nausea, headache, diarrhea, malaise, and fever, which lasted several days" (1946c: 56; 1989: 76–77). It also contradicts with account C, which was apparently derived from the USSBS report (1946a: 15)[18] and was likely added in response to ROSS' suggestions to explain how the bomb killed people in Hiroshima.[19] Account C is also consistent with his earlier accounts, such as accounts 6 and 7 in the preceding section, including its use of the USSBS report.

On the other hand, account B represents the U. S. government's position, as evidenced by President TRUMAN's statement, according to which Hiroshima was chosen because it was "an important Japanese Army base" (TRUMAN 1945: 621). According to *Time* magazine (1946: 52), ROSS feared that *Hiroshima* would be regarded as anti-American and instructed HERSEY to explain why the bomb had been dropped on the city. But if ROSS feared criticism, he would not set out to run the story in the first place. It was the magazine's intention to "wake people up" (KUNKEL 2000: 294).

HERSEY and the editors worked toward that end. The frequent references to the USSBS report attest to that effort. HERSEY apparently consulted with the USSBS report for information but also as a guideline to determine what could be said (SHIGESAWA 2012: 23–27). However, this probably led to the unfortunate absence of any allusion to the effects of lingering radiation—which may well have served to limit our general understanding and knowledge of atomic energy.

Of course, HERSEY cannot ultimately be held responsible for whatever changes may or may not have been made to the text. As we have seen, it was not possible to discuss the long-term radioactive properties of the weapon. Moreover, GROVES' censorship was unavoidable, whether HERSEY or the editors at *The New Yorker* liked it or not. There were significant constraints on what could be written about the topic, including some degree of self-censorship. There were, of course, also limits to available knowledge on the subject, and this likely also had some part to play.

One of the most significant aspects of HERSEY's work was to present, for the first time, a story as told by the *hibakusha* themselves. In doing so, HERSEY broke down the wall between former enemies, making the event a universal experience. YAVENDITTI (1974: 38) and LIFTON/MITCHELL (1994: 88) commend HERSEY's use of USSBS figures and his graphic depictions of death and suffering from radiation poisoning. BOYER (1985: 207) equally has a positive appraisal of HERSEY's careful observations concerning the exposure effects of radiation. The following sentence in *Hiroshima* perhaps best epitomizes such depictions:

> She stayed there, and later in the day, though she had no visible wounds or burns, she died. (1946b; 1946c: 24; 1989: 35)

In late August 1945, a story was globally distributed by Domei news agency concerning Hiroshima residents who were not immediately injured in the explosion but who would later lose their lives due to the delayed effects (THE UNITED PRESS 1945: 1, 5; BLAKESLEE 1945: 1–2). With HERSEY's accounts, readers were able to understand radiation effects by personalizing the suffering of the characters in the story (SHARP 2007: 135, 137–38).

Almost two decades later, HERSEY wrote, "I didn't set out to 'prove' anything [...]. I wanted to bring to readers a sense of what it must have been like to be there" (Letter to Jeffrey BAUMAN from HERSEY, 1965/11/17). In one interview, HERSEY went on to state that his "[...] hope was to have the reader enter into the characters, become the characters and suffer with them" (BONETTI 1988). HERSEY was also incredibly adept in formulating his story in very accessible prose. It would be quite a stretch, however, to claim his text "exposed" the cover-up.

6 CONCLUSION

Citing a MED expert, the USSBS report (1946a: 18) declared that "our understanding of radiation casualties is not complete". Nevertheless, radiation effects were often discounted and not only the effects but the very existence of residual radiation, among other consequences, categorically denied. But as Sean MALLOY (2012: 518) has pointed out, in spite of manifold efforts on the part of the U. S. government, "radiation effects ultimately became central to the widespread understanding of nuclear weapons as uniquely terrible and have likely contributed to the formation of a nuclear 'taboo' that has helped check their use since 1945".

Peaceful nuclear power has been introduced around the world without such checks, however. BROWN (2019: 3) has emphasized the fact that by downplaying the effects of the Chernobyl accident, mankind has inadequately prepared for the next disaster. She warns that a lack of understanding concerning events at Chernobyl will lead us to repeat its mistakes. Meanwhile, at the time of writing, military attacks are being launched against nuclear power plants in Ukraine, with Russian president, Vladimir PUTIN, threatening to use nuclear weapons.

Hiroshima and Nagasaki have so far been the only cities against which nuclear weapons have been used and those events continue to intrigue people around the world, with many focusing, in particular, on the (il)legitimacy of nuclear weapons *per se*. With its unique experience of that city, its unprecedented treatment in *The New Yorker*, the enthusiasm with which it was accepted, and the personal touches with which its characters are portrayed, HERSEY's *Hiroshima* is an unparalleled work that has generated numerous new studies. The popular attention to the heroism of its author, however, has often unwittingly perpetuated the "myth" delineated above. Contrary to popular belief, HERSEY did not describe the bomb's effects in full.

Indeed, *Hiroshima* did not shed light on any new and troubling features of the bomb. Its descriptions of the bomb's effects were already in line with the official USSBS report when it was submitted to GROVES for censorship clearance. The comments left by GROVES in the galley proofs led only to minor changes. Though some additions were most likely made based on GROVES' instructions, these did not materially change the nature, style and the power of the reportage, as its virtue lies in the experience of the bombing recreated from the point of view of the survivors.

HERSEY had to tread a fine line in order to speak as truthfully as possible while at the same time using the USSBS report as the "gold standard". However, there were limitations. With the long-term radioactive effects of the atomic bomb open to comparison with chemical weapons, both GROVES and the U. S. government had felt compelled to control information concerning Hiroshima and Nagasaki in the wake of the attacks. In terms of HERSEY's article, GROVES gave considerable importance to having a version published that could beat the USSBS to establish itself as the official narrative. His ultimate purpose of reviewing the article was to justify the bombing.

It is difficult to gauge how this manipulation of information—which occurred three quarters of a century ago—has affected humankind's responses to cases of radiation exposure at nuclear accidents in Chernobyl and Fukushima, and what influence it has exerted on the discourse surrounding atomic energy in general. Indeed, we still do not fully know the extent of what was provoked by the bombings in 1945. Insofar as we are unable to entirely distinguish between what HERSEY wrote and what he did not, and insofar as we are unable to know the reasons why, we will never escape the long shadow cast by GROVES.

NOTES

1   The experts' committee organized by the Fukushima prefectural government reported that the thyroid cancer apparently has nothing to do with the disaster because the estimated dose children in Fukushima received was far smaller than doses at the Chernobyl accident (JIJI 2022). Also, UNSCEAR stated in 2021 (2022: 87–89) that "the increased incidence rates might be due to over-diagnosis", and "the detected excess of thyroid cancers is probably unrelated to radiation exposure".
2   For discussions of the USSBS, see MACISAAC (1976), DANIELS (1981), and GENTILE (2001).
3   The word "censorship" is used herein, but official censorship had not been imposed then as the Office of Censorship had already been abolished. That is how the press recognized the ongoing practice requested by the U. S. Army. See, for example, ROSS' inquiry to his legal counsel hereinafter. The military requested that the United States' press "withhold information (beyond the official release) without first consulting with the War Department, concerning [...] the atomic bomb" (WAR DEPARTMENT BUREAU OF PUBLIC RELATIONS, 1945 / 09 / 14).
4   For OC, see SWEENEY (2001). For the press-government relationship, see KNIGHTLEY (1975) and SWEENEY (2006).
5   See LANG (1959) for his articles.
6   BLUME speculates that *Hiroshima* was subjected to censorship by GROVES based on four documents. Two of them support this hypothesis but do not make clear if GROVES had actually seen the galley proofs before publication. One is a letter from SHAWN to GROVES, dated August 15, 1946, which alludes to "proofs of a four-part article on the bombing of Hiroshima, sent to your office to look over". The other is a Letter to Jay CASSINO from John HERSEY, January 10, 1947, which states that "the article was submitted to the War Department for review", but does not specify at which point of time. Two other documents BLUME mentions offer no evidence at all. The first is GREENSTEIN's August 12 reply to ROSS—mentioned above—and is misinterpreted by BLUME as containing a suggestion by GREENSTEIN that the article be submitted to the Army for censorship clearance. The other is an entry in GROVES' diary dated August 7, 1946, in which GROVES wrote that he had called SHAWN about changing the article a little (2020: 119). This does not refer to HERSEY's *Hiroshima*,

but Lang's "Seven Men on a Problem" (titled "The Plan" at this stage), which appeared in the August 17, 1946, issue of *The New Yorker*.
7. All citations for nine accounts hereinafter are from the galley proofs (Hersey 1946b).
8. Ross had actually suggested "hurricane" instead, as tornado was a "purely American phenomenon" (Ross' notes, 1946/08/08). But "tornado" remained in the galley proof.
9. For Groves' dislike of the USSBS, see Shigesawa (2019: 320–21).
10. NavTechJap (1983 [1945]) states that 15 % of 4,030 survivors in Hiroshima, sampled randomly, reported effects of radiation (1945: 26–27), whereas the MED reported "7 per cent or less" (1946: 28). The Joint Commission concluded that radiation injuries were detected in 37.4 % and 33.7 % of injured persons in the sample groups, respectively studied in Hiroshima and Nagasaki (1951: 1, 14).
11. At the congressional hearing, Warren also said, "the gamma radiation and allied radiation effects did not add a great deal to what would have happened if the same amount of energy had been released by TNT" (1946: 510).
12. It is certain that the USSBS used "sterility" to mean "amenorrhea" because the latter is used in the report of the USSBS' Medical Division, based on which the USSBS Chairman's report (1946a) was written with slightly different, though similar, wording (USSBS Medical Division 1947: 53). Hersey could also refer to a preliminary report by the Research Commissions of the Imperial University of Kyoto, which he obtained while in Japan, and which stated that 50 % of examinees "had delayed menstruation".
13. The INS editor was "more than a little surprised" when he learned that the OC had no objection to running the story (Press Memo by Day Thorpe, 1945/08/07). When the story was distributed, it also alarmed some media outlets, urging them to ask OC officials if they had cleared the story (Press Memorandum by J. E. Warner, 1945/08/08).
14. For Groves' knowledge of the bomb's radiological effects, see Malloy (2012).
15. Hersey 1946a: 27. Ross asked, "This water is brackish [*sic*] (?) so maybe it's no wonder they vomited. This vomiting wouldn't seem to be the ordinary after-the-explosion vomiting" (Ross' notes, 1946/08/19). Ross appears to have no idea of radiation sickness from internal exposure.
16. For the reports of postwar survey missions, see Shigesawa (2019: 75–125, 193–224).
17. For the discussion of how USSBS came to include effects of radiation in such an extent, see Shigesawa (2019: 187–224).
18. It reads: "A plausible estimate of the importance of the various causes of death would range as follows: Flash burns, 20 to 30 percent. Other injuries, 50 to 60 percent. Radiation sickness, 15 to 20 percent".
19. In his notes, which customarily listed all his questions and suggestions to the author, Ross alluded to a "great interest in how this bomb kills people" (August 6), later stating that he "[...] wondered about what killed these people, the burns, falling debris, the concussion—what?" (August 8).

REFERENCES

Abbreviations:
JHP John Hersey Papers, Beinecke Rare Book & Manuscript Library, Yale University.
NACP National Archives and Record Administration, College Park, MD.
NYR New Yorker Records, New York Public Library.

Monographs and articles:
ASAHI SHINBUN 朝日新聞 (2021): "'Kuroi ame' jōkoku miokuri 「黒い雨」上告見送り]". In *Asahi shinbun* 朝日新聞, 07/27: 1.
BLAKESLEE, Howard W. (1945): "480,000 Japs Left Dead, Wounded, Homeless By Hiroshima, Nagasaki Atom Bomb Attacks". In: *Washington Post*, 08/23: 1–2.
BLUME, Lesley M. M. (2020): *Fallout: The Hiroshima Cover-up and the Reporter Who Revealed It to the World*. New York: Simon & Schuster.
BONETTI, Kay (1988): *Interview with John Hersey*. Columbia, MO: American Audio Prose Library (audio tape recording).
BOYER, Paul (1985): *By the Bomb's Early Light*. Chapel Hill, NC: University of North Caroline Press.
BRODIE, Janet Farrell (2015): "Radiation Secrecy and Censorship after Hiroshima and Nagasaki". In: *Journal of Social History* 48 (4), 842–64.
BROWN, Kate (2019): *Manual for Survival: An Environmental History of the Chernobyl Disaster*. New York: W. W. Norton & Company.
DANIELS, Gordon A. (ed.) (1981): *A Guide to the Reports of the United States Strategic Bombing Survey*. London: Royal Historical Society.
EDANO, Yukio 枝野幸男 (2011): "Edano kanbō chōkan no kaiken (yōshi) 枝野官房長官の会見(要旨)". In: *Asahi shinbun* 朝日新聞, 03/23: 4.
EDEN, Lynn (2004): *Whole World on Fire: Organizations, Knowledge, and Nuclear Weapons Devastation*. Ithaca, NY: Cornell University Press.
FORDE, Kathy Roberts (2011): "Profit and Public Interest: A Publication History of John Hersey's Hiroshima". In: *Journalism and Mass Communication Quarterly* 88 (3), 562–79.
GENTILE, Gian P. (2001): *How Effective Is Strategic Bombing?: Lessons Learned From World War II to Kosovo*. New York: New York University Press.
GLASSTONE, Samuel / DOLAN, Philip J. (eds.) (1977): *The Effects of Nuclear Weapons*. 3rd edition. Washington, DC: U.S. Government Printing Office.
GROVES, Leslie R. (1945a): *Statement in Hearings before the Committee on Military Affairs, U.S. House of Representatives, 79th Congress, 1st Session*

on H. R. 4280, October 9, An Act for the Development and Control of Atomic Energy*. Washington DC: U. S. Government Printing Office, 9–33.

GROVES, Leslie R. (1945b): *Statement in Hearings before Senate Special Committee on Atomic Energy, 79th Congress, 1st Session on S. Res. 179, November 28, A Resolution Creating a Special Committee to Investigate Problems Relating to the Development, Use, and Control of Atomic Energy*. Washington, DC: U. S. Government Printing Office, 31–79.

HASHIMOTO, Hiroki 橋本拓樹 (2021): "Fukushima Daiichi jiko futari rōsai nintei: Sagyō de hibaku inkōgan wa hatsu 福島第一事故2人労災認定：作業で被曝咽頭がんは初". In: *Asahi shinbun* 朝日新聞, 09 / 09: 29.

HERSEY, John (1946a): *Hiroshima: original manuscript of first draft*, Box 2, YCAL MSS 707, JHP.

HERSEY, John (1946b): *Galley Proofs "A Reporter at Large: Some Events at Hiroshima" attached to a letter from William Shawn to General Groves, Aug. 15, 1946*, Folder: Hiroshima + Nagasaki, Box 9, Entry 1 General Correspondence, Papers of General Leslie R. Groves, National, NACP.

HERSEY, John (1946c): "Hiroshima". In: *The New Yorker*, 08 / 31: 15–68.

HERSEY, John (1989): *Hiroshima. A new edition with a final chapter written forty years after the explosion*. New York: Vintage Books.

HOOK, Glenn D. (1991): "Censorship and Reportage of Atomic Damage and Casualties in Hiroshima and Nagasaki". In: *Bulletin of Concerned Asian Scholars* 23 (1), 13–25.

JACOBSON, Harold (1945): "Death Will Saturate Bomb Targets For 70 Years, Atomic Expert Says". In: *The Atlanta Constitution*, 08 / 08: 1, 3.

JIJI (2022): "Six people to sue Tepco over thyroid cancer after Fukushima disaster". In: *The Japan Times (online)*, 01 / 19, https://www.japantimes.co.jp/news/2022/01/19/national/fukushima-tepco-cancer-case/ (last access 2022 / 03 / 19).

JOINT COMMISSION FOR THE INVESTIGATION OF THE EFFECTS OF THE ATOMIC BOMB IN JAPAN (1951): *Medical Report of the Joint Commission for the Investigation of the Effects of the Atomic Bomb in Japan*, vol. 1. Washington, DC: Army Institute of Pathology.

KNIGHTLEY, Phillip (1975): *The First Casualty: From the Crimea to Vietnam: The War Correspondent as Hero, Propagandist, and Myth Maker*. New York: Harcourt Brace Jovanovich.

KOOP, Theodore F. (1946): *Weapon of Silence*. Chicago, IL: University of Chicago Press.

KUNKEL, Thomas (ed.) (2000): *Letters from the Editor: The New Yorker's Harold Ross*. New York: Modern Library.

LANG, Daniel (1959): *From Hiroshima to the Moon: Chronicles of Life in the Atomic Age*. New York: Simon and Schuster.

LAURENCE, William L. (1946): "Bikini 'Dud' Decried for Lifting Fears: The Underwater Atomic Blast as Pictured from the Shores of Bikini". In: *New York Times*, 08 / 04: 3.

LIFTON, Robert J. / MITCHELL, Gregg (1994): *Hiroshima in America: A Half-Century of Denial*. New York: Avon Book.

MACISAAC, David (1976): *Strategic Bombing in World War Two: The Story of the United States Strategic Bombing Survey*. New York: Garland Publishing.

MALLOY, Sean L. (2012): "A Very Pleasant Way to Die: Radiation Effects and the Decision to Use the Atomic Bomb against Japan". In: *Diplomatic History* 36 (3): 515–45.

MANHATTAN ENGINEER DISTRICT (1977 [1946]): *The Atomic Bombings of Hiroshima and Nagasaki*. Washington, DC. In: *Manhattan Project: Official History and Documents (12 microfilm collections)*, reel 1. Washington, DC: University Publications of America.

MURAKAMI, Kōichi 村上晃一 (2018): "Haigan shibō rōsai nintei: Fukushima Daiichi jiko. Sagyō de hibaku 肺がん死亡労災認定：福島第一事故　作業で被曝". In: *Asahi shinbun* 朝日新聞, 09 / 05: 26.

NAVTECHJAP (1983 [1945]): "Atomic Bombs, Hiroshima and Nagasaki, Article 1, Medical Effects". In: *Reports of the U. S. Naval Technical Mission to Japan, 1945–1946 (microfilm edition)*, reel 10. Wilmington, DL: Scholarly Resources.

OFFICE OF CENSORSHIP (1945): *A Report on the Office of Censorship*. Washington, DC: U. S. Government Printing Office.

ŌTAKI, Megu 大瀧慈 (2020): "Health Effects of Internal Radiation Exposure from 'Black-Rain' Recognized by District Court 地裁判決が認めた黒い雨による内部被曝". In: *Hiroshima Peace Science* 広島平和科学 42: 1–16.

PRIME MINISTER OF JAPAN AND HIS CABINET (2011): "Tōhoku, Kantō no kata e: Ame ga futtemo kenkō ni eikyō wa arimasen 東北、関東の方へ：雨が降っても、健康に影響はありません". In: *Prime Minister's Official Residence online*, 03 / 20, https://warp.ndl.go.jp/info:ndljp/pid/1368653/www.kantei.go.jp/saigai/20110321ame.html (last access 2023 / 02 / 08).

SAITŌ, Osamu 齋藤紀 (2018): *Hiroshima no hibaku to Fukushima no hibaku: Ryōsha wa honshitsuteki ni onaji mono ka nite hi naru mono ka* 広島の被爆と福島の被曝：両者は本質的に同じものか似て非なるものか. Kyōto: Kamogawa shuppan.

SHARP, Patrick B. (2000): "From Yellow Peril to Japanese Wasteland: John Hersey's Hiroshima". In: *Twentieth Century Literature* 46 (4): 434–52.

SHARP, Patrick B. (2007): *Savage Perils: Racial Frontiers and Nuclear Apocalypse in American Culture*. Norman, OK: University of Oklahoma Press.

SHIBATA, Yuko (2012): "Dissociative Entanglement: US-Japan Atomic Bomb Discourses by John Hersey and Nagai Takashi". In: *Inter-Asia Cultural Studies* 13 (1): 122–37.

SHIGESAWA, Atsuko (2012): "John Hersey's 'Hiroshima' Revisited: From the Vantage Point of 66 Years Later". In: *Hiroshima Journal of International Studies* 18: 19–37.

SHIGESAWA, Atsuko (2019): *Demystifying the Atomic Bomb: The U.S. Strategic Bombing Survey Goes to Hiroshima and Nagasaki*. Hiroshima: Hiroshima City University / Graduate School of International Studies, Dissertation.

SWEENEY, Michael S. (2001): *Secrets of Victory: The Office of Censorship and the American Press and Radio in World War II*. Chapel Hill, NC: University of North Carolina Press.

SWEENEY, Michael S. (2006): *The Military and the Press: An Uneasy Truce*. Evanston, IL: Northwestern University Press.

TAKAHASHI, Hiroko 高橋博子 (2008): *Fūin sareta Hiroshima, Nagasaki: Bei kakujikken to minkan bōei keikaku* 封印されたヒロシマ・ナガサキ：米核実験と民間防衛計画. Tōkyō: Gaifūsha.

TIME MAGAZINE (1946): "Without Laughter". In: *Time Magazine* 48 (11), 09 / 09: 52.

TRUMAN, Harry S. (1945): "Statement by President Truman [Aug. 6]". In: *Foreign Relations of the United States: Diplomatic Papers, 1945. The British Commonwealth, the Far East*, vol. VI. Washington, DC: U.S. Government Printing Office, 621–24.

THE UNITED PRESS (1945): "Tokyo Puts Toll of Atomic Bombs At 190,000 Killed and Wounded". In: *New York Times*, 08 / 23: 1, 5.

USSBS CHAIRMAN'S OFFICE (1946a): *The Effects of Atomic Bombs on Hiroshima and Nagasaki*. Washington, DC: U.S. Government Printing Office.

USSBS CHAIRMAN'S OFFICE (1946b): *Summary Report (Pacific)*. Washington, DC: U.S. Government Printing Office.

USSBS CHAIRMAN'S OFFICE (1946c): *Japan's Struggle to End the War*. Washington, DC: U.S. Government Printing Office.

USSBS MEDICAL DIVISION (1947): *The Effects of Atomic Bombs on Health and Medical Services in Hiroshima and Nagasaki*. Washington, DC: U.S. Government Printing Office.

UNSCEAR (2011): *Sources and Effects of Ionizing Radiation. Report to the General Assembly*, vol. II. New York: United Nations Publication.

UNSCEAR (2020 / 21): *Sources, Effects and Risks of Ionizing Radiation. Report to the General Assembly*, vol. II. New York: United Nations Publication.

WARREN, Stafford L. (1946): *Statement in Hearings before the Senate Special Committee on Atomic Energy, Feb. 15, U.S. Congress. Senate Special Committee on Atomic Energy. Atomic Energy: Hearings before the committee on S. Res. 179. A Resolution Creating a Special Committee to Investigate Problems Relating to the Development, Use, and Control of Atomic Energy, 79th Congress, 2nd Session*. Washington, DC: US GPO.

WASHBURN, Patrick S. (1988): "The Office of Censorship's Attempt to Control Press Coverage of the Atomic Bomb during World War II", https://eric.ed.gov/?id=ED295201 (last access 2023 / 02 / 08).

WASHINGTON POST (1945): "Atom Bomb's Radioactivity Fades Rapidly". In: *The Washington Post*, 08 / 09: 1–2.

WELLERSTEIN, Alex (2021): *Restricted Data: The History of Nuclear Secrecy in the United States*. Chicago, IL: The University of Chicago Press.

YAVENDITTI, Michael J. (1974): "John Hersey and the American Conscience: The Reception of "Hiroshima". In: *Pacific Historical Review* 43 (1): 24–49.

Archival Material:

A preliminary report by the Research Commissions of the Imperial University of Kyoto, Folder: Hiroshima, Box 21, Uncat Za MS 235, JHP.

Cabled message from HERSEY to William SHAWN, 1946 / 03 / 22, Folder: HERSEY, John, 1946 Editorial Correspondence, Box 434, NYR.

Invitational Travel Orders, 1946 / 05 / 21 & 24, 1946, Box 21, Uncat Za MS 235, JHP.

Letter from T. D. STAMPS to GROVES, 1946 / 12 / 31, Folder: Letters to Friends, Box 4, Entry 3, NACP.

Letter to GROVES from SHAWN, 1946 / 08 / 06, Folder: Atomic Energy-Articles (Gen. Groves) 1949–1961, Box 2, Entry 1, NACP.

Letter to H. W. ROSS from Milton GREENSTEIN, 1946 / 08 / 12, Folder: HERSEY, John, "Hiroshima", Box 50, NYR.

Letter to Jay CASSINO from John HERSEY, 1947 / 01 / 10, Folder: HERSEY, John, "Hiroshima", Box 1315, NYR.

Letter to Jeffrey BAUMAN from HERSEY, 1965 / 11 / 17, Folder: Hiroshima, Box 21, Uncat Za MS 235, JHP.

Letter to Milton GREENSTEIN from H. W. ROSS, 1946 / 08 / 01, Folder: HERSEY, John, "Hiroshima", Box 50, New Yorker Records, New York Public Library.

Memorandum by U. S. Chief of Staff to Combined Chiefs of Staff on Effects of Atomic Bomb on Hiroshima and Nagasaki, 1946 / 06 / 19, Box 3, Entry 491, RG165, NACP.

Memorandum to Secretary of War from L. R. Groves, 1946/06/19, Folder: Atomic Energy Safe File #2, Box 1, Entry 106, RG107, NACP.

Press Memo by Day Thorpe, 1945/08/07, Folder: 012–D/4 Atom Smashing, Box 481, RG216, NACP.

Press Memorandum by J. E. Warner, 1945/08/08, Folder: 012–D/4 Atom Smashing, Box 481, RG216, NACP.

Press Memorandum by Theodore F. Koop, 1945/08/08, Folder: 012–D/4 Atom Smashing, Box 481, RG216, NACP.

Ross' notes on "Reporter—Some Events at Hiroshima—Part I" by Hersey, 1946/08/06, Folder: Notes on Writings Harold Ross 1946, Box 39, NYR.

Ross' notes on "Reporter—Some Events at Hiroshima—Part II" by Hersey, 1946/08/08, Folder: Notes on Writings Harold Ross 1946, Box 39, NYR.

Ross' notes on "Reporter—Some Events at Hiroshima—Part II Revise" by Hersey, 1946/08/19, Folder: Notes on Writings Harold Ross 1946, Box 39, NYR.

Statement by Brig. Gen. T. F. Farrell, Chief, Atomic Bomb Mission, Tokyo, Japan, 1945/09/12, Folder: 10 Atomic Bomb Mission #1, Box 785-11, RG331, NACP.

War Department Bureau of Public Relations "Note to Editors", 1945/09/14, Folder: 380.01 (Security), Box 66, Entry 5, Manhattan Engineer District records, RG 77, NACP.

From Cities in Ruins to Ambassadors of World Peace: The Daily Press in Hiroshima and Nagasaki from 1945 to 1949

Marie-Christine Dreßen

1 Introduction

"They make advertisements for soap. Why not for peace?" (RESNAIS 1959: 00:28:55–00:29:02)

Quoted from the 1959 film, *Hiroshima mon amour*, by French director Alain RESNAIS, the above question is posed by the female protagonist and hints at the almost inflationary use of the word "peace" in the context of Hiroshima and Nagasaki. Following the atomic bombings on August 6 and 9, 1945, both cities were largely razed to the ground. In addition to the attention given by regional governments and city officials, it was, first and foremost, the local press in and around Hiroshima and Nagasaki that covered the physical and symbolic reconstruction of the two cities. However, the question soon arose as to how they ought best to reinvent themselves in their new roles as survivors of the atomic bomb.

As Hiroshima embarked on this colossal task, city officials soon became aware of the important international role that the city had acquired. Those who had survived the horrendous effects of the first ever atomic bomb dropped on civilians were now seen to possess significant symbolic power and were, as such, thought to carry the responsibility of spreading the idea of world peace on a global scale (ZWIGENBERG 2014: 3). Hiroshima city officials managed to have national legislation passed that sought 1) to change the city's image from that of a predominantly military town to that of one that had been subjected to a nuclear attack, and 2) to acquire the necessary financial support for its reconstruction. The law in question—the "Hiroshima Peace Memorial City Construction Law" (Hiroshima heiwa kinen toshi kensetsu hō 広島平和記念都市建設法)—was passed in 1949.[1] Article one states that: "It shall be the object of the present law to provide for the construction of the city of Hiroshima as a peace memorial city to symbolize the human ideal of sincere pursuit of genuine and lasting peace" (CITY OF HIROSHIMA 2021). As shall be examined in greater detail below, the law itself—as well as the discussions surrounding it and the level of approval among the population—was closely covered by the local media. Nagasaki was soon able to follow suit when the "Nagasaki International Culture City Construction Law" (Nagasaki kokusai bunka toshi kensetsu hō 長崎国際文化都市建設法) was passed. The legislation—which included plans for the construction of the Nagasaki Peace Park (Nagasaki heiwa kōen 長崎平和公園) and the Nagasaki Atomic Bomb Museum (Nagasaki genbaku shiryōkan 長崎原爆資料館)—entailed regional and national financial support to help rebuild the city.

Local media and newspapers gave a great deal of attention to all such legislation as centered on physical and symbolic reconstruction projects—such as those that envisioned peace parks and museums—and which aimed at turning Hiroshima and Nagasaki into peace cities. A brief glance at newspaper articles from the years immediately following the bombings shows that city officials were invested in demonstrating the new role of the cities from very early on. Relevant newspaper articles covering all print publications from the active period of the Civil Censorship Detachment (1945–49) during the Allied Occupation[2] of Japan (1945–52) are accessible via the Gordon W. Prange Collection. A detailed examination thereof gives a good overall impression of the various stakeholders—from survivors and citizens to local officials and newspapers—whose interests helped form the two cities anew. It becomes clear that certain local newspapers—such as the *Chūgoku shinbun* 中国新聞 in Hiroshima or the *Nagasaki shinbun* 長崎新聞 in Nagasaki—were more influential than others. Their coverage significantly affected discussions as to how the image of the two cities was to be reconstructed. Indeed, they essentially contributed, in a broader sense, to the peace discourse that developed around these two cities.

The first section will deal with the beginnings of the peace discourse in and around the two cities, while the second will explore the historical background of the Japanese press while it was subject to censorship under Allied Occupation. The third section will then take a closer look at the local press in Hiroshima and Nagasaki as well as five narratives that dominated the daily press in both cities.

For this paper, I consulted the table of contents database "20th Century Media Information Database" by way of which one can browse the Prange Collection with significantly greater ease. I began with about 30 keywords, such as *genbaku* 原爆 (atomic bomb) or *heiwa Nagasaki* 平和長崎 (peace Nagasaki). As my research progressed, about 20 further keywords emerged—such as *genshiryoku heiwa-teki riyō* 原子力平和的利用 (peaceful use of nuclear energy)—that I proceeded to integrate as standard keywords. I then ordered the specific articles from the National Diet Library in Tokyo either as microfilm, microfiche, or printed copy. The readings soon revealed several narratives that seemed to recur repeatedly during the occupation period. For the purposes of this paper, I elaborate on five of these predominant and recurring narratives: 1) the international scope of the atomic bombs; 2) the transition of Hiroshima to become an ambassador of peace; 3) the transition of Nagasaki to become an ambassador of peace; 4) the initially small reconstruction festivals that gradually turned into official peace ceremonies; and 5) the ensuing nationalization of atomic bomb commemoration. The aim of this paper is to examine peace discourse overall as well as the role played by local media in particular, in order to delineate the process of transition by way of which two cities that had been attacked with atomic bombs became national and international symbols for world peace. Special attention shall also be given to gaps or blank spots in the coverage, especially amidst the censorship of Allied Occupation. Placed alongside banal advertisements for soap and so on, such blank spots stand out against the wide spread of peace slogans.

2 Heiwa everywhere

Wandering through Hiroshima today, one is almost bombarded at every corner with one particular word: *heiwa* 平和, the Japanese word for peace. The initiative Hiroshima Peace Tourism—headed up by the Tourism Policy Department and the Economic Affairs and Tourism Bureau of the City of Hiroshima—even offers contemporary visitors a "Hiroshima Peace Tourism Guide Map" (City of Hiroshima 2018–23) on its website.³ Several *hibaku* 被爆 buildings are shown immersed in bright pastel colors and presented as "buildings which survived the atomic bombing" and that constitute "lasting, physical testaments to the destructive ability of the atomic bomb" (City of Hiroshima 2018–23). The first *hibaku* building to be introduced was the Atomic Bomb Dome, also known as Hiroshima Peace Memorial, or Genbaku Dome. The building—officially called Hiroshima heiwa kinenhi 広島平和記念碑—has been on the UNESCO World Heritage List since September 28, 1995. Before the bomb was dropped, the building's function corresponded to its name at the time, namely the "Hall for the Promotion of Industry of Hiroshima Prefecture" (Hiroshima-ken sangyō shōreikan 広島県産業奨励館). After "Little Boy" was dropped on August 6, 1945, the Genbaku Dome was the only building near the hypocenter to survive. Intact but significantly damaged, it was left untouched during the reconstruction of the city and came to be known as the "A-bomb Dome" (Genbaku dōmu 原爆ドーム). The Peace Park that now surrounds the monument was built between 1950 and 1964 (ICOMOS 1996: 115–17).

In the years following the atomic bomb, the Genbaku Dome also became an important visual symbol for coverage of the annual commemoration of the bombings. The walking tours described in the "Hiroshima Peace Tourism Guide Map" equally recommend other peace-related buildings and places of interest such as the Honkawa Elementary School Peace Museum, the Memorial Cathedral for World Peace, the Hiroshima Peace Memorial Museum, the Hiroshima National Peace Memorial Hall for the Atomic Bomb Victims, and the Children's Peace Monument. The names of the routes themselves equally reflect how the City of Hiroshima envisions itself. The first route is titled "Exploring Hibaku Buildings and the Legacy of Hiroshima", alluding to the city's status as the first in the world to be hit by an atomic bomb. The second route—"From the Ashes: A Look at Modern Hiroshima"—refers to its phoenix-like resurrection and examines both the city's survival of the atom bomb and its reconstruction as a modern Japanese city. The third route—"Lessons from the Past: An In-Depth Look at Peace Museums in Hiroshima"—reveals the city's vision of its future as an advocate for world peace from which the whole world can learn (CITY OF HIROSHIMA 2018–23).

The City of Hiroshima's current image campaign—predominantly oriented toward tourism—is not arbitrary, but rather the result of almost 80 years of imaginative design. Indeed, the peace declaration of 2022—which was read aloud by the mayor of Hiroshima at the annual Peace Memorial Ceremony on August 6 of the same year—sees the city: "light[ing] the way toward lasting world peace" (MATSUI 2022).

KAWAGUCHI Yūko traces this "symbol of peace" narrative back to autumn 1945 and attributes special importance in its creation to local politicians and local newspapers such as the *Chūgoku shinbun*. For KAWAGUCHI, "[t]he overall characteristic of this narrative was based on the logic that the suffering of Hiroshima was not only a local incident but also an event that bore transnational meaning" (KAWAGUCHI 2006: 233). Hence, the narrative placed the bombing of Hiroshima in the context of world history and imagined the attention of people all over the world. Soon after the bombing, especially local politicians in cooperation with local newspapers began to move away from a dark image of Hiroshima—as a city that had been bombed and destroyed and where over 100,000 people lost their lives—to a much more positive image as a peace city. This change of image can, in turn, be traced back to newspaper articles from local media. KAWAGUCHI, for example, locates the beginning of the "symbol of peace" narrative in one precise editorial in the *Chūgoku shinbun* on November 11, 1945. The text envisions the bombing of Hiroshima as an opportunity to turn the former militaristic city into a peaceful one (KAWAGUCHI 2006: 234). In her groundbreaking work *Hiroshima Traces: Time, Space, and the Dialectics of Memory* (1999), Lisa YONEYAMA also carves out the universal meaning of Hiroshima: "'Hiroshima', as master code for catastrophe in the twentieth century is apparently all-absorbing as it conflates countless particulars into a single totality in the name of world peace" (1999: 15). Since the atomic bomb was dropped, the

Chūgoku shinbun has come to establish itself as a leading actor in the creation of the "symbol of peace" narrative. The Hiroshima Peace Media Center—hosted on the *Chūgoku shinbun*'s website—gives visitors a chance to examine the front page of every morning edition since August 6, 1946. Users are invited to compare the depiction of the themes "Hiroshima / Nagasaki" and "A-bomb anniversary". The site also offers insights into the development of the peace narrative in the years following the bombing (CHŪGOKU SHINBUN).

Local news outlets, such as the *Chūgoku shinbun*, played a vital role in creating Hiroshima's "symbol of peace" narrative. Though a corollary can equally be found in the case of Nagasaki, the latter 'peace'—or rather 'culture'—narrative is significantly anchored in that of Hiroshima. As such, it makes sense to begin tracing the beginning of the "symbol of peace" narrative back to its starting point in Hiroshima.

However, to analyze the peace discourse, it is necessary to clarify which definition of 'discourse' will be used in this article. I look, specifically, to Reiner KELLER, who represents an approach to discourse theory proper to the sociology of knowledge. Amidst the wide range of approaches in discourse theory and discourse analysis, KELLER has formulated four features as the lowest denominators in the use of the term 'discourse', each of which shall support my analysis of peace discourse of Hiroshima and Nagasaki. According to KELLER, discourse theories or analyses are (1) "concerned with the actual use of (written or spoken) language and other symbolic forms in social practices". (2) They "emphasize that in the practical use of signs, meanings of phenomena are socially constructed, and these phenomena are thereby constituted in their social reality". (3) Discourse theories or analyses "claim that individual instances of interpretation may be understood as parts of a more comprehensive structure that is temporarily produced and stabilized by specific institutional-organizational contexts". (4) Lastly, they "assume that the use of symbolic orders is subject to rules of interpretation and action that may be reconstructed" (KELLER 2013: 3). Moreover, KELLER defines discourses as "more or less successful attempts to stabilize, at least temporarily, attributions of meaning and orders of interpretation, and thereby to institutionalize a collectively binding order of knowledge in a social ensemble" (KELLER 2013: 2).

Over the pages that follow, I elaborate on the peace discourse alluded to above with a view to elucidating the institutionalization that resulted from various efforts to establish or stabilize certain meanings and orders of interpretation. More concretely, I examine how it is that the Japanese daily press contributed to the formation of the discourse of the two cities as symbols of peace. However, given that the articles consulted were accessed through the Gordon W. Prange Collection—the largest collection of print publications accumulated by the Civil Censorship Detachment from the years 1945 to 1949—we must first examine the state of the Japanese press in Hiroshima and Nagasaki in the years following the bombings.

3 ALLIED OCCUPATION CENSORSHIP AND THE JAPANESE PRESS

KAWAGUCHI (2006: 223) traces the change in the *Chūgoku shinbun*'s coverage of Hiroshima—as a city devastated by the atomic bomb to that of a leading actor regarding world peace—to an edition of the paper from November 1945. The date is not arbitrary. Rather—as path-breaking studies by Monica BRAW, Glenn HOOK or KAWAGUCHI Yūko indicate—it is explicitly linked to the beginning of the regime of censorship under Allied Occupation.

Conception of the censorship apparatus did not begin with the famous radio address of Emperor Hirohito 裕仁 (1901–89), which declared the end of the Pacific War on August 15, 1945, or with the signing of the Japanese surrender agreement on September 2, 1945. As BRAW's comprehensive study—*The Atomic Bomb Suppressed: American Censorship in Japan 1945–1949* (1986)—has shown, detailed planning for the occupation of Japan can be traced as far back as 1943, "a time when most citizens there and in other countries only dared hope that the war would end [...]" (BRAW 1986: 25). In May 1944, a directive was sent from the U.S. War Department to the Commander-in-Chief of the Southwest Pacific Area, Douglas MACARTHUR (1880–1964), outlining reasons for civilian censorship. It delineated the necessity of measuring Japanese public opinion on the country's cooperation with Allied Occupation policy and claimed that censorship would serve "[...] to maintain security, protect lines of communication, detect subversive elements, prevent information from reaching the enemy and discover attempts to violate any military order or regulation in the area" (BRAW 1986: 26). Familiar though this may sound with regard to censorship, a fundamental question—formulated by BRAW (1986: 8) quite early in her study—emerges: if the Allies' stated aim for Japan was to convert it into a peaceful and democratic nation, why impose censorship?

As a possible answer to this question, Sey NISHIMURA[4] cites the state of abject devastation that engulfed Japan following its defeat in the war, a period viewed by occupation officials "as a transition from Japanese imperialism to American democracy during which authoritarian measures were necessary" (NISHIMURA: 1989: 3). He also gives a more concrete answer to the question by looking at the topics considered sensitive by the SCAP (Supreme Commander of the Allied Powers, also known as General Headquarters or GHQ) at the time of the occupation, such as those alluded to in medical journals published during the occupation and now located in the Gordon W. Prange Collection (NISHIMURA 1989: 2). As outlined by NISHIMURA (1989: 4–5), censorship performed by the Civil Censorship Detachment (CCD) of the SCAP encompassed two forms. The first was in force from autumn 1945 to late 1947 and involved pre-publication censorship, whereby two copies of each publication were submitted to the CCD for examination, of which one was kept on file. Publication, as such, was only permitted pending CCD approval. This pre-publication censorship was replaced by post-publication censorship in late 1947 and remained in force until October 1949. CCD staff consisted of roughly 80 % Japanese nationals who worked as 'readers', reading and translating material to a surprisingly reliable standard. This work was then checked by supervisors, oftentimes second-generation Japanese Americans, or other foreigners. According to NISHIMURA, censorship was very lenient and over 99 % of the checked material passed without any censorship markings. The markings that were made generally related to one of the four categories: "wartime nationalist propaganda, mention or criticism of the SCAP, reference to effects of atom bombs (which the SCAP classified as public disturbance), and allusions to censorship itself" (NISHIMURA 1989: 5–7). Against this backdrop, it might be reasonably assumed that the local press in Hiroshima and Nagasaki was well aware of what could be written and what was not favoured by the CCD.

In addition to NISHIMURA's research, HOOK (1988) and KAWAGUCHI (2006) have also carried out detailed analyses of several periods of censorship in Japan, identifying different characteristics and motivations for the practice. HOOK (1988: 1–2) divides the censorship under occupation into three periods. The first he dates from August 7 (the day after the bombing of Hiroshima on August 6) to August 14, 1945 (the day before the declaration of the end of the war on August 15), when censorship was still in the hands of the Japanese authorities. According to HOOK, the Japanese media's coverage of atomic bomb damage in this period was based on official statements made by the Japanese authorities.[5] As HOOK points out, the authorities "minimized the damage and power of the 'new type of bomb'", and emphasized instead the following two dynamics: "the possibility of adopting countermeasures against future atomic damage" as well as "the contradiction between the declared values of the enemy [...] and the injustice and inhumanity of the attacks on Hiroshima and Nagasaki" (HOOK 1988: 2). This first period was, thus, very much dominated by the war strategy of the Japanese government and did not change until Japan's surrender on August 15, 1945, followed by the publication of the Press Code around September 18, 1945. For HOOK, the end of the war marks the beginning of this second period, during which the Japanese media was free from censorship (HOOK 1988: 1). KAWAGUCHI (2006: 229–32), on the other hand, divides this latter time frame itself into two separate periods, noting—as I do—differences in how news was covered in the Japanese media before and after the arrival of the Allied Forces at the end of August 1945. For KAWAGUCHI, the period immediately following Japan's surrender was marked by a political vacuum of sorts, one which afforded the Japanese press relative freedom. In terms of the atomic bomb, articles thus appeared—some even including photographs—describing the actual destruction of the two cities as well as the cost to human life and the effects of radiation. For KAWAGUCHI, a second period is then marked by the arrival of the Allied Forces at the end of August and continues until the publication of the Press Code. She notes that as soon as the Allied Forces arrived, there was a decrease in reports on the devastation caused by the atomic bombs and the negative effects of radiation, mostly likely for fear that it could lead to repercussions in the new political environment. HOOK and KAWAGUCHI are in agreement in terms of the last period, however, which runs from the beginning of censorship under the Press Code on September 19, 1945, until its end on October 31, 1949. For KAWAGUCHI, "it cannot be denied that the coverage of the atomic destruction was effectively reduced after the Press Code" (KAWAGUCHI 2006: 232). She also states that criticism of the use of the bomb as well as criticism of the U. S. were suppressed, while "the moralistic justification of the use of the bombs [...] was permitted" (KAWAGUCHI 2006: 232).

The articles analyzed over the pages that follow are generally taken from the last period identified by HOOK and KAWAGUCHI when censorship was already in place despite the fact that it was not to be seen in the articles themselves. Before beginning with the analysis proper of the five narratives that are the focus of this paper, however, the state of the daily press in Hiroshima and Nagasaki shall briefly be explained.

4 DAILY PRESS IN HIROSHIMA AND NAGASAKI, 1945–49: FIVE NARRATIVES

As a point of departure, Ann SHERIF's article—"Who Reads Shashi? The Case of the Hiroshima Regional Newspaper" (2012)—summarizes quite well an ongoing debate regarding the role of newspapers in society. For SHERIF (2012: 1), newspapers function as "powerful institution[s] in society". Indeed, scholars have defined, as SHERIF would have it, three societal roles for newspapers. First, "newspapers are businesses that function as impartial observers of society and conveyers of information". Second, "newspapers are advocates for the public good, watchdogs of society". And third, "newspapers are servants of the state, or, since the twentieth century, agents of manipulation and control by the state and industry" (SHERIF 2012: 3). As these three possible points illustrate, the particular role assumed by the media at any given moment very much depends on the political and social circumstances. In democratic societies therefore, SHERIF (2012: 3) concludes, newspapers view themselves as promoters of freedom of expression, while working to guarantee the public's right to be informed . The role of a particular newspaper is also determined by its audience reach. SHERIF differentiates between regional block newspapers (*burokkushi* ブロック紙) such as the *Chūgoku shinbun*, and national newspapers (*zenkokushi* 全国紙) such as the *Asahi shinbun* 朝日新聞 or *Yomiuri shinbun* 読売新聞. Regional and national newspapers faced several challenges during the occupation as well as during the war. SHERIF (2012: 8) explains that from the late 1930s on, newspaper production was controlled by the Japanese government through censorship, which controlled content and rationed paper and printing materials. Viewed by the government as "mouthpieces of the state", newspapers were required to publish the specific content provided by the government. As SHERIF has pointed out, only one newspaper per prefecture was permitted, which meant that a lot of smaller regional newspapers fused with regional papers or were bought by national ones. Against this backdrop, the *Chūgoku shinbun* was the only paper in the Chūgoku region—consisting of the prefectures of Tottori, Shimane, Okayama, Hiroshima and Yamaguchi—that survived the forced consolidation of national and regional newspapers (*mochibun gōdō* 持ち分合同). From November 1944 onwards, the rationing of printing resources was reduced even further when the government decreed that newspapers were not to exceed two pages per day. The space on these two pages (front and back) was split between news and the advertisements that—along with subscriptions—provided a fundamental source of income (SHERIF 2012: 8).

As for the severe rationing of paper and resources in the publishing industry during the occupation, Chad DIEHL's study—*Resurrecting Nagasaki: Reconstruction and the Formation of Atomic Narratives* (2018)—suggests that this served to control the flow of ideas in publishing, the ultimate aim of which was to turn Japan from an enemy to an ally (2018: 95–96). The United States, which suffered no shortage of paper at home, used the resource to exert control over the Japanese publishing industry. Indeed, the United States disseminated millions of leaflets and newsletters—so-called 'paper bombs'—over Japan and Southeast Asia that conveyed information about the atrocities of the Japanese government and military as well as ideas that were favorable to the United States. According to DIEHL (2018: 96), the "SCAP hoped that control of paper allocation and printing would help shape ideas in Japanese society and mould the nation in ways that suited the interests of the United States". This also meant taking control over the atomic narratives that emerged in the aftermath of the bombings. Instead of completely censoring all discussion of the atomic bombings, the SCAP instead tried to control the discussion. This meant permission and, indeed, paper were given for literature that was in line with the politics of the U.S. government, such as NAGAI Takashi's 永井隆 (1908–51) "The Bells of Nagasaki" (*Nagasaki no kane* 長崎の鐘, 1946). A central figure for atomic bomb literature (*genbaku bungaku* 原爆文学), NAGAI was favoured by the SCAP. Rather than criticizing the U.S. deployment of the atomic bomb, he "expressed gratitude for it; he professed anti-communist ideas; [...] he praised the future of atomic energy and related it to world peace" (DIEHL 2018: 96–97). For DIEHL (2018: 98), the utility of such statements in controlling and shaping the flow of information about the war and the atomic bombings is clear. It is no wonder, he claims, the publishing industry was one of the first commercial sectors to recover from the war. However, it was clear to the leaders of the SCAP that they would not be able to control the discussion eternally. Hence, "they selectively approved books to cultivate a narrative of Nagasaki and Hiroshima that did not explicitly indict the Americans" (DIEHL 2018: 101).

For the local news media in Hiroshima and Nagasaki, being part of the publishing industry controlled by the SCAP brought difficulties of its own. It meant navigating the interests of the local public such as the reconstruction of the two cities and their self-image, as well as those of occupation officials and the censorship authorities, upon which the survival of the newspapers depended ultimately. Furthermore, in both Hiroshima and Nagasaki, the headquarters of various local newspapers were severely damaged by the bombings. As direct survivors of the bomb, they needed to reconstruct themselves while also participating in the reconstruction of their cities.

There is less literature available on the history of the *Nagasaki shinbun* and the other Nagasaki newspapers that it eventually absorbed[6] as there is for the *Chūgoku shinbun*. As such, the latter shall serve to illustrate the condition of the local news media in the cities. In her detailed study on the Hiroshima Peace Memorial Museum, Stefanie SCHÄFER elaborates on the comprehensive recording of the *Chūgoku shinbun*'s own history. According to SCHÄFER (2018: 42), the *Chūgoku shinbun* has dedicated an entire archive to coverage of the atomic bomb. It appears to see itself first and foremost as a chronicler of the city of Hiroshima and attaches great importance to factual reporting. While political interventions or investigative journalism seem to be the exception for the *Chūgoku shinbun*, the newspaper has put great emphasis on the atomic bomb commemoration (SCHÄFER 2018: 42). Commemoration of the atomic bomb has played an important role in the newspaper's self-perception to this day. MIYAZAKI Tomomitsu—who directed the Hiroshima Peace Media Center—introduces the newspaper as follows: "The Chugoku Shimbun has always stood beside the people of Hiroshima as a newspaper company that also endured the tragedy, and it worked hard to support the city's reconstruction in the aftermath of the atomic bombing"

(MIYAZAKI 2015: 527). If we relate the *Chūgoku shinbun* to SHERIF's three roles for newspapers outlined above, the paper would seem to regard itself as part of the local community in Hiroshima, serving the city as an advocate for peace. In this respect, the newspaper has seen an image-change in comparison to its prewar role. As SHERIF explains (2012: 5–6), the *Chūgoku shinbun* had a strong media presence as only one of three local newspapers in the 1930s. By conforming to the demands of the imperial government, it became the dominant source of news in the Chūgoku region and an important stakeholder during the war. Yet, as the newspaper itself explains in its own company history (*shashi* 社史), this changed after the atomic bombing and the end of the war. Even though the newspaper lost employees and its headquarters were destroyed by the atomic bomb, located, as they were, only about 1,000 meters from the hypocenter, the newspaper managed to resume printing in the form of special editions a few days after the bombing, owing to a contingency plan that provided alternate printing facilities. Within several months, regular publication was once again secured (SHERIF 2012: 9–10). The extent to which the *Chūgoku shinbun* sees its own history as intertwined with that of the atomic bombing and the reconstruction of Hiroshima can clearly be seen in the specific content of the paper's online presence. This includes a manga called "Manga story of the newspaper of the city hit by the atomic bomb"—which was created by the *Chūgoku shinbun* on the occasion of the 130th anniversary of the company's founding[7]—as well as a 28-minute long program describing the state of newspaper after the bomb was dropped as well as its new start in the bomb's wake.

The following sections will deal with five narratives that were featured prominently in the daily press in the years following the atomic bombing (1945–49), not only by the *Chūgoku shinbun* but also by several other newspapers in Hiroshima and Nagasaki. A look at dominant narratives in: 1) coverage about the atomic bombings; 2) the reconstruction of the two cities; 3) the festivities commemorating the bombings; and 4) the nationalization of the atomic bomb commemoration all offer insights into topics and themes that were seemingly accepted by the censorship department and that shape the image of Hiroshima and Nagasaki up to this day.

The most prominent theme dominating the daily press in the two cities is the international scope of the atomic bombings. As alluded to above, "the most comprehensive archive in the world of Japanese print publications issued during the early years of the occupation of Japan, 1945–1949" (UNIVERSITY OF MARYLAND) can be found in the Gordon W. Prange Collection.[8] As Yukako TATSUMI (2019) has pointed out, it took some time until the more than 500 wooden crates containing the comprehensive print collection were finally opened to the public in the 1960s, even though they had arrived at the University of Maryland in the early 1950s.⁹ One reason for this was that the SCAP had been intent on "protecting the confidentiality of its censorship operation and keeping the general public unaware of SCAP's censorship of Japanese publications". Ultimately, it was a Japanese scholar who brought most attention to the great amount of postwar Japanese print publications held at the libraries. Cooperative work with the National Diet Library of Japan (NDL) did not begin until the 1990s (TATSUMI 2019a: 7–8). Today, the Prange Collection is accessible via an online database.¹⁰

Navigating the database and scrolling through newspaper and magazine articles that deal with the bombings reveals various writings both in the database entries as well as the related print articles. The different forms of Hiroshima and Nagasaki in print language have been discussed in detail by Hiroko OKUDA (2011: 12–15). The names of the two cities have been rendered in different ways since the bombs were dropped, sometimes with Kanji (Chinese characters), Hiragana (Japanese syllabary used for Japanese words) or Katakana (Japanese syllabary used especially for foreign loanwords). According to OKUDA, the different writings represent different memories (OKUDA 2011: 12).

There are two different ways of writing Hiroshima in Kanji and the difference between the two "show[s] the historical discontinuity between the military city hit by the atomic bomb and the present representative of world peace" (OKUDA 2011: 12). The older method—廣島—is reminiscent of the city's history as a military capital (*gunto* 軍都) and calls to mind imperial colonialism and Japanese military expansionism. The modern version—広島—does not have these negative implications given that the two characters stand simply for 'wide' (*hiroi* 広い), and 'island' (*shima* 島). Nowadays, the Hiragana for Hiroshima—ひろしま—is especially used for municipal publications though it also evokes childish and nostalgic images owing to the fact that Hiragana syllabary is the first form of writing learned by native Japanese speakers. The final form of syllabic writing in Japanese is Katakana, which is mostly used to transliterate foreign words. According to OKUDA (2011: 13), when written in Katakana, Hiroshima evokes an entirely different image: "While symbolizing Japan's recovery from the defeat in the war in a national context, in a global context it characterizes post-nuclear Hiroshima for the peace and antinuclear movements". OKUDA here refers to the established English slogan 'No more Hiroshimas' (*Nō moa Hiroshimazu* ノーモアヒロシマズ).

OKUDA (2011: 13–14) lists three different forms of writing for Nagasaki. The Kanji version—長崎—refers to Nagasaki's international and Christian past, as well as the city's history as having prospered from the construction of warships. It thus "stands for both its international influence and its atomic destruction". The Katakana version ナガサキ, in contrast, is closely connected to the image of Nagasaki as an 'atomic wasteland' as well as to the slogans 'peace begins in Nagasaki' (*Heiwa wa Nagasaki kara* 平和はナガサキから) and 'let Nagasaki be the last city to suffer an atomic bomb' (*Nagasaki o saigo no hibakuchi ni* ナガサキを最後の被爆地に). The Hiragana version ながさき "[...] encourages the nostalgic notion of child-like innocence, and at the same time transforms the city's 'ordeal' into its 'redemption'" (OKUDA 2011: 14). In the narratives that follow, a decisive role is also played by the different local, national and global identities that were formed through the various ways of spelling the name of the two cities following the bombings (OKUDA 2011: 14).

4.1 The international scope of the atomic bombings

Even today, over 75 years after the atomic bombings, their legacy is still deeply engraved into both cities. Indeed, the mayors of both Hiroshima and Nagasaki deliver peace declarations each year at the annual commemoration of the bombings. In these speeches, Nagasaki's task is clearly defined: to "achieve permanent world peace" (TAUE 2022). About 75 years ago, this self-imposed role was already imminent in the daily press of the two cities. One article titled "Towards the promotion of world peace" (*Sekai heiwa no sokushin e* 世界平和の促進へ), for example, covers NAGAI Takashi's call for a commemorative peace tower in the city to demonstrate the city's peace efforts (NNS 1949/04/25). Through the publication of articles, such as "Echoing to the world 'No More Hiroshimas'" (*Sekai ni hibike 'Nō moa Hiroshimazu'* 世界に響け「ノー・モア・ヒロシマズ」; NNS 1949/08/07), local newspapers contributed to the cities' self-image as important international actors for world peace. Yet, not only local newspapers but also newspapers from other prefectures, such as the Hiroshima's neighbor prefecture Ehime, picked up on the city's new role, as an article on the fourth anniversary of the bombing indicates. The article—"Resounding Peace Bell: From Hiroshima, the land of the atomic bomb, to the whole world" (*Hibike heiwa no kane: Genbaku no chi Hiroshima kara zen-sekai e* 響け平和の鐘：原爆の地広島から全世界へ)—covers the festivities that took place in Hiroshima to commemorate the fourth anniversary of the bombings under the banner 'No more Hiroshimas' (SES 1949/08/07). Furthermore, international visits, such as those made by the American writer Helen KELLER (1880–1968) to Nagasaki, further highlight the international role now played by both cities. Commenting on her visit, the *Nagasaki min'yū* 長崎民友 published the following article: "Nagasaki Station where the light of love remains: Ms Keller on her way home yesterday" (*Ai no hikari nokoru Nagasaki eki: Kinō Kerā joshi kikoku no to e* 愛の光残る長崎駅：きのうケラー女史帰国の途へ; NM 1948/10/20). However, as is shown by the article "Finally appearing in the world: 'The Bells of Nagasaki'" (*Iyoiyo yo ni deru 'Nagasaki no kane'* いよいよ世に出る「長崎の鐘」; SS 1948/07/31), the international publication of "The Bells of Nagasaki" by Nagai Takashi was closely monitored by local press. A fundamental change in self-perception from the two cities as militaristic locations to symbols of world peace is particularly evident in the SCAP's favorable reception of this work, which included an appendix on Japanese wartime atrocities in Manila (Diehl 2018: 96–98).[11] This change was equally indicated by articles such as "What is UNESCO? Movement to overcome illiteracy and ignorance and invoke permanent world peace: UNESCO and Japan" (*Yunesuko to wa nani ka? Monmō to muchi o kokufuku shi: Kōkyū-teki sekai heiwa no shōrai e no undō: Yunesuko to Nihon* ユネスコとは何か?文盲と無知を克服し:恒久的世界平和の招来への運動:ユネスコと日本; NM 1948/09/02), which refers to the early UNESCO influence in the two cities.

4.2 Towards peace: Hiroshima in transition

That Hiroshima has become a peace city is self-evident. Even tourists visiting the city incorporate the city's peace principles into their online reviews, thus becoming "spontaneous conveyor[s] of that memory" (Van der Does / Kawano 2020: 527). In the first years after the bombing, tourism was a fundamental necessity to securing the reconstruction and financial stability of the city. According to Schäfer (2018: 68–78), atomic bomb tourism played a vital role for the reconstruction of the city and was one of the reasons for the conservation of the many relicts and sites of atomic devastation. Indeed, the city of Hiroshima was aware that the anticipated visitors would expect to see traces of the bomb and its effects, which made it necessary to preserve these despite the fact that this would lead to conflicts over reconstruction and preservation. Schäfer notes that the combination of atomic bomb commemoration and financial interests did not pose a contradiction for the city's most influential newspaper, with the *Chūgoku shinbun* equally promoting atomic bomb tourism (Schäfer 2018: 74). After all, the Hiroshima Peace Memorial City Construction Law was passed shortly after the war and sought to transform the location into a peace city of international importance. Yet it also functioned to ensure financial support from the national government (Schäfer 2018: 83). At the same time, local actors such as the press played an important role in shaping the public Hiroshima commemoration from the beginning (Schäfer 2018: 83). First and foremost, it was the *Chūgoku shinbun* that—alongside the Hiroshima City Hall and the Hiroshima Peace Memorial Museum— "fashioned a remarkably consistent narrative of the city's history and identity, [...] of deep involvement in the empire's project of modernization and its wars, of the continuity from Hiroshima's development as a modern city to the bombing" (Sherif 2012: 6). The transition from a city in ruins to a modern "peace city" (*heiwa toshi* 平和都市) reverberates in the reportage of the local press. Take, by way of example, the article "August 6 is world peace day: Humanity reflects on Hiroshima's sacrifice" (*8/6 wa sekai no heiwabi: Hiroshima no gisei o jinrui ga hansei* 8・6は世界の平和日：廣島の犠牲を人類が反省; CS 1948/07/13), which reinterprets August 6 as an international world peace day and imagines Hiroshima in a universal role. Moreover, it refers to the atomic bombing as a "sacrifice", thus drawing on interpretations such as those asserted by Nagai Takashi. Meanwhile, Hiroshima's transition is closely monitored by the Nagasaki newspapers, as is shown by an exemplary article that includes a photograph of the reconstructed and modern city landscape of Hiroshima: "This is how Hiroshima has been rebuilt. Calling out to the whole world for the 'No more Hiroshimas' movement: The two cities only have one wish" (*Hiroshima wa kō shite fukkō shita: Zen-sekai ni yobikakeru Nō moa Hiroshimazu undō: Ryōshi no negai wa tada hitotsu* 広島はこうして復興した：全世界に呼びかけるノーモア・ヒロシマズ運動：両市の願いは唯一つ; NM 1948/08/09).

4.3 Towards culture: Nagasaki in transition

The 'one wish' ostensibly shared by both cities, both now and then, is for 'world peace'. This message was spread to the world by the local press in Hiroshima and Nagasaki very early on. For evidence of such, one needs look no further than the article: "Peace from Nagasaki: Two messages flying across the sea: From Nagasaki to global citizens" (*Pīsu furomu Nagasaki: Umi o tobu messēji futatsu ni: Nagasaki kara sekai shimin e* ピース・フロム・ナガサキ：海を飛ぶメッセージ二つに：長崎から世界市民へ; NM 1949/05/25). The petition that sought the above-mentioned Nagasaki International Culture City Construction Law featured significantly in the local press in Nagasaki, as can be seen in articles such as "Nagasaki as a symbol of peace: Petition for the 'Special law concerning the Peace Memorial City Nagasaki'" (*Genbaku Nagasaki o heiwa no shinboru ni: 'Heiwa kinen toshi Nagasaki ni kan suru tokubetsuhō' seigan* 原爆長崎を平和のシンボルに：『平和記念都市長崎に関する特別法』請願; NM 1949/04/24). Yet, as a look at newspaper articles in the early years after the bombings reveals, the transformation into an international city of peace and culture did not only take place on paper or in the imagination of its various local actors, but also very physically through a change in the city landscape. Again, atomic bomb tourism played a significant role in the reconstruction of the city, as visitors to—or more accurately readers of—the *Nagasaki min'yū* are invited to "Please take a look: 'Tourist Nagasaki Information Tower' in front of Nagasaki Station (*Ichimoku go-ran kudasai: Nagasaki ekimae ni 'kankō Nagasaki annaitō'* 一目ご覧下さい：長崎駅前に『観光ナガサキ案内塔』; NM 1948/09/06).

Yet, as Tomoe OTSUKI (2016: 396–400) has pointed out, the transformation of the city has come at the cost of minorities such as the Christian community of Nagasaki and the *buraku* community[12] who used to inhabit the Urakami area, the part of Nagasaki that was most heavily damaged by the atomic bomb. In contrast to the Christian community and its most prominent representative NAGAI Takashi, however, the *buraku* community is not at all represented in local atomic bomb commemoration. Moreover, as OTSUKI emphasizes, Nagasaki is "one of the pivotal cities in the Japanese war effort [...] [and] remains one of the largest manufacturing sites for military materials under the Mitsubishi Heavy Industrial Corporation" (OTSUKI 2016: 396). OTSUKI (2016: 402–09) also uncovers NAGAI Takashi's imperial past as a surgeon in the Japanese imperial army, as well as that of KITAMURA Seibō 北村西望 (1884–1987), who used to produce statues of prominent military figures during the war. After the war he was commissioned to create the Statue for Peace Memorial (Heiwa kinenzō 平和祈念像) in the Nagasaki Peace Park.

4.4 From small reconstruction festivals to official peace ceremonies

The fourth narrative consists of the ongoing transformation of the small reconstruction festivals that took place in the cities in the first years after the bombing to the official peace ceremonies that are being held today on a national scale and find their way into international coverage.[13] As SCHÄFER (2018: 58) notes, the first small anniversary festival was held 1946 in Hiroshima in light of growing demands to commemorate the fallen. Specifically, a "memorial tower for the fallen" (*senbotsu kuyōtō* 戦没供養塔) was erected on the grounds of today's Hiroshima Peace Memorial Park (Hiroshima heiwa kinen kōen 広島平和記念公園). Looking at the Kanji used for Hiroshima—広島—the modern writing does not evoke negative implications of Hiroshima as a military capital but stands instead for the modern peace city, as OKUDA (2011: 13) has shown. According to SCHÄFER (2018: 70), the first "Peace reconstruction festival" (*heiwa fukkōsai* 平和復興祭) in 1946 had the character of a public mourning ceremony. However, the peace trope had already found its way into the media reports and the local press was simultaneously shaping the atomic bomb commemoration that was forming in the years immediately following the bombing. A look at the *Chūgoku shinbun* title page from August, 1946, reveals that the coverage of the first anniversary of the atomic bombing of Hiroshima was full of references to "peace". The word "peace" appears in most articles and headlines in the form of: "Peace conference" (*heiwa kaigi* 平和会議), "Hiroshima peace construction" (*heiwa Hiroshima kensetsu* 平和廣島建設), or "Peace reconstruction festival". A look at the older Kanji writing for Hiroshima—廣島—indicates that while the city was moving forward in creating its new peace-driven self-image, it was still rooted a past enshrouded by war. The atomic bomb, on the other hand, is presented on the title page of August 6, 1946, in the following terms: "The atomic bomb is humanity's greatest gunpowder" (*Genbaku wa jinrui saidai no kayaku* 原爆は人類最大の火薬; CS 1946/08/06). Surprisingly though, the title page from the following day—on which one would expect pictures of the first "Peace reconstruction festival"—is rather plain (CS 1946/08/07) (see Figure 1). One year later, the title page from August 7, 1947, already shows a crowd at a "Peace festival" (*heiwasai* 平和祭) with the Genbaku Dome partially visible in the background. It also prominently features photographs of the first official peace ceremony held in Hiroshima. These show visitors attending the ceremony as well as depicting speeches given by the Supreme Commander for the Allied Powers, General Douglas MACARTHUR, together with the mayor of Hiroshima, HAMAI Shinzō 浜井信三 (1905–68). In addition to several peace-related articles, the edition of August 7, 1947, also features a message from General Douglas MACARTHUR to the Japanese nation. At the end of the page, the slogan "The bell of peace rings" (*Heiwa no kane ga naru* 平和の鐘が鳴る; CS 1947/08/07) is surrounded by several advertisements. The latter exemplifies a point made by SHERIF (2012: 8) with regard to the daily press after the war. To wit, the newspapers depended on advertisements as a fundamental source of income. At the same time, they were only able to print two pages per day, which

Fig. 1: Title page of the *Chūgoku shinbun* on August 7, 1946 © Courtesy of The Chugoku Shimbun Hiroshima Peace Media Center.

Fig. 2: Title page of the *Chūgoku shinbun* on August 7, 1947 © Courtesy of The Chugoku Shimbun Hiroshima Peace Media Center.

leads to the rather odd combination of atomic-bomb commemoration content and advertisements (see Figure 2).

The *Chūgoku shinbun* title page of August 7, 1948, shows almost the same scenery and motifs as one year earlier. In this instance, however, the Genbaku Dome is fully visible and framed by a banner with the slogan "No more Hiroshimas". Printed next to the photograph is the "Peace declaration" (*heiwa sengen* 平和宣言) (CS 1948/08/07). The edition from August 7, 1949, shows almost the same content including the speeches, the bell, and the peace declaration. Again, peace slogans like "Peace forever" (*heiwa yo eien ni* 平和よ永遠に) are placed alongside advertisements (CS 1949/08/07).

Unfortunately, there were no such online collections available as would provide a point of comparison for the front pages of Nagasaki newspapers from a comparable period. Interestingly, the title page of *Nagasaki min'yū* from August 9, 1948—the anniversary of the atomic bombing on Nagasaki—is, in fact, focused on Hiroshima. The headline reads: "This is how Hiroshima has been reconstructed" (*Hiroshima wa kō shite fukkō shita* 廣島はこうして復興した). The first photograph of the Nagasaki city landscape appears on the second page with the description "Nagasaki City revived for the 3rd Anniversary of Nagasaki" (*Nagasaki sanshūnen o mukae yomigaetta Nagasaki-shi* 長崎三周年を迎えよみがえった長崎市; NM 1948/08/09).

These examples show that in the years after the bombings, both cities began to ritualize their commemoration in the form of peace festivals that evolved into grand national ceremonies. An important role in this process was played by local newspapers and by the fact that each city provided coverage of the other's commemorative activities.

4.5 Japan remembers: the nationalization of atomic bomb commemoration

KAWAGUCHI Yūko's article on newspaper reports of the atomic bombing of Hiroshima in the early postwar years has pointed out that according to the nationwide coverage the bombing of Hiroshima and Nagasaki was not initially considered a national event. Rather, the memories of Hiroshima were transformed by a process of nationalization (KAWAGUCHI 2006: 241). SCHÄFER (2018: 54) also notes that it was local actors who shaped public atomic bomb commemoration in the first years even before the national government began to take an interest. In the case of Hiroshima and Nagasaki, the nationalization of the atomic bomb commemoration as seen through the daily press in both cities took place, first and foremost, through royal visits. On May 29, 1949, the *Ehime shinbun* 愛媛新聞 covered Emperor Hirohito's visit to NAGAI Takashi: "His Majesty is paying a visit to Dr. Nagai" (*Heika, Nagai-hakase o o-mimai* 陛下、永井博士をお見舞; ES

1949 / 05 / 29). The Emperor's visit with one of the most prominent advocates of Nagasaki atomic bomb commemoration signalled the beginning of a process of nationalization, though in a certain sense, this reflects the attitudes and goals of the Allied Occupation given that NAGAI's literature was in line with American policy (DIEHL 2018: 96–97). Not necessarily in the first years after the bombings, but at the latest from 1949 on, the Emperor's visits to the two cities began to function as significant demarcations of commemoration while making the Emperor himself an important figure in the movement. Accordingly, on May 28, 1949, the *Chūgoku shinbun* covered his visit to Hiroshima in the article titled: "His Majesty, supporting the citizens of the atomic bomb city Nagasaki" (*Heika, bakuto Nagasaki shimin go-gekirei* 陛下、爆都長崎市民ご激勵; CS 1949 / 05 / 28).

As OKUDA (2011: 15) writes: "[...] Tokyo has integrated A-bomb memories of Hiroshima and Nagasaki and the national feeling against nuclear weapons, which is well known as Japan's 'nuclear allergy', into a shared sense of being a *heiwaaikoo* (peace-loving) nation". Carol GLUCK describes this as "the realm of national history and public memory, where the past is collectively constructed, disputed and perpetuated" (GLUCK 1993: 65). To return to KELLER's (2013: 3) formulations on discourse, the sociologist writes "that in the practical use of signs, meanings of phenomena are socially constructed and these phenomena are thereby constituted in their social reality". The peace ceremonies can be seen to have socially constructed meaning constituted in their social reality in the local communities as well as the local and national commemoration ceremonies.

5 CONCLUSION: NUCLEAR GAPS

Hiroshima and Nagasaki would appear to have reached their goal of becoming national and international symbols. While they integrate the memory of the atomic bomb into their self-representation to different degrees, the landscapes of both cities are living witnesses to events that occurred almost 80 years ago. They incorporate peace parks, museums, walking tours, annual commemoration ceremonies and much more into their cities' public image. Shortly after the bombings, both cities enacted laws—the Nagasaki International Culture City Construction Law and the Hiroshima Peace Memorial City Construction Law—to ensure financial support for reconstruction and to manifest their new roles as international symbols of peace. As SCHÄFER (2018: 292) has shown, for the respective administrative bodies of each city, this served, first and foremost, to ensure the future of the cities especially regarding funding for its reconstruction. The word "peace" was thus used in an almost inflationary fashion in all kinds of festivities, slogans (such as 'No more Hiroshimas'), urban planning, and even in laws. This incredibly broad field of application attests to the discursive transformation of the two cities. For KELLER, this is exemplary of a type of discourse that is "more or less successful [in its] attempts to stabilize, at least temporarily, attributions of meaning and orders of interpretation, and thereby to institutionalize a collectively binding order of knowledge in a social ensemble" (KELLER 2013: 2). Actors such as the city administration, the annual peace declarations of the respective mayors, authors of atomic bomb literature such as NAGAI Takashi, and ultimately the national government through their royal visits, made the transformation from two cities in ruins into international ambassadors of world peace possible.

However, as a look at the local newspapers and their coverage in the years following the bombings has shown, the new self-image of the two cities was demonstrated to a significant extent by local newspaper coverage concerning their architectural transformation and legislation, about visits by national and international guests, and about local and national commemoration. As SHERIF (2012: 3) has indicated, in the wake of the bombing, local newspapers such as the *Chūgoku shinbun* saw themselves as "advocates for the public good, watchdogs of society", a role which they continue to embody to this day. As the foregoing has sought to demonstrate, however, their interests do appear to have been in line—at least to a certain extent—with the policies and goals of the U. S. during its occupation of Japan. This can clearly be seen in the categories of censorship employed by NISHIMURA (1989: 5–7), namely any "mention or criticism of SCAP, reference to effects of atom bombs (which SCAP classified as public disturbance), and allusions to censorship itself". Although the materials accessible through the Gordon W. Prange collection occasionally show censorship markings, self-censorship of the press in advance of publishing may also have played an important role in the articles that were ultimately published and the direction taken by public discussions in the press. This may also have been a reason for the frequent mention of the author NAGAI Takashi, who was favored by the SCAP given his apparently neutral stance on the U. S. decision to use the atomic bomb in Japan. NAGAI instead related the bombing to the end of war and to world peace (DIEHL 2018: 96–97). The predominance of the peace narrative—which was fundamental for the image change of the two cities—also occurred in accordance with U. S. hopes given that it shifted the focus away from the destruction caused by the atomic bombs toward a vision of a bright future of world peace.

It is worth noting, as OTSUKI (2016: 397) has shown, that the voices of minorities such as the *buraku* community—who used to inhabit the Urakami area in Nagasaki that was most heavily damaged by the atomic bomb—have been silenced in the process of the cities' transformation into "symbols of peace". In contrast to the Christian community, they are absent from the atomic bomb commemoration discourse. Similarly, the imperial past of prominent figures such as NAGAI and KITAMURA has been concealed. These are just a few blank spots that can be identified in the reportage of the local press. After all, censorship control, paper shortage and the financial situation in which local newspapers found themselves after the atomic destruction are likely to have played an important role in directing their coverage. As for financial difficulties, newspaper dependence on advertisements sometimes resulted in consumer advertisements placed in rather odd contexts, such as directly alongside peace messages. Nonetheless, what has endured after 80 years of transformation are the bright pastel colors of tourism advertisements seen in Hiroshima and Nagasaki today as well as the cities' apparently never-ending road to world peace. Contrary to what is stated by the protagonist in *Hiroshima mon amour*, there may well be more to see in Hiroshima—and, indeed, Nagasaki, than just soap advertisements, it seems that peace itself reigns.

NOTES

1 The primary objective of the law was to use Hiroshima's special status to receive governmental aid for reconstruction. The law enabled a different approach to Hiroshima and broader financial support for the city compared to others that were destroyed by conventional air raids but which also received support from the government.
2 One generally speaks of 'Allied Occupation', but in the case of Japan it was almost exclusively carried out by the United States.
3 The guide maps on the Hiroshima Peace Tourism's website are introduced with the words: "Proud to announce that the first ever English language Hiroshima Peace Tourism Guide Map is now available!". The website offers three different walking tours featuring "iconic landmarks, museums, and more". It also claims to offer a multimedia experience through QR codes and mobile sites such as Twitter or Instagram, encouraging its readers to "[b]e sure to get your copy today!" (CITY OF HIROSHIMA 2018–23).
4 In contrast to the reference of Japanese authors publishing in Japan, where the surname is given before the given name. Since Sey NISHIMURA publishes in the United States, the order of his name is adopted in the order taken from his publication, not in the usual way which involved writing the surname before the given name. This will apply to all Japanese authors in this English-language article.
5 Censorship in Japan did not begin when the Allied forces arrived. Rather, the Japanese military government had already used censorship as a means to control mass culture (YAU / WONG 2020: 27–28). Some of the Japanese censorship personnel went on to be employed by the CCD (BRAW 1986: 64–66).

6   In contrast to the *Chūgoku shinbun*, the *Nagasaki shinbun* does not seem to build its self-image as much on the atomic bombing and peace efforts. The internet presence of the *Nagasaki shinbun* does not have an extensive section on the atomic bombing and its aftermath or the company's history against this background. Navigating the newspaper's website, one has to look under the subheading "projects and special features" (*kikaku, tokushū* 企画・特集) to find a relatively plain "peace site" (*pīsu saito* ピースサイト) containing articles related to the atomic bomb (NAGASAKI SHINBUN).

7   The Manga, created in 2022, was published as a one-volume booklet and is currently sold at the *Chūgoku shinbun*'s head office as well as at the Rest House near Hiroshima Peace Memorial Park. The manga consists of four parts: "August 6—I'll never forget that horrifying day", "Viewfinder blurred with tears", "The newspaper of the city hit by the atomic bomb that worked to help the city orally" and "Newspaper published by the Chugoku Shimbun from amidst the ruins" It can be accessed by CHŪGOKU SHINBUN (2022).

8   The Prange Collection holds every single Japanese-language print publication issued between 1945 and 1949. Among that are about 71,000 books, 18,000 newspaper titles, 13,800 magazine titles and 10,000 news agency photographs (TATSUMI 2019b: 263).

9   The Prange Collection is named after its donor, Dr. Gordon W. PRANGE (1910–80), history professor at the University of Maryland. In 1945, he became Chief of the Historical Branch of the SCAP's Intelligence Section in Japan. Since the Civil Censorship Detachment of the SCAP required a copy of all potential publications before release for review, massive amounts of copies were accumulated. When the SCAP began lifting its censorship operation in November 1949, PRANGE ordered the CCD collection to be brought to the University of Maryland and archived there, since he was aware of the historical significance of the material (TATSUMI 2019b: 263–64).

10 Since 2000, the NPO Institute of Intelligence Studies (Hōjin interijensu kenkyūjo 法人インテリジェンス研究所) at Waseda University has been working on a table of contents database to make research of the Prange Collection easier to access. The database today is called "20th Century Media Information Database" (*20-seiki media jōhō dētabēsu* ２０世紀メディア情報データベース) (YAMAMOTO 2013). The database contains 3,226,180 entries, about 1,964,900 belonging to magazine articles and 1,261,280 newspaper articles (FLACHE 2020).

11 The appendix was not initiated by NAGAI himself but had been demanded by the SCAP as a condition for publication (DIEHL 2018: 107).

12 The "buraku people" (*burakumin* 部落民) have been excluded from mainstream Japanese society for centuries for performing work that was stigmatized as 'dirty' such as leather processing and butchery. The stigmatization against them continues to this day (OTSUKI 2016: 396).

13 For example, Germany's most prominent newscast covers the commemoration ceremonies every year (TAGESSCHAU 2022a and 2022b)

REFERENCES

BRAW, Monica (1986): *The Atomic Bomb Suppressed—American Censorship in Japan 1945–1949* (Lund Studies in International History 23). Malmö: Liber Förlag.
DIEHL, Chad (2018): *Resurrecting Nagasaki—Reconstruction and the Formation of Atomic Narratives*. Ithaca / London: Cornell University Press.
GLUCK, Carol (1993): "The Past in the Present". In: GORDON, Andrew (ed.): *Postwar Japan as History*. Berkeley / Los Angeles / London: University of California Press, 64–95.
HOOK, Glenn (1988): "Roots of Nuclearism: Censorship and Reportage of Atomic Damage in Hiroshima and Nagasaki". In: *Working Paper* 16 (First Annual Conference on Discourse, Peace, Security, and International Society): 1–23.
ICOMOS (1996): "World Heritage List—Hiroshima: Advisory Body Evaluation"; https://whc.unesco.org/en/list/775/documents/ (last access 2023 / 03 / 19).
KAWAGUCHI, Yūko 川口悠子 (2006): "Newspaper Reports of the Atomic Bombing of Hiroshima in the Early Postwar Years: Local, National, and Transnational / *Sengo shoki no genbaku hōdō: Rōkaru, nashonaru, toransu nashonaru* 戦後初期の原爆報道：ローカル、ナショナル、トランスナショナル". In: *Pacific and American studies / Amerika taiheiyō kenkyū* アメリカ太平洋研究 6: 227–42.
KELLER, Reiner (2013): *Doing Discourse Research. An Introduction for Social Scientists*. Translated by Bryan JENNER. Los Angeles et al.: Sage.
MIYAZAKI, Tomomitsu (2015): "The view from under the mushroom cloud: The Chugoku Shimbun newspaper and the Hiroshima Peace Media Center". In: *International Review of the Red Cross* 97 (899): 527–42.
NISHIMURA, Sey (1989): "Medical Censorship in Occupied Japan, 1945–1948". In: *Pacific Historical Review* 58 (1): 1–21.
OKUDA, Hiroko (2011): "Remembering the atomic bombing of Hiroshima and Nagasaki: Collective memory of post-war Japan". In: *Acta Orientalia Vilnensia* 12 (1): 11–28.
OTSUKI, Tomoe (2016): "Reinventing Nagasaki: the Christianization of Nagasaki and the revival of an imperial legacy in postwar Japan". In: *Inter-Asia Cultural Studies* 17 (3): 395–415.
RESNAIS, Alan (1959): *Hiroshima, mon amour*. Argos et al.
SCHÄFER, Stefanie (2018): *Das Atombombenmuseum Hiroshima—Erinnern jenseits der Nation (1945–1975)*. Bielefeld: transcript Verlag.
SHERIF, Ann (2012): "Who Reads Shashi? The Case of the Hiroshima Regional Newspaper". In: *The Journal of Japanese Business and Company History* 1 (1): 1–18.

Tatsumi, Yukako (2019a): "Making a Case for Local Relevance: Strategic Exhibition Planning for the Gordon W. Prange Collection". In: *Journal of East Asian Libraries* 168: 4–20.

Tatsumi, Yukako (2019b): "Toward a Comprehensive Collection on the Allied Occupation of Japan: A Partnership between the University of Maryland Libraries and the National Diet Library of Japan". In: Luckert, Yelena / Inge, Lindsay (eds.): *The Globalized Library: American Academic Libraries and International Students, Collections, and Practices*. Chicago: ACRL Press, 261–74.

Van der Does, Luli / Kawano, Noriyuki (2020): "Online tourist reviews and accidental conveyors of memories of the atomic bomb". In: *Journal of Tourism and Cultural Change* 18 (5): 514–31.

Yoneyama, Lisa (1999): *Hiroshima Traces—Time, Space, and the Dialectics of Memory*. Berkeley / Los Angeles: University of California Press.

Yau, Hoi-yan / Wong, Heung-wah (2020): "A brief history of censorship in Japan". In: Yau, Hoi-yan / Wong, Heung-wah (eds.): *Censorship in Japan*. London: Routledge, 22–43.

Zwigenberg, Ran (2014): *The Origins of Global Memory Culture*. Cambridge: Cambridge University Press.

Internet Sources:

City of Hiroshima (2018–23): "Hiroshima Peace Tourism Guide Map"; https://peace-tourism.com/en/news/entry-191.html (last access 2023 / 06 / 11).

City of Hiroshima (2021): "The Hiroshima Peace Memorial City Construction Law and Commentary"; https://www.city.hiroshima.lg.jp/uploaded/attachment/151843.pdf (last access 2023 / 06 / 10).

Chūgoku shinbun (2022): "Manga story of the A-bombed city newspaper"; https://www.hiroshimapeacemedia.jp/?page_id=127116 (last access 2023 / 06 / 03).

Chūgoku shinbun: "Hiroshima 1945: The A-bombing and the Chugoku Shimbun"; https://www.hiroshimapeacemedia.jp/?page_id=25648 (last access 2023 / 06 / 03).

Flache, Ursula (2020): "Neu lizenziert: '20th Century Media Information Database'"; https://blog.crossasia.org/neu_lizenziert-20th-century-media-information-database/ (last access 2023 / 06 / 04).

Matsui, Kazumi (2022): "Peace Declaration"; https://www.city.hiroshima.lg.jp/site/english/158103.html (last access 2023 / 06 / 04).

Nagasaki shinbun: "Pīsu saito ピースサイト"; https://www.nagasaki-np.co.jp/feature/peace-site/ (last access 2023 / 06 / 04).

TAUE, Tomihisa (2022): "Nagasaki Peace Declaration"; https://www.city.nagasaki.lg.jp.e.jc.hp.transer.com/heiwa/3070000/307100/p036984.html (last access 2023/06/04).
TAGESSCHAU (2022a): "Gedenkfeier in Hiroshima—'Die Menschheit spielt mit einer geladenen Waffe'"; https://www.tagesschau.de/ausland/asien/hiroshima-gedenken-113.html (last access 2023/06/11).
TAGESSCHAU (2022b): "9. August 1945—Nagasaki gedenkt der Opfer des Atombombenabwurfs"; https://www.tagesschau.de/ausland/asien/nagasaki-gedenken-atomwaffen-101.html (last access 2023/06/11).
UNIVERSITY OF MARYLAND: "Postwar Japan"; https://www.lib.umd.edu/collections/special/japan (last access 2023/06/03).
YAMAMOTO, Taketoshi 山本武利 (2013): "Hon dētabēsu no tokushoku to mokuhyō 本データベースの特色と目標"; http://20thdb.jp/greeting (last access 2023/06/04).

Newspaper Articles:
Chūgoku shinbun 中国新聞 (CS)
CS (1946/08/06): "Genbaku wa jinrui saidai no kayaku 原爆は人類最大の火薬": 1.
CS (1947/08/07): "Heiwa no kane ga naru 平和の鐘が鳴る: 1.
CS (1948/07/13): "8/6 wa sekai no heiwabi: Hiroshima no gisei o jinrui ga hansei 8・6は世界の平和日：廣島の犠牲を人類が反省": 1.
CS (1948/08/07): "Heiwa sengen 平和宣言": 1.
CS (1949/05/28): "Heika, bakuto Nagasaki shimin go-gekirei 陛下、爆都長崎市民ご激勵": 2.
CS (1949/08/07): "Heiwa yo eien ni 平和よ永遠に": 1.

Ehime shinbun 愛媛新聞 (ES)
ES (1949/05/29): "Heika, Nagai-hakase o o-mimai 陛下、永井博士をお見舞": 1.

Nagasaki min'yū 長崎民友 (NM)
NM (1948/08/09): "Hiroshima wa kō shite fukkō shita: Zen-sekai ni yobikakeru Nō moa Hiroshimazu undō: Ryōshi no negai wa tada hitotsu 広島はこうして復興した：全世界に呼びかけるノーモア・ヒロシマズ運動：両市の願いは唯一つ": 1.
NM (1948/08/09): "Nagasaki sanshūnen o mukae yomigaetta Nagasaki-shi 長崎三周年を迎えよみがえった長崎市": 2.
NM (1948/09/02): "Yunesuko to wa nani ka? Monmō to muchi o kokufuku shi: Kōkyū-teki sekai heiwa no shōrai e no undō: Yunesuko to Nihon ユネスコとは何か？文盲と無知を克服し：恒久的世界平和の招来への運動：ユネスコと日本": 2.

NM (1949/09/06): "Ichimoku go-ran kudasai: Nagasaki ekimae ni 'kankō Nagasaki annaitō' 一目ご覧下さい：長崎駅前に『観光ナガサキ案内塔』": 2.

NM (1948/10/20): "Ai no hikari nokoru Nagasaki eki: Kinō Kerā joshi kikoku no to e 愛の光残る長崎駅：きのうケラー女史帰国の途へ": 2.

NM (1949/04/24): "Genbaku Nagasaki o heiwa no shinboru ni: 'Heiwa kinen toshi Nagasaki ni kan suru tokubetsuhō' seigan 原爆長崎を平和のシンボルに：『平和記念都市長崎に関する特別法』請願": 2.

NM (1949/05/25): "Pīsu furomu Nagasaki: Umi o tobu messēji futatsu ni: Nagasaki kara sekai shimin e ピース・フロム・ナガサキ：海を飛ぶメッセージ二つに：長崎から世界市民へ": 2.

Nishi Nippon shinbun 西日本新聞 (NNS)

NNS (1949/04/25): "Sekai heiwa no sokushin e 世界平和の促進へ": 1.

NNS (1949/08/07): "Sekai ni hibike 'Nō moa Hiroshimazu' 世界に響け「ノー・モア・ヒロシマズ」": 1.

Saga shinbun 佐賀新聞 (SS)

SS (1948/07/31): "Iyoiyo yo ni deru 'Nagasaki no kane' いよいよ世に出る「長崎の鐘」": 1.

Shin Ehime shinbun 新愛媛新聞 (SES)

SES (1949/08/07): "Hibike heiwa no kane: Genbaku no chi Hiroshima kara zen-sekai e 響け平和の鐘：原爆の地広島から全世界へ": 1.

Depictions of the Nuclear in Children's Manga from the Occupation Period

Katharina Hülsmann

1 Introduction

Japanese society's relationship to the topic of the nuclear has been the subject of many studies, some of which are showcased in this edited volume. The complicated nature of this relationship can be understood by examining various means of meaning-making such as public policy, news coverage, literature, museification, art and popular media. What I mean by "the nuclear" are the inextricably linked properties of nuclear fission as harnessed for its power of mass destruction on the one hand, and for its use as an energy source for industry and households on the other. The splitting of these nuclear properties into 'good' and 'evil' in popular discourse is a process that began almost immediately after the United States dropped the atomic bombs on Hiroshima and Nagasaki in 1945. It was invoked at that time by President EISENHOWER in his 1953 speech "Atoms for Peace", and has continued to influence Japanese public policy by rendering the nation largely dependent on nuclear energy for everyday life, a process that has not undergone significant change despite the Fukushima nuclear disaster in 2011.

In this paper, I will contribute to a more comprehensive understanding of how the nuclear issue was perceived in the popular discourse of the postwar period in Japan. I examine Japanese children's manga (comics), highlighting the nuclear issue as an emerging trope in children's manga during the occupation period (1945–52). Most manga research, especially in western languages, focuses on manga after World War II and after TEZUKA Osamu 手塚治虫 (1928–89) revolutionized the medium in the 1950s with the creation of his long-running, dramatic stories in the so-called "story manga". Some existing research also analyzes wartime manga magazines, identifying manga aimed at young boys as an effective way of spreading propaganda (SKABELUND 2014; CHENG CHUA 2015; SEO 2021). One of the first long-running series of illustrated storytelling worth mentioning in this context is SHIMADA Keizō's 島田啓三 (1900–73) "Adventurous Dankichi" (*Bōken Dankichi* 冒険ダン吉), which was published in the magazine *Shōnen kurabu* 少年倶楽部 (1914–62) between 1933 and 1939. The narrative centers the young boy Dankichi and very much glorifies Japan's colonialism. SHIMADA's depiction of native people as intellectually inferior goes hand in hand with his depiction of white foreigners as evil colonizers, while Japan, personified by the story's hero Dankichi, is presented as a benevolent conqueror. The more detailed examination of wartime manga in chapter 2 will serve to provide more extensive context of this trend.

The works that I focus on are less well-known than the aforementioned wartime classic, but they are of interest here insofar as they center the powerful properties of the nuclear in their stories of young male protagonists. This paper looks, in particular, at "Atomic Genkichi" (*Genshi no Genkichi* 原子のゲン吉) by SHIMADA Keizō, which ran in the newly founded magazine *Manga shōnen* 漫画少年 (1947–55), as well as the work "Superhuman Power Nuclear Ball" (*Kairiki genshi bōru* 怪力原子ボール) by TAKEDA Shinpei 竹田慎平, which was published as a book by Yashio shobō in Tokyo in 1948. These two works were selected because they prominently feature objects imbued with the power of the nuclear, which, as will be shown, became a trope in children's manga during the occupation period. Special attention will also be paid to the context in which these works emerged. SHIMADA's work will serve here as an example of children's manga that emerged from a manga magazine while TAKEDA's exemplifies children's manga published directly in book form. We thus cover two popular publication channels for children's media of that time.

Anything published during the occupation period was subject to censorship by the Allies. It is thus especially interesting to see which new narrative tropes appeared during the occupation period as well as which tropes from wartime manga can now be found that were apparently not subject to censorship. The censorship process in operation at that time will be examined in detail in chapter 3, with a focus on which children's manga were censored and how.

For this study, I have used a very important corpus of Japanese print publications from the occupation period that is available through the Gordon W. Prange Collection at the University of Maryland. The Gordon W. Prange Collection contains all print publications of that time, including 71,000 books, 13,800 magazines, 18,000 newspapers, 10,000 news agency photos as well as a lot of other material. Of particular interest here is the Collection of Children's Books, which spans 8,000 Children's books, of which only 47 were censored.[1]

In chapter 4, I give a brief overview of nuclear themes that appear in the Collection of Children's Books, as the nuclear issue does not only appear in children's manga but in non-fiction books and novels as well. Having provided ample contextualization for the background of these works, I will then delve into the qualitative analysis of the two works selected for this study in chapter 5.

My conclusions do not only illuminate the emergence of the nuclear issue in Japanese children's manga, they also point to the complicated project of reeducation that the Allies had envisioned for Japan. I will show that even in publications for popular consumption aimed at a very young audience, the reframing of the nuclear as a power for good can be found as early as 1947, just two years after the atomic bombs were dropped. However, the depictions examined here still contain caveats to this nuclear power, insofar as control—or the idea of power falling into the wrong hands—also takes center stage. Thus, as we shall see, even in simple illustrated storytelling aimed at children, the destructive power of nuclear energy as well as a subtle discursive preparation for the Cold War was reframed as something positive.

2 WARTIME MANGA

Early forms of manga have existed since the Meiji period (1868–1912) and as Stephan KÖHN (2005: 200) points out, they were themselves influenced by earlier forms of humoristic and satirical depictions from the Edo period (1603–1868), during which time they were still distributed as pamphlets (*kawaraban* 瓦版). Various sorts of short text / picture storytelling become common and popular in magazines during the Meiji period, amongst them "joke pictures" (*odokee* 戯絵) or "Toba pictures" (*tobae* 鳥羽絵, named after the famed picture scroll artist TOBA Sōjō 鳥羽僧正, who was active at the beginning of the 11th century). Meiji period manga were also influenced by American-style newspaper supplements. The satirical magazine *Japan Punch* (1862–87) founded by Englishman Charles WIRGMAN in Yokohama, for example, greatly influenced the Japanese satirical newspapers that would follow it (KÖHN 2005: 201). Jean-Marie BOISSOU points out that in Japan large publishing houses were publishing manga as soon as they began operating, and this stands in stark contrast to Western nations such as the United States and France, where comics and bande dessinées were normally associated with small publishing houses. Japan's biggest publisher Kōdansha 講談社 (then named Dai-Nippon yūbenkai 大日本雄辯會) was a notable exception. It published the popular magazines *Shōnen kurabu*, as well as *Shōjo kurabu* 少女俱楽部 (1923–62)—written for a young female readership and published some time later—as well as *Yōnen kurabu* 幼年俱楽部 (1926–46), aimed at older children (BOISSOU 2010: 23). *Shōnen kurabu* achieved widespread success from the 1920s, circulating up to 700,000 copies per new issue in the 1930s (CHENG CHUA 2015: 3), peaking at 750,000 copies in 1935 (SEO 2021: 52). These magazines were comprised of various formats, including early forms of manga, as well as illustrated articles, serial novels, and columns of reader contributions. During the Pacific War, the publishing industry suffered from paper shortages and many magazines that published manga were forced to cease operations. It is no coincidence that the few which remained active the longest served as a popular platform for military propaganda. In his 1995 monograph "The year 1945 as Seen in Manga" (*Manga ni miru 1945nen* 漫画にみる1945年), SHIMIZU Isao 清水勲 examines manga (here: satirical illustrations and short sequential illustrated storytelling) in newspapers and newspaper supplements during the year 1945. One recurring character created by *mangaka* YOKOI Fukujirō 横井福次郎 (1912–48) is the eponymous Riki-san 力さん, a hard-working factory employee, whose everyday life is depicted in short and humorous 4-panel stories, which appeared in *Asahi Graph* アサヒグラフ (1923–2000)—a weekly illustrated supplement to the national newspaper *Asahi shinbun* 朝日新聞 (1879–). Manga such as *Riki-san* are a typical example for the style of storytelling in so-called "panel pictures" (*komae* コマ絵) that became popular at the end of the 19th century (KÖHN 2005: 201–02). YOKOI had been deployed to the Philippines but was sent home after contracting Malaria (MANGASEEK / NICHIGAI ASSOCIATED 2003: 413). YOKOI was also a member of the "New Manga group" (Shin mangaha shūdan 新漫画派集団) (SHIMIZU 1995: 16; MANGASEEK / NICHIGAI ASSOCIATES 2003: 415), which was founded in 1932 by SUGIURA Yukio 杉浦幸雄 (1911–2004), KONDŌ Hidezō 近藤日出造 (1908–79) and YOKOYAMA Ryūichi 横山隆一 (1909–2001). This group brought new trends to manga, such as humorous non-sensical manga (*nansensu manga būmu* ナンセンス漫画ブーム), for which SUGIURA has been credited (MANGASEEK / NICHIGAI ASSOCIATES 2003: 202).

In one strip from a January 1945 issue of *Asahi Graph*, YOKOI's Riki-san is pictured dreaming about joining a queue of ordinary people to assault a helpless Franklin D. ROOSEVELT, who is shown tied to a pole (SHIMIZU 1995: 17). In another strip from March 1945, he first finds a tin of chocolates on the ground, then some biscuits, then a tire and finally the wreck of a B-29 bomber, the strategic bomber used by U. S. armed forces against Japan (SHIMIZU 1995: 33). Riki-san is frequently depicted as focused on his factory work, so much so that when he wakes up at night during an air raid, his first thought is to run to the factory (SHIMIZU 1995: 39). In one strip from June 1945, a fellow factory worker and colleague of Riki-san sees that the latter's house has burned down and is concerned for his safety—only to find Riki-san alive and well at the factory, working as diligently as ever (SHIMIZU 1995: 75). Factory work was seen as essential to Japan's war efforts and Riki-san can be seen as a model citizen, tirelessly focused on his work. In the short manga strips featuring Riki-san, we can see a vision of Japan winning the war, with B-29 planes shot down and Franklin D. ROOSEVELT captured, allowing the Japanese public to take revenge on him. These comic strips were primarily aimed at newspaper readers. However, in Kanji, the main character's name is spelled with the *furigana*, the reading aid for the Kanji in the Hiragana syllabary, also given in the title of each strip. This suggests that even small children were invited to read about the exploits of Riki-san. Below, I offer a closer examination of what is very likely the most striking example of illustrated reading material aimed specifically at children during World War II: *Shōnen kurabu*. Of the illustrated magazines for children from the time, there is little doubt that *Shōnen kurabu* has received the most scholarly attention thus far. As such, it provides us with important context for the works examined in this paper.

In his 2015 *Japan Forum* article, Karl Ian Uy CHENG CHUA examines the depiction of non-Japanese foreigners in *Shōnen kurabu*. He highlights the recurring images of the South Sea islander (the *nan'yō* 南洋) as an exotic other, as well as depictions of China as an inferior partner (CHENG CHUA 2015: 3, 8). As CHENG CHUA points out, this specific depiction of foreigners is not limited to manga storytelling but is rather a form of colonialist ideology that can equally be found in other forms of articles published in *Shōnen kurabu*. The first depiction of South Sea islanders in *Shōnen kurabu* was drawn by TAGAWA Suihō 田河水泡 (1899–1989) in a double-page spread image titled "New Year in the South Sea" (*Nan'yō no o-shōgatsu* 南洋のお正月) in January 1930. Notably, TAGAWA's depiction of South Sea islanders celebrating New Year draws on stereotypical depictions of black people in U. S.-American popular culture, while the festivities in the picture depict an interesting mixture of Japanese New Year traditions (such as mochi) and exotic pastimes (such as riding an elephant) (CHENG CHUA 2015: 5–6). As CHENG CHUA points out, other narratives produced by the magazine and which justify Japan's colonialist pursuits can be found in published maps and travel diaries (CHENG CHUA 2015: 14).

Shōnen kurabu's most well-known depiction of the exoticized inhabitants of South Sea islands is the illustrated story "Adventurous Dankichi" by SHIMADA Keizō, one of the two authors whose later work will also be discussed here. "Adventurous Dankichi" ran from 1933 to 1939 in the magazine and tells the story of a young boy, Dankichi, who is swept away by a flood and ends up on a South Sea island, where he encounters natives and dangerous animals, eventually himself becoming the leader of the local tribe. The depictions of the island inhabitants amount to racist caricatures of African people: Their only clothes consist of grass skirts (until Dankichi gifts them ponchos that feature a different number on the chest of each inhabitant), they have no hair and rather large and round lips. CHENG CHUA notes that this depiction demonstrates Dankichi's superiority over the natives. Moreover, he is seemingly obliged to give them individual numbers as they are otherwise depicted as essentially identical in appearance (CHENG CHUA 2015: 8). In the later story, Dankichi also comes up against Western foreigners, depicted as "white devils" and "bad foreigners", intent on enslaving the islanders and exploiting their natural resources (CHENG CHUA 2015: 8). CHENG CHUA summarizes the depiction of the South Sea islanders as "inferior, and thus both conquerable and in need of Japan's civilizing help" (CHENG CHUA 2015: 4).

Racist caricatures aside, "Adventurous Dankichi" also features the typical protagonist of the emerging shōnen manga genre: a young boy with an animal companion. Dankichi has a mouse named Calico, which occasionally helps him out by cleverly outwitting the natives. Another very popular motif to be found in "Adventurous Dankichi" is the setting of an exotic jungle world, with exotic animals to either be overcome or who are occasionally helped by the protagonist. The story features all kinds of animals that are not native to Japan but that equally do not share a real-life habitat. Among these are giant snakes, hippos, ostriches, giraffes, elephants, and camels. This strange mix was already apparent in TAGAWA Suihō's piece about New Years in the South Sea, as CHENG CHUA points out:

> Of course, there are neither elephants nor gorillas in the Pacific islands—these animals are present simply to heighten the general sense of the exciting otherness of a nan'yo composed of stereotypical features of an assortment of exotic places where uncivilized people live—not only the south Pacific itself, but also Africa, India and Southeast Asia. (CHENG CHUA 2015: 7)

It thus becomes very apparent that the world depicted in "Adventurous Dankichi" is a perfect example of the colonialist fantasy of an exotic other that Japan was trying to conquer at the time. The exotic island jungle worlds promise treasure—a motif that later reappeared in TEZUKA Osamu's "New Treasure Island" (*Shin Takarajima* 新宝島, 1947)—and make for an appealing and adventurous setting featuring many animals that young readers were unlikely to have ever seen.

Another successful wartime manga in *Shōnen kurabu* is TAGAWA Suihō's "Norakuro" (1931–41). "Norakuro" follows the exploits of an anthropomorphic black dog who—as the hero of the story—joins the army to fight a nation of pigs on the mainland, an obvious allusion to the Chinese (BOISSOU 2010: 24; GUILBERT 2018: no pagination). "Norakuro" was the most successful manga of its time and was also one of the first manga to be reprinted in hardcover (*tankōbon*) after its serialization in the magazine. SKABELUND points out that TAGAWA was an avant-garde artist influenced by jazz music, socialism, and American cartoonists. His popular character Norakuro, for example, resembles Felix the Cat, a 1919 cartoon character created by Pat SULLIVAN and Otto MESSMER (SHIMIZU 1989: 21). However, SKABELUND also remarks that the manga clearly "familiarized children with the rhetoric of empire and pan-Asianism" (SKABELUND 2014: 11). As with "Adventurous Dankichi", "Norakuro" features much adventure and sees its main character employ many a clever ruse. It also depicts various animals among its cast of characters, often representing different nations as was common in war-time media for children (LAMARRE 2008: 76). While there is some conflict in "Adventurous Dankichi", "Norakuro" focuses more overtly on war. Indeed, Dankichi is still a young boy, but Norakuro is an enlisted soldier. As a character, Dankichi is thus closer to the intended readership of *Shōnen kurabu*, while Norakuro—as Gijae SEO (2021: 54) points out—depicts a life in which the young readers themselves do not participate.

Amongst other content published in *Shōnen kurabu*, "Adventurous Dankichi" and "Norakuro" are lively examples that show images of a global hierarchy of nations, which was bolstered by public discourse in Japan. As SEO points out, the magazine was frequently supported by subscription-paying parents as well as teachers and schools who endorsed the pedagogical content of the maps printed in the magazine (SEO 2021: 52). Examining the influence of *Shōnen kurabu* on the Japanese attitude to war, SEO comes to the following conclusion:

> Since children could not take part in the war because of their young age, there was a need for a tool to help them experience war without actual participation. Children could prepare themselves as soldiers for the future through the romanticizing of war. Through such efforts, war could be transformed into something that could inspire them with cheer and excitement. (SEO 2021: 54)

SEO observes that although many children were too young to experience the war, it did become a reality for many others, with boys as young as 15 recruited as the war dragged on. In sum, *Shōnen kurabu* was perhaps the most important vehicle for wartime children's manga, featuring the most popular manga "Norakuro", as well as "Adventurous Dankichi". Both illustrated stories can be seen as an amplification of state propaganda, praising Japan's colonialist pursuits in the South Seas and glorifying the war in mainland China. The scenarios depicted—adventurous jungle settings and exotic animals, as well as animal protagonists and companions—draw on popular tropes to appeal to the school children that made up their audience. After Japan's surrender in 1945, however, the Allies assumed administrative power and the Civil Censorship Detachment (CCD) was formed to examine all Japanese publications, either prior to printing or—in the later occupation period—thereafter.

3 JAPANESE PUBLICATIONS IN THE OCCUPATION PERIOD

3.1 The Censorship Process

In *The Atomic Bomb Suppressed* (1986), Monica BRAW gives a comprehensive overview of the Allied censorship apparatus, paying special attention to its relationship with nuclear issues of that period, first and foremost the atomic bombs and the news coverage that surrounded them. On September 19, 1945, a press code was issued to "educate the press of the Japanese in the responsibilities and meaning of free press" (CIVIL CENSORSHIP DETACHMENT 1945: no pagination). The point of this press code was ostensibly to facilitate the printing of true and factual news as opposed to propaganda and rumors. However, news that might disturb public tranquility or any criticism of the Allied Powers was not permitted. In addition, key logs containing forbidden topics were periodically issued as deemed necessary by the Allied Powers.

According to BRAW's research (1986: 98), these key logs make no mention of the bombs dropped on Hiroshima and Nagasaki. However, in September 1945, the *Asahi shinbun* was suspended for two days following the publication of a statement declaring the bombings a war crime (BRAW 1986: 89, 97). We can thus assume that damage caused by the atomic bomb fell within the purvey of topics considered likely to bring about civil unrest (BRAW 1986: 40).

Overseen by the Deputy Chief of Intelligence, the CCD came into operation at the beginning of the occupation and was responsible for the censorship of all print publications. It consisted of 8,734 civilian and military personnel, 90 of them officers. The main censorship work was carried out by over 8,000 Japanese and Korean censors that provided translations and highlighted questionable passages (BRAW 1986: 46). Censorship lasted until October 1949 when the CCD operation was brought to a close. By 1948, most magazines and newspapers had moved from pre-censorship to post-censorship, by which process their content was examined post-publication rather than via the submission of galley proofs prior to printing (BRAW 1986: 80). If infractions were found, however, publishers could find themselves returned to pre-censorship status. Furthermore, "trustworthy" media outlets enjoyed privileges in terms of the allocation of paper, which was a scarce resource at the time (BRAW 1986: 89–90).

BRAW also states that the CCD had initially intended to rely on the old Japanese censorship apparatus, even availing of the same physical facilities for their operations (BRAW 1986: 38). Indeed, it could be argued that little had changed for the Japanese press, which was equally subject to censorship during the war. SEO, for example, points to a certain continuity in the case of the popular magazine *Shōnen kurabu*, claiming that after the war it "continued to be published with no change of its identity, only reducing the business in size" (SEO 2021: 55). Despite *Shōnen kurabu*'s obvious contribution to military propaganda, the magazine was allowed to continue publication after the war. They ostensibly signaled a change in the times by writing the second half of the magazine title in Katakana (少年クラブ) from 1946 on. However, there were some new themes that appeared in manga, as well as in other publications of the time. Writing on this topic, SEO claims that:

[i]mmediately after the defeat, the greatest change in *Shonen Kurabu* was an increase in the description of the US forces with whom the Japanese had to cooperate, and an increase of articles written in the Roman alphabet. "Masao's Picture Diary," a cartoon about the friendship between the US soldiers and Japanese children, was printed in both Japanese and English. Japanese pronunciation was added to anything written in English along with a glossary [...]. The language of the enemy had now become a medium of cooperation. (SEO 2021: 63)

This theme can be found more widely in manga of the occupation period, for example in "Hello Jeep" (*Harō, jīpu* ハロー・ジープ), a short comic strip that was equally subject to censorship at the time. As SEO and BRAW both point out, there was very little allusion to war guilt or regret in publications of the time (BRAW 1986: 139; SEO 2021: 66). It is thus particularly interesting to delineate, on the one hand, the construction of the nuclear within Japanese children's manga and, on the other hand, point out shared themes with manga of the wartime period. I will thus highlight and contextualize a part of the nuclear dispositive that would develop during the occupation period and continue to influence thinking about the nuclear issue in postwar Japan.

3.2 Censored Children's Manga

Of all publications printed during the occupation period, only two percent or so show signs of deletion or modification (THE GORDON W. PRANGE COLLECTION 2013). What sort of children's manga would warrant deletion or modification by the Allies? In this sub-chapter, I briefly draw on my examination of the Collection of Children's Books in the Gordon W. Prange Collection, which is digitized and available to the public at the library of the University of Maryland. The Collection of Children's Books contains 8,000 children's books, of which only 47 were censored or received censorship interventions. There are about 2,000 manga amongst these books, of which only 25 were censored. I examined the entirety of censored children's manga and found that the vast majority of them depicted samurai.

In some cases, the reasons for the censorship were given explicitly, such as with the samurai and ninja story "Pommel Horse Tengu" (*Kurama Tengu* 鞍馬天狗, 1948) by TŌGE Teppei 峠哲兵 (dates unknown), which features a stamp on the cover declaring it a violation, as well as scribbled remark reading "Rightist [sic!] Propaganda". Even anthropomorphic samurai such as the cute rabbit that fights other woodland creatures in "Mimisuke's Pleasure Trip Diary" (*Mimisuke man'yūki* 耳助漫遊記)—a 1948 manga by TANAKA Masao 田中正雄 (1927–2014)—were subject to disapproval on the part of the censors. All scenes involving drawn swords were subject to intervention. Many of these interventions happened in the later years of the occupation, when newspapers, magazines and books were already operating on a post-censorship basis.

The censorship documents for the illustrated book "Mysterious Kuroshio Island" (*Kaii Kuroshiotō* 怪異黒潮島) by KISHIDA Seiichi 岸田靖一 (dates unknown), for example, indicate that the book was published in October 1948 but was not censored until December 10, 1948. According to the documents, the copy was received by the CCD on December 8, reviewed by the censor on December 9, and officially censored as "Rightist [sic!] Propaganda" on December 10. As with the other censored children's books containing samurai, all passages containing sword fights were crossed out.

There is another example of an early manga title from an unknown author, which passed censorship with the deletion of one 'false' fact. It is called "Hello Jeep" and depicts an American MP and his jeep. The jeep has amazing technical capabilities such as driving extremely fast or over water. In the panel that was marked for deletion however, the driver states "I got this jeep for being in the American Army". In the margin, the panel was marked as 'false' by the censors and thus had to be removed. This presence of this single CCD stamp suggests that the manga otherwise passed censorship without any further deletion.

This particular manga exemplifies the change in children's manga following Japan's surrender, as SEO observes (2021: 63). Depictions of benevolent military personnel became very common, as did the use of the Roman alphabet and efforts to render the English language with Katakana, as can be seen from the title *Harō, jīpu*.

It is clear from the foregoing analysis of the Collection of Children's Books in the Gordon W. Prange Collection that there was very little censorship of children's manga. What censorship did exist focused mainly on depictions of samurai sword fights. Other instances of violence, such as drawn guns and shootings (whether of people or animals)—which can be seen in "Nuclear Ball" (*Genshi bōru* 原子ボール, 1949) by TAKEDA Shinpei for example—were not censored. Though neither of the manga in the following, more in-depth analysis, were subject to any intervention on the part of the censors, it certainly cannot be claimed that the authors enjoyed total freedom of expression. Censorship practices were not made known to the public at large, yet individual authors, publishing houses, and newspaper agencies were well aware of them as they were obliged to provide censorship copies directly to the CCD.

Thus, all publications from the occupation period allude to the influence of censorship, albeit a sort of self-censorship enacted by authors wary of the censor's pen. In terms of the construction of the nuclear issue with which we are concerned, we would do well to bear media censorship in mind when viewing these publications, despite the lack of specific censorship of the individual publications.

4   NUCLEAR THEMES IN NON-FICTION CHILDREN'S PUBLICATIONS

In the Collection of Children's Books in the Gordon W. Prange Collection, there are seven non-fiction books, the titles of which contain the words nuclear (*genshi* 原子), nuclear energy (*genshiryoku* 原子力) or atomic bomb (*genshi bakudan* 原子爆弾). None of the books containing these themes were subject to any direct censorship. All seven of the titles are science textbooks for the education of children on the function of nuclear elements, though occasionally basic chemistry and physics knowledge—such as the formation of atoms from neutrons, protons and electrons—is also present. They contain references to the research of Albert EINSTEIN and other well-known western physicists.

A particularly interesting example is the title "The Atomic Bomb" (*Genshi bakudan* 原子爆弾, 1948) by IIDA Yukisato 飯田幸郷 (*1918). This book contains several chapters that deal with the destructive potential of the atomic bomb and the power of nuclear energy. Chapter 22 is titled "The Atomic Bomb" (*genshi bakudan* 原子爆弾) and mostly explains the technological functioning of the atomic bomb, referencing Albert EINSTEIN as a physicist elementary to the discovery of nuclear power. This is followed by chapter 23—"The History of Nuclear Fission Research" (*genshi kakubunretsu kenkyū no rekishi* 原子核分裂研究の歴史)—which similarly focuses on technological content. Chapter 24—"The Power of the Atomic Bomb" (*genshi bakudan no iryoku* 原子爆弾の威力)—discusses atomic bomb tests carried out by the U. S. in New Mexico and in the Bikini atoll, even featuring an illustration on page 129 of an atomic explosion in the latter location, drawn from the perspective of an airplane. The birds-eye view in the image can be classified as a depiction from the perspective of the perpetrator and coheres with many other early depictions of the mushroom clouds from the atomic explosions over Hiroshima and Nagasaki. Chapter 25 is then titled "The Control of Nuclear Energy" (*genshiryoku no kanri* 原子力の管理) and states that it is Japan's duty to advocate against the use of atomic bombs given the damages that were suffered in Hiroshima and Nagasaki (damages which are referenced only briefly in passing). Chapter 26—"The Age of Nuclear Energy" (*genshiryoku jidai* 原子力時代)—goes on to detail the great power that can be harnessed from nuclear energy, stating that this will bring peace and prosperity to humankind. In the space of just 14 pages (IIDA 1948: 119–33), the book touches on the subject of the destruction unleashed by atomic bombs in World War II before immediately switching to advocate the great power that can be harnessed in this new nuclear age.

"The Atomic Bomb" is of particular interest, as it was published in Tokyo by Shōheisha 昌平社 in 1948. In 1949, however, an identical manuscript was published in Tokyo by the same author with publisher Jiyū shoin 自由書院 under the title "Nuclear Energy Research" (*Genshiryoku no kenkyū* 原子力の研究). The content of these two publications is identical, right down to the page numbers. The book was printed for a second time with a different publisher within a year, suggesting that the nuclear topic was a popular one and that many people—children included—were interested in reading about it. There is also the matter of the title, which was changed from *Genshi bakudan* to *Genshiryoku no kenkyū*. This change may well have been made in order to sell the same book twice. It is possible, however, that the new title—"Nuclear Energy Research"—was chosen as it better represents the subject matter of the book, namely technological descriptions. In any case, the author clearly puts more emphasis on the industrial uses of the nuclear than making any overt reference to the atomic bomb as is suggested by the former title.

Other titles that deal with nuclear physics in particular or with chemistry and physics in a more broadly educational manner include "Atomic Story" (*Genshi monogatari* 原子物語, 1948) by SHINOHARA Ken'ichi 篠原健一 (1905–95) and "Dance of the Atoms" (*Genshi no odori* 原子の踊り, 1948) by KIMURA Tsuneyuki 木村恒行 (1910–73). As with the aforementioned titles, these books also explain complex scientific and technological matters in simple terms for a young audience. They occasionally contain straightforward and light-hearted drawings of atoms or ions, which are depicted with little faces and limbs, to make chemical reactions easier to understand.

Overall, references to nuclear issues are certainly present and not just in fiction but also in science textbooks aiming to educate children about nuclear physics. These textbooks generally focus on purely scientific issues, describing the release of energy for example. Naturally, they make no allusion to the destruction caused to Hiroshima and Nagasaki by the atomic bombs. In the works of IIDA Yukisato, even children are asked to consider the issue of "the control of nuclear energy", while also learning about the great power and prosperity that nuclear energy will bring. While these scientific texts purportedly remain faithful to pure scientific fact, the depiction of the nuclear in popular fictional texts actually demonstrates a rather creative vision.

5 THE NUCLEAR IN CHILDREN'S MANGA

The following manga examined in this chapter are accessible in the Gordon W. Prange Collection. They are amongst a small number of stories that make direct reference to the nuclear issue (*genshi* 原子 or げんし) in the title. The titles are significant insofar as they show similar motives: The nuclear is presented as something adventurous that must be controlled by proper authorities. Beyond that, they make use of typical storytelling tropes common in manga and can be seen as representative for children's manga of the time. Both authors—TAKEDA Shinpei and SHIMADA Keizō—were quite prolific and it can be argued that their works are a representative example for popular children's manga. While SHIMADA's work here represents the manga published in magazines, TAKEDA's work serves as an example for children's manga published in books.

5.1 Takeda Shinpei's "Superhuman Power Nuclear Ball" (1948)

The works of TAKEDA Shinpei can be found in the Gordon W. Prange Collection of Children's Books in which he is listed as the author of 21 books. To put this number into perspective, the Prange Collection lists, for example, 27 books by SHIMADA Keizō, who can certainly be considered a prolific author of the same period. TAKEDA cannot be found in reference works such as the *Mangaka Directory* by MANGASEEK AND NICHIGAI ASSOCIATES, INC (2003), and his date of birth and background, as such, could not be verified for this paper. Though there are some publications from the time that may be lost, the Prange Collection's digital collection catalogue serves as a very good inventory with which to learn about early manga and children's literature. It is TAKEDA's use of the nuclear as a prominent feature in at least two of his manga, that brought his work to my attention. I selected TAKEDA Shinpei's work "Superhuman Power Nuclear Ball", in particular, because it is a relatively long story of 64 pages that fundamentally deals with the titular 'nuclear ball' as an object of nuclear power.

TAKEDA tells humorous and exciting adventure stories for children that often involve a young male protagonist with an animal companion. Exotic animals, such as lions and elephants in the case of "Demon Jungle" (*Akuma no mitsurin* 悪魔の密林, 1948), make frequent appearances in his works and are featured on the cover. Some of his stories also contain elements of science fiction, such as "Superfast Rocket Missile" (*Kaisoku rokettodan* 怪速ロケット弾, 1948), which shows flying rockets on its cover.

"Superhuman Power Nuclear Ball" is a science fiction story. The *genshi bōru* of the title is a small ball that is identical in appearance to a baseball: white with a characteristically curved seam. In the story, the ball is invented by a bespectacled professor. When a button on the ball is pressed, it emits a blast of air that will enlarge any living being that comes into contact with it. If pressed again, it will emit a blast of air that does the opposite—thus alternating in its function and frequently leading to slapstick humor.

At the beginning of the story, the professor receives a letter by an ominous group called the "QQ group", which informs him that spies are trying to lay their hands on his invention. Indeed, while the professor is distracted reading this letter, a thief (as indicated by a stereotypical black and white striped top, reminiscent of a prison uniform) enters his laboratory and steals the ball. The professor then sends his young son to retrieve it. The young baseball-cap-wearing boy and his canine companion thus become the heroes of the story. The boy pursues the group of thieves and finally returns the ball to his father, turning the thieves over to the police.

In a number of scenes throughout the story, the nuclear ball is used by the different parties. Having stolen the ball, the thief returns to his gang and tries pressing the button for the first time. A blast of air erupts and hits the leader, causing him to fall out of a tree. He is picked up by his underlings, now measuring only 20 centimeters or so (TAKEDA 1948: panel 59).[2] In awe of this power, the thieves decide to return to their hideout. Meanwhile, the protagonist seeks out the stolen ball, aided by his pet Pochi. Many chase scenes follow before Pochi and the boy are themselves captured by the thieves.

The ball is once again used in the scene that follows. Pochi is brought with its legs bound before the leader of the thieves. The leader asks the dog to show him how to return to normal size. The dog cleverly outwits the leader, telling the leader to point the ball at him and press the button. A blast of air is thus issued but instead of hitting the leader and enlarging him back to his normal size, the blast hits the dog, enlarging him to giant proportions (TAKEDA 1948: panel 83).

Pochi's ruse allows him to break free, retrieve the ball, and free the protagonist, who then uses the ball to return the dog to his normal size (TAKEDA 1948: panel 93). In their subsequent pursuit of the gang leader, the boy and his dog don diving suits and dive deep into the sea. Again, the ball saves them from a dangerous situation when Pochi is almost caught by an octopus. The boy throws the ball at the octopus but, strangely, rather than this causing the octopus to swell in size, the animal instead develops an enormous growth on the side of its head, leading it to release the dog. The protagonist ties a rope around this growth, saying "Maybe we can stop any more damage this way" (TAKEDA 1948: panel 114), as tears stream down the octopus's face and it cries for help (TAKEDA 1948: panel 115).

Having faced various other adventures (a great black sea monster, which also appears on the book cover), the protagonist and his dog finally find the gang leader who then begs them to return him to his normal size, promising them treasure in return. The story concludes with the protagonist finding treasure on an island and in the final panel, returning the ball to his father, and handing the gang leader over to the police.

"Superhuman Power Nuclear Ball" is a fairly typical adventure story told in many amusing vignettes that are loosely connected by the underlying quest of the protagonist. Good and evil are clearly separated in this story, though the antagonists are never terribly menacing, instead continuously proven to be inept. In particular, the gang leader, shrunk to miniature size for most of the story, is presented as funny and almost cute. This is achieved by contrasting him with his taller henchmen as well as with the larger objects he uses. When Pochi is taken captive and brought to him, the tiny leader is shown smoking his pipe (enormous in comparison to the shrunken figure), thus creating a very amusing image (TAKEDA 1948: panel 82).

The nuclear ball is an interesting storytelling device. It looks like an everyday object, well-known to children even today. Yet it also carries this strange and wondrous power to enlarge and shrink living beings. The story offers no explanation as to why this property is useful, but it is assumed that the antagonists have a motive for obtaining this ball. The letter received by the professor makes explicit mention of espionage as a motive (TAKEDA 1948: panel 10).

Furthermore, clever ruses in which the nuclear ball plays a key part are the main way the protagonist outwits his foes. The ball and its powers are generally used without lasting consequence given that those affected are always returned to their original state (TAKEDA 1948: panel 93, panel 141). The ball's effects are thus generally presented as reversible. The octopus, by contrast, constitutes an ominous exception. In this case, the ball is thrown at the attacker, as compared with all other instances when it is activated by pushing the button, ostensibly the intended use method.

The bodily harm suffered by the octopus can thus be seen as a result of misuse. Be that as it may, the octopus is afforded no pity in the story—having first attacked the protagonist, it receives its just deserts.

In this story, the nuclear ball is an everyday item imbued with a strong, wondrous power. The protagonist can easily use it to achieve his goals, while the other parties do not understand its capabilities and end up with unintended results. Thus, one interpretation is that nuclear power can be used to do good in the right hands. Of course, this is far from a realistic depiction of nuclear power given that the power it provides—the ability to shrink and enlarge living creatures—is purely fictional. Furthermore, TAKEDA Shinpei has more stories involving a ball with these properties. Two months after "Superhuman Power Nuclear Ball" was published, TAKEDA published "The Adventure of Gen-chan" (*Gen-chan no bōken* げんちゃんの冒険), in which a ball with the same properties appears. There is, however, nothing nuclear about the ball in this instance. In sum, nuclear power in this story is simply used as a property to signify that an item has special powers. The use of everyday items seemingly magically imbued, so to speak, with a nuclear property, can also be found in the manga examined below.

5.2 Shimada Keizō's "Atomic Genkichi" (1947–48)

SHIMADA Keizō's "Atomic Genkichi" first appeared in the *Manga shōnen*, a magazine founded during the occupation period and published monthly by Gakudōsha学童社 from December 1947 to October 1955. The editor, KATŌ Ken'ichi 加藤謙一, had been the editor in chief for *Shōnen kurabu* from 1921 to 1935 (SHIMIZU 1989: 18). The publication is noteworthy because it featured many well-known *mangaka* of the time and especially those that had been active in *Shōnen kurabu*, such as TAGAWA Suihō and HARA Kazushi 原一司 (1915–57). The popular manga strip *Sazae-san* サザエさん by HASEGAWA Machiko 長谷川町子 (1920–92) was also featured on page two, next to the table of contents in certain issues of *Manga shōnen* in 1947 and 1948. However, as could already be observed with *Shōnen kurabu*, the cover illustrations during the occupation period differed strikingly from the imagery on manga magazines during the war. The artist responsible for most of the *Manga shōnen* covers was SAITŌ Ioe 斎藤五百枝 (1881–1966), who was famous for drawing the covers of *Shōnen kurabu* during the war (HOLMBERG 2013: 183). *Manga shōnen*'s stated mission was to "fill children's hearts with cheer and happiness" (*Manga shōnen*, vol. 1 (1947): 2) and its covers feature laughing boys, occasionally accompanied by exotic animals such as the giraffe on number three. The back cover of the first issue, in particular, is interesting insofar as it shows a color illustration depicting a bird's eye view of a baseball field. There is an ongoing game involving Dankichi from "Adventurous Dankichi", Norakuro from TAGAWA Suihō's eponymous manga, Popeye and Minnie Mouse—cheered on by the South Sea islanders from "Adventurous Dankichi"—a yellow duck and a pink rabbit, vaguely resembling Donald Duck and Bugs Bunny.

The way in which U. S.-American cartoon characters are intertwined with popular manga characters on this back cover points to the influence of the Allies on Japanese popular culture at the time. It should also be noted that the characters are partaking in a friendly game of baseball, a U. S.-American sport that was popular in Japan even before the war and which continues to be popular even today. Indeed, chief editor KATŌ Ken'ichi had tried his hand at publishing the baseball-themed magazine *Yakyū shōnen* 野球少年 some eight months prior to starting publication of *Manga shōnen* (SHIMIZU 1989: 29). The first issue of *Manga shōnen* featured the manga *Batto-kun* バット君 (1947–50) on its first pages after the table of contents along with a message from the editor. This spot is usually reserved for the most popular series in the magazines. *Batto-kun*, drawn by INOUE Kazuo 井上一雄 (1914–49), with FUKUI Eiichi 福井英一 taking over after INOUE's sudden death in 1949, is said to be the first baseball manga (HOLMBERG 2013: 190), a genre which would become very popular in postwar Japan. A brief glance at the front and back cover of this issue is enough to show just how palpable the influence of U. S.-American pop culture on *Manga shōnen* was.

Manga shōnen also featured other serial manga, such as "Atomic Genkichi" by SHIMADA Keizō. The title is distinctly reminiscent of "Adventurous Dankichi" and the protagonist, Genkichi, resembles Dankichi as well in that he is a light-skinned young boy. However, while "Adventurous Dankichi" can still be classified as a picture story (*emonogatari*) and not quite a manga, the style of sequential storytelling used in "Atomic Genkichi" is much closer to the style of storytelling found in later manga. As with TAKEDA Shinpei's "Superhuman Power Nuclear Ball", the panels in the first chapter of "Atomic Genkichi" are marked with their proper numbers indicating the intended reading sequence. In later chapters, this was omitted, indicating that the readers were quite familiar with the intended reading sequence. Furthermore, the panels are designed to resemble an unfurling roll of pictures, almost like a film reel. Speech bubbles and symbols like gusts of wind and stars, indicating movement and energy are used much more frequently than in "Adventurous Dankichi", where they are only used occasionally. Thus, the visual style of sequential storytelling in "Atomic Genkichi" appears much more refined and dynamic than in "Adventurous Dankichi". The narrative, however, is told in a similar way to "Adventurous Dankichi" and also shows similarities to TAKEDA Shinpei's work: Each chapter centers on an amusing vignette, which—given the use of slapstick humor—can be enjoyed without knowing the entire story. It begins when Genkichi is sent a parcel by his researcher uncle who lives overseas. The parcel contains the gift of 'nuclear shoes', which give Genkichi the ability to fly. However, once he puts those shoes on, he finds their power too great to control and immediately flies out of the window, ending up on an exotic island. At this early stage, the premise of the story is similar to "Adventurous Dankichi". Indeed, Genkichi encounters many exotic animals and the story contains similar racist caricatures of South Sea islanders.

Chapter two of "Atomic Genkichi" is titled "The Hippo that is Flying in the Sky" (*sora o tobu kaba* 空を飛ぶ河馬). It was published in January 1948 and sees Genkichi helping out a hippopotamus that begins to fly because one of Genkichi's "nuclear rocket slippers" (*genshi rokettokutsu* 原子ロケット靴; SHIMADA 1948: 11) becomes stuck in its nostril. The narration for the first three panels of chapter two makes special reference to the power of the slippers: "What terrific nuclear powers! In the blink of an eye, the hippo is propelled with rocket speed over the water surface... and just like that it flies into the sky" (SHIMADA 1948: 11). The first panel depicts the moment in which the errant slipper gets stuck in the nostril of the hippopotamus. The animal is wide eyed and visibly shocked with two drops of sweat coming off its forehead. Its eyebrows are creased upwards, expressing mild pain, an impression that is amplified by three stars and energy lines being emitted from its nostril. The slipper is carried on an energy line punctuated by little round puffs of steam as it becomes lodged. Genkichi looks on in the background, a question mark over his head as he holds his cap in one hand. In the next panel, the hippopotamus is pushed backwards through the water, rendered with wavy lines. Perspiration continues to fly from the hippopotamus' mouth and its eyes remain wide open, eyebrows furrowed in an expression of fear and pain. All the while, the slipper emits the energy line and little puffs of steam. In the third panel, the hippo is then pictured flying backwards through the sky with energy lines, perspiration following, along with little puffs of steam coming steadily from the slipper. In this panel, the hippopotamus is depicted smaller and in the lower left corner of the image, the tips of some palm trees are visible, showing the hippopotamus's rapid ascent. Here the hippopotamus emits its first cries (*fuwa fuwa*), mouth and eyes still wide open (SHIMADA 1948: 11).

Great attention is paid to the depiction of a very human-like expression on the hippopotamus's face in this little scene. The fact that the slipper remains stuck in its nostril, propelling it backwards through the water and sky has a comedic effect. The readers get a glimpse here of what can go wrong if the power of the nuclear slippers evades control: In the blink of an eye, even a heavy and seemingly slow creature such as a hippopotamus can be propelled into the sky. Immediately, the protagonist Genkichi feels compelled to help this poor creature. With the power of the one slipper remaining on his foot, he pursues the flying hippopotamus, gets a hold of it, and brings it back to earth. However, they both crash into a hut belonging to the chief of a South Sea tribe (see Figure 1).

The islanders in "Atomic Genkichi" are portrayed in exactly the same way as those in "Ad-

Fig. 1: SHIMADA: "Genshi no Genkichi: Sora o tobu kaba", p. 11 © Gakudōsha. Photo taken by the author, original copy available at the Gordon W. Prange Collection, University of Maryland Libraries.

venturous Dankichi". They are drawn completely black with large, round eyes and lips, wearing nothing but grass skirts and bracelets on their wrists. To set himself apart from the otherwise identical islanders, the chief of the tribe wears feathers or leaves attached to his head. Once Genkichi—aided, of course, by the nuclear slippers—has helped the islanders and taken away the hippopotamus, they kneel in front of him, believing him a god, owing to his having descended from the skies. He is at once invited into their home where they prepare him an exotic meal: a whole cooked ostrich. Genkichi then finds a diamond ring in the ostrich meat, inadvertently eaten by the animal while in the company of pirates burying treasure. Genkichi is able to use the ring to pursue the pirates. This simple story unfolds over four pages and 32 panels, and principally focuses on Genkichi helping the hippopotamus and the islanders. By the time the following chapter begins, the reader is already primed for a confrontation with the pirates.

Genkichi is a good-natured protagonist, who comes to the aid of various creatures, first with his nuclear shoes then with his super strong 'nuclear hand'. His pursuit of thieves and the riches they have stolen shares much with the storyline in the "Superhuman Power Nuclear Ball". Amongst the stolen goods, Genkichi finds his uncle's ring and is thus impelled to travel abroad in order to find the professor and, indeed, inventor of the nuclear slippers and nuclear hand—a quite similar bespectacled character as the one in TAKEDA's work.

The humor in "Atomic Genkichi" is primarily derived from situations in which things go awry because the great powers of the nuclear objects have been underestimated, with various characters flying around and crashing as a result. The presence of the nuclear in this manga can also be seen as a strange and wondrous property that can imbue everyday objects with inexhaustible power. Again, children are educated—albeit in a very playful fashion—about 1) the potential to harness the nuclear in everyday life, and 2) the need to control this power. Just as in "Superhuman Power Nuclear Ball", the nuclear never exhibits any long-term negative effects on those who use it yet the need for control in terms of who uses those powers is at the core of both stories. The researcher in both stories is presented as well-intentioned and competent in their use of nuclear devices, while it is incumbent on the protagonist to safeguard the devices against improper use.

6 CONCLUSION: THE NUCLEAR AS AN ADVENTUROUS TROPE IN CHILDREN'S MANGA

It was not until the late 1950s and 1960s that realistic depictions of the horrors of the atomic bombs and their consequences would appear in manga. An overview of postwar manga dealing with the war and the atomic bomb can be found in MI-

Yabe et al. (2013). A collection of manga dealing specifically with the bombings was published in four volumes titled "Atomic Bomb Manga Collection" (*Gensuibaku manga korekushon* 原水爆漫画コレクション), published in 2015 by Heibonsha. Notable examples include "Bikini Ashes of Death" (*Bikini shi no hai* ビキニ死の灰), published as a stand-alone volume in 1954 by Hanano Kaoru 花乃かおる (dates unknown), "Disappearing Girl" (*Kieyuku shōjo* 消えゆく少女), published as two stand-alone volumes in 1959 by Shirato Sanpei 白土三平 (1932–2021), "Tragedy of a Planet" (*Aru wakusei no higeki* ある惑星の悲劇), published 1969 in *Gekkan shōnen magajin* by Asaoka Kōji 旭丘光志 (1938–) and Kusaka Tatsuo 草河達夫 (dates unknown), and the well-known works of Nakazawa Keiji 中沢啓治 (1939–2012), namely "I saw it" (*Ore wa mita* おれは見た), published 1972 in *Gekkan shōnen janpu*, and "Barefoot Gen" (*Hadashi no Gen* はだしのゲン), published 1973–87 in *Shūkan shōnen janpu*, and later in *Shimin*, *Bunka hyōron* and *Kyōiku hyōron*.

Nuclear power as we find it in occupation period manga is a wondrous element of adventure-style stories. In these stories, the nuclear can be used safely in everyday life. In this paper I have pointed to certain connections to be made to wartime manga. Many of the occupation period manga make use of the same tropes that we find in wartime manga such as "Adventurous Dankichi" and "Norakuro". The protagonists are often good-hearted young boys with which young readers can easily identify. They go on adventures and are often aided by their animal companions.

Notable racist tropes that appear—unsurprisingly—in both wartime and occupation period manga by Shimada Keizō, for example, are the South Sea islanders that are depicted in exactly the same way, that is, as primitives who are quick to worship the light-skinned protagonist, in veritable awe of his superior intellect and capabilities. Takeda Shinpei's manga, while it does not contain racist caricatures, contains a similar reference to the exotic as a destination of sorts, one that promises riches if conquered, as is demonstrated by the protagonist at the end of his voyage to a treasure island at the end of the story. In this way, both Shimada and Takeda perpetuate colonialist depictions of the South Sea, depictions which the CCD did not apparently deem worthy of intervention during the censorship process. This shows how colonialist ideology remained unquestioned, as long as the negative depictions only concerned Black, Indigenous and People of Color.

A new trope that emerged in these occupation period manga was, however, the nuclear as an adventurous trope. In the two stories examined, the nuclear serves to empower the protagonist. The nuclear is depicted as a property that can imbue everyday objects (such as balls or shoes) with great, inexhaustible powers. The protagonists use the items in combination with their superior intellect to cleverly overcome their foes, while the other characters in the story do not usually know how to properly make use of the powers of the nuclear. However, even handling these items incorrectly does not generally lead to any lasting effects or damages.

This is another important property in the narrative of the nuclear in occupation period manga: The objects imbued with nuclear power must be handled correctly and this can only be done by the right people. In Takeda Shinpei's story, the leader of the thieves experiences undesirable effects, while in Shimada Keizō's story the flyaway slipper causes mayhem when it gets stuck in the nose of a hippopotamus. In each case, it is up to the protagonist to remedy the situation. Both stories also feature a very similar figure: a benevolent, bespectacled professor who invents these wondrous nuclear devices and can competently handle them. The antagonists in each case are thieves while the explicit danger—in Takeda's story in particular—is that the nuclear devices might be stolen.

Other than depictions of nuclear power, the manga examined in this paper do not differ from typical popular manga of the time. We can thus logically conclude that nuclear power during the occupation period came simply to constitute another trope among the many used in adventure manga. The nuclear signifies an inexhaustible supply of energy that can enhance everyday objects but which must, at the same time, be treated with care, and it is the responsibility of the protagonists to ensure this proper use. Through this depiction, the promise of prosperity supplied by the supposedly unlimited energy provided by nuclear power is very clearly visible. Nuclear power is reframed as a power for good, accessible even to children. Indeed, these simple stories for children's consumption already prepare the audience to face the issues of nuclear control that would come to be so important during the coming Cold War.

Notes

1   The catalogue for the Prange Digital Children's Book Collection can be accessed online via this URL: https://digital.lib.umd.edu/prange (last access 2023 / 06 / 01). The advanced search includes a setting indicating whether the publication was subject to censorship action or not. All censored children's books can thus easily be displayed in one step.
2   Similar to other early manga, such as the above-mentioned *Riki-san*, Takeda numbers the panels of his manga to guide readers to the intended reading sequence. Most pages contain two rectangular panels, arranged on top of one other, but some pages may contain more panels and feature a slight variation in the arrangement. The pages themselves are not numbered. As a point of reference in this analysis, I instead indicate the panel number.

References

ANONYMOUS (1946): *Harō, jīpu* ハロー・ジープ. Fukuoka: Suzuki Tokutarō.

BOUISSOU, Jean-Marie (2010): "Manga. An Historical Overview". In: JOHNSON-WOODS, Toni (ed.): *Manga. An Anthology of Global and Cultural Perspectives*. New York: Continuum International, 17–33.

BRAW, Monica (1986): *The Atomic Bomb Suppressed*. Malmö: Liber Förlag.

CHENG CHUA, Karl Ian Uy (2015): "Boy meets world: the worldview of *Shōnen kurabu* in the 1930s". In: *Japan Forum* 28 (1): 74–98.

CIVIL CENSORSHIP DETACHMENT (1945): *Code for Japanese Press*; https://prangecollection.wordpress.com/2013/07/21/sample-ccd-documents/censor-docs_presscode/ (last access 2023 / 01 / 06).

GUILBERT, Xavier (2018): "Norakuro". In: *du9, l'autre bande dessinée*; https://www.du9.org/dossier/norakuro/ (last access 2023 / 01 / 06).

HOLMBERG, Ryan (2013): "Manga Shōnen: Katō Ken'ichi and the Manga Boys". In: *Mechademia* 8: 173–93.

IIDA, Yukisato 飯田幸郷 (1948): *Genshi bakudan* 原子爆弾. Tōkyō: Shōheisha.

IIDA, Yukisato 飯田幸郷 (1949): *Genshiryoku no kenkyū* 原子力の研究. Tōkyō: Jiyū shoin.

KÖHN, Stephan (2005): *Traditionen visuellen Erzählens in Japan. Eine paradigmatische Untersuchung der Entwicklungslinien vom Faltschirmbild zum narrativen Manga* (Kulturwissenschaftliche Japanstudien 2). Wiesbaden: Harrassowitz.

LAMARRE, Thomas (2008): "Speciesism, Part I: Translating Races into Animals in Wartime Animation". In: *Mechademia* 3: 75–95.

MANGASEEK まんがseek / NICHIGAI ASSOCIATES, INC. 日外アソシエーツ編集部 (2003): *Mangaka jinmei jiten* 漫画家人名事典. Tōkyō: Nichigai Associates, Inc.

MIYABE, Seiichi 宮部精一 et al. (2013): *Mangakatachi no sensō* 漫画家たちの戦争. Tōkyō: Kinnohoshisha.

SEO, Gijae (2021): "Shonen Kurabu and the Japanese Attitude Toward War". In: *Children's Literature in Education* 52: 49–67.

SHIMADA, Keizō 島田啓三 (1947): "Genshi no Genkichi 原子のゲン吉". In: *Manga shōnen* 漫画少年 1: 10–14.

SHIMADA, Keizō 島田啓三 (1948): "Genshi no Genkichi: Sora o tobu kaba 原子のゲン吉:空を飛ぶ河馬". In: *Manga shōnen* 漫画少年 2: 11–14.

SHIMIZU, Isao 清水勲 (1989): "Manga shōnen" to akahon manga: Sengo manga no tanjō 「漫画少年」と赤本マンガ:戦後マンガの誕生. Tōkyō: Zōonsha.

SHIMIZU, Isao 清水勲 (1995): *Manga ni miru 1945 nen* 漫画にみる1945年. Tōkyō: Yoshikawa kōbunkan.

SKABELUND, Aaron (2014): "Leading Dogs and Children to War". In: *The Journal of the History of Childhood and Youth* 7: 5–13.

TAKEDA, Shinpei 竹田慎平 (1948): *Kairiki genshi bōru* 怪力原子ボール. Tōkyō: Yashio shobō.

TAKEDA, Shinpei 竹田慎平 (1949): *Genshi bōru* 原子ボール. Tōkyō: Ōkawaya shoten.

THE GORDON W. PRANGE COLLECTION (2013): "The Process of Censorship—Sample Documents"; https://prangecollection.wordpress.com/2013/07/21/sample-ccd-documents/ (last access 2023 / 01 / 06).

Depicting the Atomic Bombings of Hiroshima and Nagasaki: History and Continued Significance

Christopher P. Hood[1]

1 Introduction

The atomic mushroom cloud has marked our minds, histories, and cultural consciousnesses for over 75 years. It is a curiously multivalent image, for people can look at it and know, with no words, to what it refers. Present and absent in the same glance, the image takes us to key debates about symbols and their power; we see the results of possessing a powerful weapon while the actual annihilation remains unseen, hidden by the cloud. This paper considers the way in which the atomic attacks on Hiroshima and Nagasaki are depicted and how this conveys meaning beyond the image itself.

To date atomic weapons have only been used in wartime attacks on another country on two occasions: the bombings of Hiroshima and Nagasaki in August 1945. In addition to documentary footage of these bombings, the story of what happened has been told in many ways and numerous times since. However, there is an apparent gap in the analysis of how the bombings have been dramatized and documented. While there are books and articles which, from their titles, would appear to be about the explosions or mushroom clouds, the contents do not consider these parts of the attack. For example, *Producing Hiroshima and Nagasaki* (SHIBATA 2018) has no discussion on the explosion and cloud. While Markus NORNES (2003: 247) comments that "[t]he final atrocity of the war, the strategic bombing attacks on civilians, has been elided by the iconic spectacle of the mushroom cloud", it is the only mention of the "iconic spectacle" in the book and it appears in a footnote. For my own paper, as will be demonstrated, the words of Tomomitsu MIYAZAKI (2015: 541) that "[w]e must maintain our perspective as human beings, and should not merely view the aftermath of the atomic bombing from above the mushroom cloud or from a distance, which is the perspective of nation" are very pertinent. But his article, like many others, is on the impact of the bombs on the *hibakusha* 被爆者 (atomic bomb victims) rather than any analysis of "the view from under the mushroom cloud" and what is significant about this viewing point. Seemingly it is only Mick BRODERICK

(1996: 211) who gives any mention of the importance of the point of view in relation to the mushroom cloud, as he notes that IMAMURA Shōhei 今村昌平, director of "Black Rain" (*Kuroi ame* 黒い雨), "intended to show the mushroom cloud from underneath, that is from the point of view of the victims of the bomb, for up to now we have been accustomed to seeing it from the safe point of view of an American aircraft".

On this basis, I hypothesize that the way TV movies, documentary-dramas, and films / movies (all of which are hereafter referred to as "dramatizations") handle the depiction of the atomic attacks on Hiroshima and / or Nagasaki reflect the difference between the position of the ones doing the bombing and those being bombed, and how the presented image may influence its interpretation by the viewer. To understand this variation in depiction, the paper reviews the relevant literature on how directors construct, or "frame", images. Second, the paper considers the actual documentary footage of the bombings themselves, suggesting that this footage will have influenced how the dramatizations will visualize the atomic attacks through a desire of directors for their work to be seen as authentic and historically accurate. Having established the framework of influences for depicting the bombings, I then analyze 16 dramatizations from 1947 to 2022 that show the bombings of Hiroshima and / or Nagasaki. The study does not include those dramatizations that solely use actual footage from 1945, but those which have their own recreations of at least some part of the bombings.

This paper is significant for scholars of atomic history and those working on the complex ways in which cultural, political and technical forces shape images according to time, space and political context. This paper speaks specifically to the positions of 'perpetrator' and 'victim' and the way they shape both the images in question and their evolving reception. While the paper considers the depictions from the positions of 'perpetrator' and 'victim', these two words are used merely to reflect the status of those who dropped the bombs and those on whom the bombs were dropped. They are not intended to reflect any judgement about the justification or otherwise of the usage of the bombs, which is a subject beyond the scope of this study.

2   DEVELOPING A FRAMEWORK TO INTERPRET DEPICTIONS OF THE ATOMIC BOMBINGS OF HIROSHIMA AND NAGASAKI

At the heart of this paper is an endeavour, based on the work of Roland BARTHES, to understand how atomic bombs are depicted in images. BARTHES embraces the mobility of the text and its meaning, underlining the impossibility of fixing absolute meaning. BARTHES (1977: 147) contends that "[o]nce the Author is removed, to claim to decipher a text becomes quite futile". BARTHES makes clear that it is the reader who has the responsibility to interpret the text. He thus opens the door to a plurality of shifting, different meanings, and subjective readings of the same source. BARTHES' work on text extends persuasively to the images this paper evaluates, an extension BARTHES himself makes in the interactive relationship of his written pieces with their accompanying photographs in *Roland Barthes by Roland Barthes* (1994). It also offers us a powerful framework via which to read the images of the atomic bombings of Hiroshima and Nagasaki, images whose meaning shifts depending on the time, cultural moment and political stance of the viewer in intriguing and unexplored ways.

In terms of the film-making aspect of the imagery, what this paper is primarily concerned with is "framing" and, more specifically, issues relating to the camera angle and how this constructs the image and the point of view (POV) that is employed. Jennifer VAN SIJLL (2005: 160) suggests that a high angle, i. e., the camera is in a high position and we, the audience, are looking down, "makes the subject appear small and vulnerable". However, David BORDWELL, Kristin THOMPSON, and Jeff SMITH note the dangers of trying to "assign absolute meanings to angles, distances, and other qualities of framing", questioning whether "filming from a high angle always renders the character as dwarfed and defeated" (2017: 190). Michael RABIGER and Mick HURBIS-CHERRIER (2020: 186) agree with this position, noting that "high-angle shots emphasize the vulnerability of a subject—and though this may be true in some circumstances, you should look beyond such facile associations". Indeed, within a dramatization, dialogue and character development, for example, also have significance. However, this paper works with the theoretical frameworks on camera angle in order to understand better the political, cultural, temporal and technological forces shaping both the construction and reception of the atomic bombings of Hiroshima and Nagasaki.

Blain BROWN (2012: 64) points out, with a high camera angle "we seem to dominate the subject. The subject is reduced in stature and perhaps in importance". This may be connected to our experiences as children, looking up to adults, and as adults, looking down on children, for example, while still needing to look up to things that dominate and are bigger than us (VIDEOMAKER 2019). Consequently, when considering the depictions of an atomic bombing, a high camera angle of a mushroom cloud would be expected to be used to show how it, and by extension those doing the bombing, is dominating the city being attacked. On the other hand, a low camera angle would be used to show how the victims are being dominated. BROWN (2012: 33) notes: "Each camera angle has a point-of-view" and the two are used in tandem. POV is different to the camera angle in that it is about from whose perspective we are seeing the image unfold.

Director Oliver STONE (cited in TIRARD 2002: 137) says that POV is "[t]he most important thing" and that "[t]he rest is just [...] scenery. Even the script." BROWN (2012: 10) suggests that "POV shots tend to make the audience more involved in the story for the simple reason that what they see and what the character sees are momentarily the same thing—in a sense, the audience inhabits the character's brain and experiences the world as that character is experiencing it". VAN SIJLL (2005: 158) and Christopher P. HOOD (2022b) also show how changes in the POV allow the director to show the narrative in different ways.

However, similar to the criticisms about camera angles, James ZBOROWSKI (2015: 2–3) provides a warning when thinking about POV in that we may be encouraged "to conceptualise the topic in particular and limited ways" and "that many theories of character and point of view rest, implicitly or explicitly, on a model of human experience that is similarly powerful but incomplete". In terms of the bombings of Hiroshima and Nagasaki, the POV will matter as it will show whether, in that moment, we are seeing them from the perspective of those doing the bombing or those being bombed. It would be expected that this, in turn, would link to the camera angle being used. However, just because the bomber's POV is being shown, for example, does not necessarily mean that the dramatization is wanting us to understand the motivations of those doing the bombing, which would need to be done through additional dialogue throughout the dramatization.

Although the relevant literature related to the technical aspects of filming provides a sound basis for studying the depictions of the atomic bombings of Hiroshima and Nagasaki in dramatizations, there is one aspect that it cannot take account of: the actual situation. While this paper considers recreations of the attacks on Hiroshima and Nagasaki, it is also cognisant of the fact that documentary footage of both attacks exists, and we need to engage with this footage. It does so ever mindful of the fact that the documentary footage of both explosions adheres to many of the conventions established above, revealing themselves to be as shaped by space, time and political moment as their more fictional counterparts, as is demonstrated further below.

When considering the atomic bombings of Hiroshima and Nagasaki, the pervasive images are of the two mushroom clouds. However, these are just one element of the bombing. The detonation is accompanied by a blinding flash, followed by the sound of the explosion and the release of an enormous pressure wave. These are, in turn (but in rapid succession), followed by the creation of the mushroom cloud, within which there will be a fireball (or "thunderball"). After this, there may be other features, such as fallout and black rain. In all cases, there is also the non-visible, but longer lasting and sometimes more impactful, aspect of radiation. This paper is concerned with the first 3 elements: 1) the flash; 2) the sound and accompanying pressure wave; and 3) the mushroom cloud.

The bombings of Hiroshima and Nagasaki were both filmed from planes that flew with the bombers. In the case of the Hiroshima bomb, while the bomb was dropped from the Enola Gay, the attack was filmed from a plane subsequently named after the mission as Necessary Evil, while the Nagasaki bomb was dropped from the Bockscar and the attack was filmed from Big Stink (The Great Artiste was part of both missions, taking scientific measurements) (REED 2020: 392, 407). The nature of the attacks meant that it was inevitable that the POV of the bombings would be from the U.S. perspective as there would have been no practical way to have camera crews at ground level, even at a distance. Furthermore, as the whole process of the bombings was filmed, this meant that predominantly the imagery had a high, top-down, camera angle. Although there were pictures looking up at the Enola Gay as the bomb-bay doors were opened, the detonation and creation of the mushroom cloud happened below the altitude of the planes and so ensured the high, top-down, angle. As the mushroom cloud grew to an altitude above the planes themselves, the angle did change to one that was less obviously top-down, but in the main images that linger on the cloud, the inclusion of land still ensures that the predominant angle is high (see Figure 1). We can also see that the cloud that formed over Hiroshima was, in mushroom terms, elongated like a shiitake rather than the button-mushroom shape that may be more typically associated with an atomic explosion. Overall, however, we are left in no doubt that the bombing of Hiroshima is filmed from a high camera angle that represents the perpetrator's POV.

Fig. 1: The Mushroom Cloud Over Hiroshima (ATOMIC ARCHIVE 2023a).

As with the Hiroshima bombing, the Nagasaki bomb detonated below the altitude of the planes and the angle of the growing mushroom cloud is much steeper with the cloud almost seemingly directly below the plane. As the cloud grew, the plane moved away to a greater distance. Compared to the most commonly used images of the Hiroshima bomb (such as Figure 1), the image of the Nagasaki mushroom cloud that appears to be used most frequently, has a less obviously high camera angle. The high angle becomes obvious due to being able to see islands in the distance and clouds over which the mushroom cloud towers (see

Figure 2). There is no doubt, however, that the POV, as expected, is again that of those who have dropped the bomb.

In terms of the stages of the bombings, neither of the pictures show the initial explosion, nor, in the case of the video footage, is the pressure wave and sound of the explosion apparent. Furthermore, as one of the crew of the Enola Gay noted (*Hiroshima*, directed 2005 by Paul WILMSHURST, 0:52:11), from their altitude they could not see the people on the ground or what was happening. The mushroom cloud becomes the focus and it is dehumanizing. The POV and high camera angles have ensured that we have seen the attacks from the relatively safe perspective of those in the planes which have dropped the bombs. We do not see the attacks from those on the ground. We do not get to see the flash, the effects of the pressure wave, or even the mushroom cloud from below. We cannot understand from these images the level of suffering that is being experienced at ground level. Watching the actual video footage of the Nagasaki bombing (ATOMIC ARCHIVE 2023c) becomes quite surreal as it gives the appearance of the cameraman just having an interest in an unusually-shaped cloud, while there is no indication at all of the destruction that has been wrought and is still unfolding beneath the foot of the cloud.

Fig. 2: The Mushroom Cloud Over Nagasaki (ATOMIC ARCHIVE 2023b).

On this basis of the literature reviewed and the documentary footage, we can hypothesize how a dramatization may show the bombings depending on the perspective—i. e., whether showing from a POV of those doing the bombing or a POV of those being bombed—that the director wishes to convey. If the director wishes the audience to see the attack from the perspective of those dropping the bomb, the POV will be from the sky and a high (up-to-down) camera angle will be used. Conversely, and although no documentary footage for the attacks on Hiroshima and Nagasaki exist to support this, we would expect the director to use a land-based POV with low (down-to-up) camera angle to emphasize the bomb being dropped down on us, the viewers, as victims. In the next section, this paper turns to analyzing a range of dramatizations using this hypothesis as a basis.

3 Depicting the Bombings of Hiroshima and/or Nagasaki in Dramatizations

Having established the basis by which we can analyze the dramatizations that include the bombings of Hiroshima and/or Nagasaki, this section focuses upon 16 such dramatizations (see Table 1). Due to space constraints, this section will not go into detail about each dramatization, but focuses on five so that greater depth can be provided. It should be noted here that four of the 16 dramatizations are animated (anime), but the principles of camera angle remain the same in relation to the direction that the atomic attacks are shown.

Title	Director	Country	Format	Year
The Beginning or The End	Norman TAUROG	USA	Film	1947
"Children of the Atomic Bomb" (*Genbaku no ko* 原爆の子; a. k. a. "Children of Hiroshima")	SHINDŌ Kaneto 新藤兼人	Japan	Film	1952
Hiroshima ひろしま	SEKIGAWA Hideo 関川秀雄	Japan	Film	1953
"Pica-don" (*Pikadon* ピカドン; a. k. a. "Atomic Bomb")	KINOSHITA Renzō 木下蓮三	Japan	Anime	1978
Enola Gay, The Men, The Mission, and The Atomic Bomb	David Lowell RICH	USA	TV movie	1980
"Barefoot Gen" (*Hadashi no Gen* はだしのゲン)	MASAKI Mori 真崎守	Japan	Anime	1983
"Leave These Children" (*Kono ko o nokoshite* この子を残して; a. k. a. "Children of Nagasaki")	KINOSHITA Keisuke 木下惠介	Japan	Film	1983
"The Diary of Yumechiyo" (*Yumechiyo nikki* 夢千代日記)	URAYAMA Kirio 浦山桐郎	Japan	Film	1985
"Black Rain" (*Kuroi ame* 黒い雨)	IMAMURA Shōhei 今村昌平	Japan	Film	1989
Hiroshima Out of the Ashes	Peter WERNER	USA	TV movie	1990
Hiroshima	Paul WILMSHURST	UK	TV movie	2005
"Hiroshima 8 August 1945" (*Hiroshima, Shōwa nijūnen hachigatsu muika* 広島・昭和20年8月6日)	FUKUZAWA Katsuo 福澤克雄	Japan	TV movie	2005
"Nagasaki 1945: The Angelus Bells" (*Nagasaki 1945—Anzerasu no kane* NAGASAKI 1945 アンゼラスの鐘)	ARIHARA Seiji 有原誠治	Japan	Anime	2005
"Barefoot Gen" (*Hadashi no Gen* はだしのゲン)	NISHIURA Masaki 西浦正記 / MURAKAMI Shōsuke 村上正典	Japan	TV	2007
"In This Corner of the World" (*Kono sekai no katasumi ni* この世界の片隅に)	KATABUCHI Sunao 片渕須直	Japan	Anime	2016
"Seaside Cinema—Labyrinth of Cinema" (*Umibe no eigakan—kinema no tamatebako* 海辺の映画館—キネマの玉手箱)	ŌBAYASHI Nobuhiko 大林宣彦	Japan	Film	2019

Tab. 1: List of dramatizations included in the study (ordered chronologically and using the standard English title for Japanese releases)

In terms of selecting the dramatizations, while I was already familiar with a number of the titles, I wanted to expand the study to capture as many of the relevant dramatizations as possible. Many dramatizations were covered or mentioned in books and articles that are referred to in this chapter even though, as noted in the introduction, they did not contain analysis on the depiction of the explosions or mushroom clouds within them. It was possible to find some through internet search, including on specific sites such as IMDb, for dramatizations that related to the bombings of Hiroshima and/or Nagasaki. Often by looking at one title, recommendations for similar ones would also be provided. Many titles, of course, were considered in other texts that discuss the atomic bombings (as discussed above). Others came about through suggestions from people who heard or read about my research.

As I gathered the list of titles, I then had to set about obtaining them so that the analysis could be done. Unfortunately, this was not possible for every dramatization that I wanted to include in the study. For example, I had wanted to include *All that Remains* (directed by Dominic HIGGINS and Ian HIGGINS in 2016), which is based on the diary and accounts of *hibakusha* NAGAI Takashi 永井隆, but it seems that there is no DVD available. This was a particular disappointment as it would have been one of the few films that exclusively focus upon the bombing of Nagasaki. For the reality is that the bombing of Nagasaki is often over-looked, and even of the Japanese dramatizations contained in this study there are only two out of eleven that feature the Nagasaki bombing.

While ultimately this study includes 16 dramatizations, additional ones were studied but not ultimately included due to factors such as 1) only using actual documentary footage for the bombings themselves (e.g., *Oppenheimer*, directed by Barry DAVIS in 1980, and *Day One*, directed by Joseph SARGENT in 1989); 2) the bombings themselves are not included even when the dramatization is seemingly about the bombings (e.g., *Fat Man and Little Boy*, directed by Roland JOFFÉ in 1989); or 3) the dramatization includes a depiction of the bombing in a way that could not have happened in reality (e.g., *Empire of the Sun*, directed by Steven SPIELBERG in 1987).

Considering the dramatizations, in relation to technical aspects of showing an atomic explosion and a mushroom cloud, we need to acknowledge the changes in film-making during the period being studied. While early films may have had limited options, with the development and use of computer-generated-imagery (CGI), the possibilities have become seemingly limitless. Indeed, these temporal considerations are something that need to be carefully considered in relation to how authentic the mushroom cloud may look compared to the actual atomic attacks.

Of the 16 dramatizations, four were British or American, and the other 12 were Japanese. It may be expected that the use of POV and camera angle, given that these nations reflect the nations that dropped the bombs and had the bombs dropped on them, would be correlated. However, analysis of the dramatizations themselves finds that this is not the case. As noted earlier, the depictions of the atomic bombings will be just one part of the whole dramatization. *Hiroshima Out of the Ashes* (Peter WERNER, 1990) only presents the victim's POV of the attacks and predominantly appears to be wanting the audience to be sympathetic to the victim's, i. e., Japanese, position rather than that of the bomber's, i. e., the American, position, despite the country of production being the USA. Therefore, rather than differentiating the dramatizations on the basis of country of production, the analysis focuses upon whether the dramatization shows the bomber's POV, the POV of those being bombed, or whether it includes both. There were no dramatizations which could be considered to have a purely neutral POV, whereby we would see, for example, the plane dropping the bomb and city in the same shot with a camera angle that was neither high nor low.

Of the 16 dramatizations ultimately included in the study, only two solely had the bomber's POV, with both *Enola Gay, The Men, The Mission, and The Atomic Bomb* (directed by David Lowell RICH in 1980) and *The Beginning or The End* (directed by Norman TAUROG in 1947) being U. S.-made. Given that it was made only two years after the end of the war and may have had the potential to further influence subsequent directors making a dramatization about the bombings, it is appropriate to focus upon *The Beginning or The End* in this part of the paper.

The Beginning or The End features the development of the atomic bomb and its ultimate usage. Consequently, and like many other dramatizations, it also includes the first use of an atomic bomb, the Trinity Test in New Mexico (USA) on July 16, 1945. Like a number of other dramatizations, it does not include the Nagasaki bombing. The Hiroshima bombing comes about 100 minutes into the 112 minute film. In terms of the depiction of the bombing, we see a view of the city of Hiroshima (which appears to be made from a model) from the sky as the bomb is released, a bright flash (keeping in mind that the film is black and white) (see Figure 3a, 1:40:29), the sound of an explosion, and then see a rising mushroom cloud. The view of the mushroom cloud is very much looking down on it growing until, much like the original documentary footage,

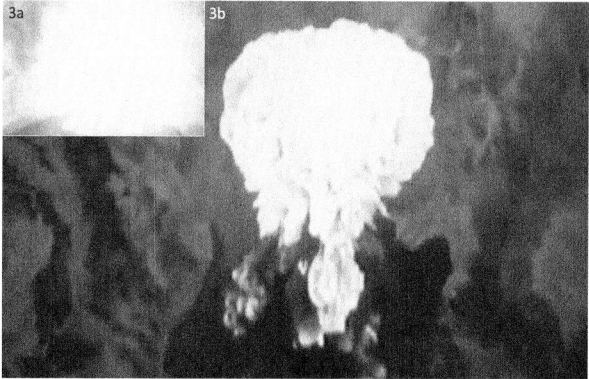

Fig. 3 (a–b): The Mushroom Cloud in *The Beginning or The End* (1947): The view of the explosion over Hiroshima and a downwards view of the mushroom cloud developing © Metro-Goldwyn-Mayer

the top of it is at the viewer's own altitude. After a view of some of those in the plane who are struggling due to the brightness of the flash, the view changes to look back down on the cloud, the base of which is spreading out, enveloping the city (see Figure 3b, 1:40:46). Technically, the depictions are noteworthy due to the relatively limited technology that would have been available at the time. While recreating the flash would not be overly difficult, the cloud looks quite authentic in terms of both the shape in relation to the actual Hiroshima bomb and how a mushroom cloud tends to grow. Yet the film's cloud was not a cloud at all, but was done by blowing up some sacks of material in water. This had to be done as the U. S. military would not give permission for the official footage to be used, but, ironically, the dramatization's results were considered so good by the U. S. miliary that they ended up using the footage themselves (KUZNICK 2022).

In terms of the analysis of the depiction, we find that *The Beginning or The End* contains all three elements—the flash of the explosion, the sound of the explosion, and the mushroom cloud. Further, the POV is very much from those in the plane or a position close to them, with us, as viewers, also looking down, with a high camera angle, at the target, Hiroshima. While the words spoken in the film itself—the full analysis of which are beyond the scope of this study—aim to provide additional context and justification for the use of the bomb, the depiction of the bombing itself leaves the viewer in no doubt about which is the dominant force and which has been overpowered.

Of the 16 dramatizations included in the study, two had both the POV of those doing the bombing and those being bombed. Of these, one, *Hiroshima* (directed by Paul WILMSHURST in 2005), was made in the UK and the other, "Barefoot Gen" (*Hadashi no Gen* はだしのゲン, directed by MASAKI Mori 真崎守 in 1983), was made in Japan. Both of these are worth considering in some detail for differing reasons. *Hiroshima* (2005) is worthy of study as a docu-drama, in that it both contains documentary elements, including actual footage of the Hiroshima bombing, and interviews with relevant people, as well as its own dramatizations (including the bombing itself), and so ensures that this study covers a range of styles of dramatization. "Barefoot Gen", on the other hand, is worth studying as one of the most well-known Japanese dramatizations and stories relating to the bombing of Hiroshima, as well as being an example of the popular anime genre. "Barefoot Gen" was so popular that there was a follow-up film, "Barefoot Gen 2" (*Hadashi no Gen* はだしのゲン 2, directed by HIRATA Toshio 平田敏夫 in 1986), which was not included in this study as it contained no new imagery of the attack. There have been other versions, such as a TV adaptation, "Barefoot Gen" (*Hadashi no Gen* はだしのゲン, directed by NISHIURA Masaki 西浦正記 and MURAKAMI Shōsuke 村上正典 in 2007), that was also included in the study.

Depicting the Atomic Bombings of Hiroshima and Nagasaki   115

In *Hiroshima* (2005), the bombing is shown by switching between actual documentary footage, interviews with survivors and flight crew, a voice over that explains the technical aspects of how the bomb worked, and recreations that use a mixture of sets and CGI for the parts relating to the explosion itself. In terms of the explosion, in a sequence that lasts about one minute (0:47:54–0:48:50), it is depicted several times, with the mushroom cloud itself being shown clearly (during the stage of it developing) four times. All three elements—the flash, sound, and the cloud—are shown, with the flash being shown more times than the cloud. In terms of the angle of showing the cloud, the first is a high angle shot that also shows the Enola Gay flying away from Hiroshima (see Figure 4a, 0:48:09). In that it also includes the plane means that it may not strictly be the bomber's POV, although it may have been similar to that from one of the other planes that flew with the Enola Gay. The next view of the cloud is from ground level, and although the positioning switches, there is no doubt that this is the POV of those being bombed and that the camera angle is low (see Figure 4b, 0:48:19). The third view is looking down on the city, albeit from a lower angle than the first one and perhaps not from the bomber's POV, and the focus is as much on the way that the foot of the cloud envelops the city as on the growth of the mushroom cloud and the fireball above the city (see Figure 4c, 0:48:29). Indeed, the view changes before much of the mushroom shape becomes apparent. The final view is another from ground level and a little further away than the second, so like the third view it is a shallower angle than the previous one from that POV, but still has an upwards angle (see Figure 4d, 0:48:38). The cloud itself, however, is soon obscured by the dirt being blown along at ground level by the pressure wave. An additional view of the mushroom cloud comes another 90 seconds later, by which time some of the previous sequences have been shown again with additional explanations and discussion, in a much more distant fashion which links to the view of one the survivors being interviewed (see Figure 4e, 0:50:12). The additional distance means that the angle is somewhat level for the explosion and initial development of this cloud, but this soon switches to a more pronounced low, upward angle (see Figure 4f, 0:50:14). With further explanation from the survivor, this explosion sequence is shown again, but with additional footage of the cloud developing. This time we see the upwards motion of the cloud as it develops, but the distance still means that there is a relatively level camera angle. Even though the survivor refers to the mushroom shape of the cloud, the mushroom-like-shape is on screen only a few seconds (see Figure 4g, 0:50:39), and, as noted earlier, this shape is not consistent with the shape of the actual cloud that developed over Hiroshima. Further repeats and extensions to previous views of the explosion and the cloud developing are shown. While one, an extension of the third view (see Figure 4i, 0:51:35), shows the cloud as relatively linear, one that extends the fifth view, shows the cloud continuing to be mushroom in shape (see Figure 4h, 0:51:44). Soon after this (0:51:57), now four minutes after the sequence began, the actual documentary footage of the cloud, which clearly looks very different to what was depicted only a few seconds previously, is shown.

Overall, the use of different camera angles and POV appear to have been an attempt to frame *Hiroshima* (2005) as being an objective docu-drama rather than presenting one standpoint in favour of or against the bombings. However, given one of the hypotheses of this paper was that the depiction of the mushroom cloud would be influenced by the actual shape of the Hiroshima mushroom cloud, it is important to note that this is not the case with *Hiroshima* (2005). Indeed, it does not appear to be the case for any dramatization. Rather, the depiction may have been

Fig. 4 (a–j): The various mushroom clouds of *Hiroshima* (2005) © BBC

more influenced by what people would understand to be the shape of a cloud after an atomic explosion. This dramatization was made in 2005, by which time viewers and those making the dramatization alike may have had expectations of what a mushroom cloud looks like due to images from numerous nuclear tests or how the cloud has been depicted in fictional dramatizations. This may mean that historical accuracy is not as significant as was hypothesized. This is further supported by the cover of the DVD of *Hiroshima* (2005) (see Figure 4j). Rather than either use one of the CGI depictions from the film itself or a picture of the actual Hiroshima cloud with its somewhat disconnected top portion (see Figure 1), a picture of the Nagasaki cloud with its more button-mushroom shape (see Figure 2), and closer to what people may have associated as being an atomic mushroom cloud, is used. *Hiroshima* (2005) is not the only dramatization to do this, as *Day One*, which was ultimately not included in this study (see above), also used the Nagasaki mushroom cloud for the Hiroshima bomb.

Another important point to note about *Hiroshima* (2005) is, that the focus is much more on the initial flash and explosion, rather than the cloud. This is not surprising, and is consistent with many other dramatizations (as discussed further below), given that the film contains survivor accounts from those who were in and around Hiroshima at the time of the bombing. In English, the ways to refer to the bombings of Hiroshima and Nagasaki tends to use words such as "atomic" and "nuclear". While these words may also be used in Japanese, for example the best-known building for memorializing the bombing of Hiroshima is the "Atomic bomb dome" (Genbaku dōmu 原爆ドーム) (see Figure 5), it is more common to refer to the bombings as *pika* ピカ or *pikadon* ピカドン. *Pika* literally means "flash", while *don* is an onomatopoeia referring to the sound of the explosion. While both words may be used in other contexts in Japanese, there are times when *pika* may be used to refer to an atomic explosion and *pikadon* has no other meaning and usage and, if anything, may be used solely to refer to the bombings of Hiroshima and Nagasaki rather than, for example, an atomic test, due to the way that the word conveys the victim's perspective of the bomb.

Fig. 5: A photograph of the Atomic Bomb Dome in Hiroshima with a memorial plaque showing what the building, the Hiroshima Prefectural Industrial Promotion Hall, looked like prior to the bombing. Photograph by Christopher P. Hood.

Let us now turn to "Barefoot Gen" (1983). As one of the most well-known Japanese narratives about the bombing of Hiroshima, it may come as a surprise to some that it also contains the bomber's POV. It is the only Japanese dramatization in the study to do so. However, in reality, these parts (there is some switching between the bomber's and those-about-to-be-bombed POVs) of the dramatization are very short, being about 60 seconds (0:29:37–0:29:48 and 0:31:06–0:31:56) in total within the 86 minutes of the film. The bomber's POV parts of the dramatization show the situation in the Enola Gay and includes dialogue in English, with Japanese subtitles included on the screen. Once the bomb explodes, with the city effectively disappearing within the flash and initial development of the mushroom cloud as we look down on it (see Figure 6a, 0:31:52), the bomber's POV comes to an end. With the switch to the POV of those being bombed, the *pika* is shown multiple times, turning the color images to black and white to emphasize the flash. Color returns to show victims being killed very graphically (see Figure 6b, 0:32:42), and buildings—including the Hiroshima Prefectural Industrial Promotion Hall which is now known as the Atomic bomb dome (see Figure 6c, 0:33:05)—being destroyed in the explosion, with colors moving up the screen in a way that suggests the process of things being sucked up into the mushroom cloud. That that is the intention becomes clearer about two-and-a-half minutes after the initial explosion, when we can see the colors going up into the base of the mushroom cloud (see Figure 6d, 0:34:17). We then follow the cloud up as it billows with a second bulge developing. We never see the top. Instead, the screen goes completely red before being replaced by a still picture of a view over the city (see Figure 6e, 0:34:34) which is seemingly based on a photograph taken of the city (see Figure 7), probably some time after the bombing, in which the largest cloud was likely as much due to the fires raging in the city as the original atomic blast (BROAD 2016). Certainly, this cloud is not one that bears any resemblance to the bomber's POV soon after the bombing.

Fig. 6 (a–e): The stages of the Hiroshima bombing in "Barefoot Gen" (1983) © Gen Production

Considering these images in more detail, "Barefoot Gen" (1983) has all three elements of the atomic bombing, but although the streaks of colored lines end up being revealed as part of the formation of the mushroom cloud, the mushroom cloud as a whole is of much less significance than the *pikadon* and destruction that is happening on the ground. This, in many respects, should not be a surprise since those within Hiroshima would have been too close to really see the cloud in its entirety. Consequently, "Barefoot Gen" (1983) is typical of most of those dramatizations which focus on a POV of those actually in the cities of Hiroshima or Nagasaki themselves, as is discussed further below. In "Barefoot Gen" (1983) it is not really clear from whose viewpoint or where we are seeing the mushroom cloud as shown in Figure 6d develop or why it is included. We have to assume it is included to provide a familiar frame of reference for those viewers who may equate an atomic attack with a mushroom cloud. Overall, however, "Barefoot Gen" (1983) clearly uses low camera angles and a focus on the POV of those in the city to emphasize the viewpoint of them being victims.

Fig. 7: The Cloud Over Hiroshima (Atomic Archive 2023d).

Let us now turn to those dramatizations that only consider the victims' POV. Of these, as noted above, there is only one (*Hiroshima Out of the Ashes*, directed by Peter WERNER in 1990) that is not Japanese, while the other ten are all Japanese. To help illustrate the typical elements amongst these dramatizations, this paper will now focus upon two, chosen, in part, as they were made in different time periods.

The first of the two films, and one of the first Japanese dramatizations about the bombings, is *Hiroshima* ひろしま (directed by SEKIGAWA Hideo 関川秀雄 in 1953). Stylistically, there are some overlaps between this film and one that was made the previous year, "Children of the Atomic Bomb" (*Genbaku no ko* 原爆の子, directed by SHINDŌ Kaneto 新藤兼人 in 1952), but it is *Hiroshima* (1953) that has gained greater recognition and is still more readily available on DVD and Blu-ray, for example. All of the dramatizations that depict a victim's POV contain the *pika* and although the way this is achieved varies somewhat, technically, even for a film made in the 1950s (and earlier as we saw with *The Beginning or the End*) it was not difficult to replicate a bright flash. *Hiroshima* (1953) not only contains the *pika*—albeit with there seemingly being multiple flashes rather than a single one even in one view. It also shoes, like many dramatizations, the reactions of those exposed to the *pika*. For example, one of the teachers is looking up to the sky with her pupils. This is further emphasized by a low camera angle when she tries to cover her face (see Figure 8a, 0:27:39). The film then also depicts the *don*, as the sound and accompanying pressure wave bring destruction to the city (see Figure 8b, 0:27:39). However, with other dramatizations contained in this study, we do not see people reacting to the sound by covering their ears to protect themselves from the loud sound, in the way that we see people cover their eyes with their arms or hands to protect them from the *pika*, but rather the focus is on the destruction brought by the accompanying pressure wave.

As one of the earliest dramatizations made about the atomic bombings, *Hiroshima* (1953) is very effective in showing the impact of the *pikadon*, and, although beyond the scope of this study, the graphic horrors of what happened to people in the city. As noted above, this is also an aspect in "Barefoot Gen" (1983). Such outputs may have also led to greater acceptability of graphically showing the dead and dying in Japan (see also HOOD 2013). In relation to the mushroom cloud, Figure 8d (0:27:53) shows that an image of a mushroom cloud does appear in the film, albeit briefly. However, Figure 8c (0:27:43) is more significant as it represents

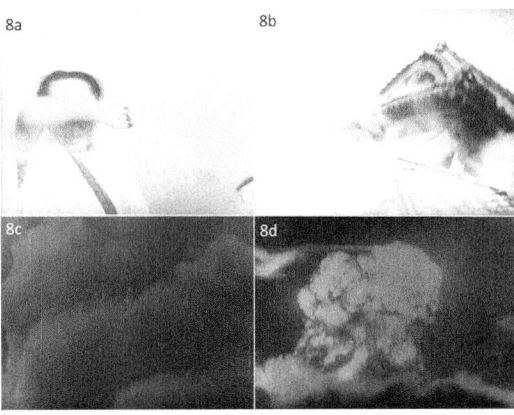

Fig. 8: The stages of the Hiroshima bombing in *Hiroshima* (1953) © Shin Nippon Films

how the cloud, where no clear shape is discernible, is shown for a longer period of time. This billowing tower, where the overall shape of the cloud is not discernible, reflects the reality of what it would have been like for many of the victims—i. e., they were too close to the cloud itself to see its shape. Figure 8d, in that respect, is a jarring image as this view means that we are suddenly taken to a position further away from the city. A similar effect is used in Figure 6e in "Barefoot Gen" (1983), before we are returned to the centre of Hiroshima and the horrors in the next scene. This mushroom cloud in Figure 8d is itself more reflective of what people have come to understand an atomic mushroom cloud to look like rather than what the Hiroshima cloud itself looked like, as discussed in relation to *Hiroshima* (2005) above.

The second dramatization that only considers a victim's POV is the anime "The Bells of Angelus" (*Nagasaki 1945—Anzerasu no kane* NAGASAKI 1945 アンゼラスの鐘, directed by ARIHARA Seiji 有原誠治 in 2005). While, arguably, "Black Rain" by IMAMURA Shōhei (1989) is one of the better-known dramatizations that could be included here, using this film would have meant that there would be no analysis of any dramatizations that include or focus upon the bombing of Nagasaki. It is important that this paper does not overlook the bombing of Nagasaki. The title "The Bells of Angelus" refers to the bells in the Immaculate Conception Cathedral in Nagasaki, a city which is well-known for being one of the places in Japan where Christianity survived since its introduction in the late 16th century, despite being outlawed for over 200 years (DIEHL 2018: 2). In terms of the dramatization, as with others, we see many *pika* and *don*, and, unsurprisingly, the impact on the Cathedral of both the *pika* (Figure 9a, 0:14:39) and the *don* (Figure 9b, 0:14:44) are a feature of some of these. One of the main views of the mushroom cloud is from outside the city itself (Figure 9c, 0:14:50). While this is comparable in some ways to *Hiroshima* (1953), the more distant view is less jarring than in *Hiroshima* (1953) as some of the char-

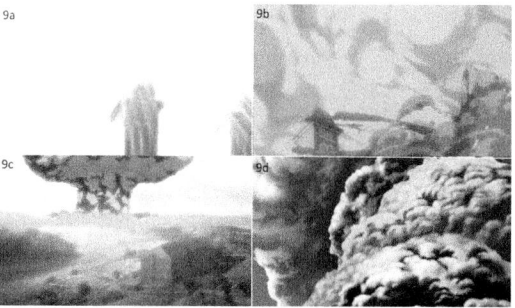

Fig. 9 (a–d): The stages of the Nagasaki bombing in "The Bells of Angelus" (2005) © Mushi Production

acters in "The Bells of Angelus" are based outside of the city and so the view is in keeping with their experiences. The "Bells of Angelus" also contains a close-up view of the mushroom cloud itself (Figure 9d, 0:15:21), which is comparable to what was seen in Figure 8c in *Hiroshima* (1953) and Figure 6d in "Barefoot Gen" (1983). Later in the film, other characters also speak of their experiences of the bombing (0:27:04 and 0:44:00), but in both cases depictions of their recollections focuses on the *pika* and do not include the mushroom cloud, while the *don* element is just a sound and does not include the impact of the pressure wave as is done in many other dramatizations. This further underlines the fact that in the victims' experiences, due to being so close to the bomb itself, the *pika* (and, to a lesser degree, the *don*) is of greater significance than the mushroom cloud. Overall, the POV and angles are consistent with what would be expected when the focus is on the victims. However, in relation to the depiction of the mushroom cloud itself, "The Bells of Angelus" (2005)—as with *Hiroshima* which was released in the same year—the preference appears to be to show what modern audiences may expect a mushroom cloud to look like rather than something more similar to the documented mushroom cloud.

4 Conclusions

At the time of writing in 2022, with the Russian invasion of Ukraine and continued North Korean missile tests, the possibility of atomic weapons being used had perhaps not been as high since the end of the Cold War at the start of the 1990s. While a broad understanding of the dangers of atomic weapons may exist amongst the populations of many countries, the knowledge may not be uniform or as great as it was during the Cold War. To date, only Japan has experienced atomic weapons being used against it in a war with the bombings of Hiroshima and Nagasaki in 1945. This paper has considered the way in which dramatizations have depicted these bombings. The hypothesis was that conventional filming techniques in relation to Point-Of-View (POV) and camera angle, together with the influence of the official documentary footage of the bombings, would dictate how the bombings are shown.

The paper ultimately considered 16 dramatizations that showed the atomic attacks on Hiroshima and/or Nagasaki which did not solely use documentary footage of the bombings. The study found that POV and camera angle are indeed used in tandem, while noting that there was no strict divide based on the national origin of the dramatization. It was further noted that the depictions are only one part of the complete narrative, so one may not be able to surmise the overall message the director is seeking to convey by studying the portrayal of the atomic attacks alone. As was demonstrated by analyzing five of the dramatizations in more detail, the combination of POV and camera angle was used to emphasize the position of power of those dropping the bomb and the helplessness of the victims having the bomb dropped on them. The POV from the sky and high camera angle leaves the viewer in no doubt of being the do-er and that the aggrieved party is someone in a weaker position. A POV from land with a low angle, on the other hand, emphasizes the position of being a victim.

However, the study found, contrary to the hypothesis, that the actual shapes of the mushroom clouds in 1945 did not seem to influence the shape of the mushroom clouds in the dramatizations. Factoring in that the first film made only two years after the actual bombings managed to recreate quite an authentically shaped cloud depicting the attack on Hiroshima, technical reasons do not appear to have been the limiting factor here. Rather, when considering the years in which the dramatizations were made, it seems that mushroom clouds that were more consistent with the shape of those after many nuclear tests and which had appeared in fictional depictions, were adopted in preference over historical authenticity, particularly in the case of the Hiroshima mushroom cloud. Indeed, for those dramatizations focusing on the victim's POV in the city being attacked, the shape of the whole mushroom cloud was relatively unimportant in comparison to seeing the base of the cloud and, even more importantly, depicting the actual flash (*pika*) and the sound of the explosion (*don*) with its accompanying pressure wave.

The extant literature, such as BORDWELL, THOMPSON, and SMITH (2017), warns of the dangers of directors being overly reliant on conventional usage of POV and camera angles to show power relationships. Yet, the reality is that people often do, naturally, interpret power relations vis-à-vis view, size and angles. The study of the depictions of the atomic attacks on Hiroshima and Nagasaki has clearly shown that this dynamic is a central element to how the bombings are presented.

Not all dramatizations that include a nuclear explosion are about the bombings of Hiroshima and/or Nagasaki. Indeed, those covering Hiroshima and/or Nagasaki are in the minority, with the author having identified over 50 fictional dramatizations that contain an atomic explosion (Hood 2022a). With this paper having established that the POV and camera angle are a significant element of how an atomic attack is portrayed in determining how it is interpreted, what needs to be researched further is the degree to which this factor is also true in fictional dramatizations. Based on the analysis, the expectation would be that a sky-based POV with a high camera angle being used for the perpetrator and a land-based POV and low camera angle being used for the victim. Further research should also be done about the shape of the mushroom cloud in such dramatizations. If fictional dramatizations use button-mushroom-shaped clouds, this could further support an unexpected finding of this paper that dramatizations about the bombings of Hiroshima and/or Nagasaki were not overly concerned with the historical accuracy of the shape of the mushroom cloud and that it may have been other documented clouds and fictional depictions that were of greater significance. In other words, directors may have a preference to use imagery that they believe the audience expects to see rather than present more authentic imagery.

The story of the atomic bombing of Hiroshima, and, to a lesser extent, Nagasaki, and how they are shown in dramatizations does not end. In 2023 the movie *Oppenheimer* (Christopher Nolan, 2023) will be released and potentially (at the time of writing it is not known what the movie will show) will become yet another dramatization which the findings of this analysis can be tested against. In the meantime, this paper is significant as it adds to the body of literature that helps our understanding of the importance of POV and camera angles in the making of dramatizations and how the images are interpreted.

Notes

1   The author would like to thank the following people in relation to the development of this paper: all of the participants at the conference "Hiroshima—Nagasaki—Fukushima: Articulations of the Nuclear. The Case of Japan", University of Cologne, May 2022, where an early version of this paper was presented, Sarah Buddery (film critic), Kate Griffiths (Cardiff University), Sheldon Hall (Sheffield Hallam University), and Nick Hodgin (Cardiff University).

REFERENCES

ATOMIC ARCHIVE (2023a): "Atomic Bomb Cloud over Hiroshima"; https://www.atomicarchive.com/media/photographs/hiroshima/image-1.html (last access 2023/03/03).

ATOMIC ARCHIVE (2023b): "Atomic Bomb Cloud over Nagasaki"; https://www.atomicarchive.com/media/photographs/nagasaki/image-1.html (last access 2023/03/03).

ATOMIC ARCHIVE (2023c): "Atomic Bombing of Nagasaki"; https://www.atomicarchive.com/media/videos/nagasaki.html (last access 2023/03/03).

ATOMIC ARCHIVE (2023d): "Atomic Bombing over Hiroshima"; https://www.atomicarchive.com/media/photographs/hiroshima/image-2.html (last access 2023/03/07).

BARTHES, Roland (1977): *The Death of the Author*. London: Fontana.

BARTHES, Roland (1994): *Roland Barthes by Roland Barthes*. Santa Cruz: University of California Press.

BORDWELL, David / THOMPSON, Kristin / SMITH, Jeff (eds.) (2017): *Film Art: An Introduction*. 11th edition. New York: McGraw-Hill Education.

BROAD, William J. (2016): "The Hiroshima Cloud That Wasn't"; https://www.nytimes.com/2016/05/24/science/hiroshima-atomic-bomb-mushroom-cloud.html (last access 2022/09/02).

BRODERICK, Mick (1996): *Hibakusha Cinema: Hiroshima, Nagasaki, and the Nuclear Image in Japanese Film*. London: Routledge.

BROWN, Blain (2012): *Cinematography: Theory and Practice*. 2nd edition. Waltham, MA: Focal Press.

DIEHL, Chad R. (2018): *Resurrecting Nagasaki: Reconstruction and the Formation of Atomic Narratives*. Ithaca: Cornell University Press.

HOOD, Christopher P. (2013): "Visualisation of Death in Japan: The Case of the Flight JL123 Crash". In: AARON, Michele (ed.): *Envisaging Death: Visual Culture and Dying*. Cambridge: Cambridge Scholars Publishing, 120–39.

HOOD, Christopher P. (2022a): "Nuclear Imagery and Looking for Movies and TV Programmes"; https://hoodcp.wordpress.com/2022/03/24/nuclear-imagery-and-looking-for-movies-and-tv-programmes/ (last access 2022/03/24).

HOOD, Christopher P. (2022b) "Filming Points of View and Angles—The 'Jaws' Way"; https://hoodcp.wordpress.com/2022/11/22/filming-points-of-view-and-angles-the-jaws-way/ (last access 2022/11/22).

KUZNICK, Peter (2022): *Interview and comments at the conference "Hiroshima—Nagasaki—Fukushima: Articulations of the Nuclear. The Case of Japan"*, University of Cologne, 22 May 2022.

MIYAZAKI, Tomomitsu (2015): "The View from under the Mushroom Cloud". In: *International Review of the Red Cross* 97 (899), 527–42.

NORNES, Markus (2003): *Japanese Documentary Films: The Meiji Era Through Hiroshima*. Minneapolis: University of Minnesota Press.

RABIGER, Michael / HURBIS-CHERRIER, Mick (2020): *Directing: Film Techniques and Aesthetics*. 6th edition. Abingdon: Routledge.

REED, Bruce C. (2020): *Manhattan Project: The Story of the Century*. Cham: Springer.

SHIBATA, Yuko (2018): *Producing Hiroshima and Nagasaki: Literature, Film, and Transnational Politics*. Honolulu: University of Hawai'i Press.

TIRARD, Laurent (2002): *Moviemakers' Master Class*. London: Faber & Faber.

VAN SIJLL, Jennifer (2005): *Cinematic Storytelling: The 100 Most Powerful Film Conventions Every Filmmaker Must Know*. Studio City: Michael Wiese Productions.

VIDEOMAKER (2019): "Camera Angles—a complete guide"; https://www.videomaker.com/how-to/shooting/composition/your-complete-guide-to-camera-angles/ (last access 2022 / 08 / 15).

ZBOROWSKI, James (2015): "Filmic Point of View and the Representation of Character Interaction"; https://www.academia.edu/14359457/Filmic_point_of_view_and_the_representation_of_character_interaction (last access 2022 / 08 / 09).

Looking for Possibilities: Urban Planning in Manchuria and Tange Kenzō's Plan for Hiroshima

Chantal Weber

1 Introduction

The urban planning that began in Japan with the arrival of the Meiji period (1868–1912) was visionary. Highly influenced by Western concepts, Japanese architects and planners were constantly on the lookout for a traditional—or uniquely Japanese—aesthetic with which to express their designs for the modern city. They could not achieve their goal of blending Japanese and Western design features, however, without giving due consideration to the political restrictions and regulations in effect at the time. Moreover, postwar planners were confronted with the peculiar situation that many cities had been utterly destroyed, and those in need of housing had constructed temporary dwellings, leading to widespread urban sprawl. Visionary planning had to take the population's basic needs into account while at the same time building cities that could create and represent the new, democratic identity of the battered country.

One star of this period of postwar architecture was Tange Kenzō 丹下健三 (1913–2005), proclaimed the best-known Japanese architect—a "world architect" indeed—by Seng Kuan and Yukio Lippit (2012: 13). During the war, Tange had worked as assistant to Maekawa Kunio 前川国男 (1905–86), student of Le Corbusier (1887–1965), but returned to university for further research and founded his own research laboratory, yet his real glory days began after the war. As to the origins for Tange's fame, any and all research inevitably brings one into contact with the Peace Memorial Park in Hiroshima, which includes the Peace Memorial Museum and the cenotaph that was erected to the victims of the atomic bomb of 1945. The reconstruction of Hiroshima—including the planning of the Peace Memorial Park—was Tange's first postwar project. In point of fact, though his original plans were only partially implemented, it was the only project of Tange's to involve urban landscape design on such a grand scale. Tange received considerable international recognition in 1951 when he was invited to the eighth meeting of the Congrès Internationaux d'Architecture Moderne (CIAM) where he presented his proposal for the reconstruction of Hiroshima. Over the course of his life, TANGE would construct many other buildings and engage in various urban planning projects, providing the plan for Tokyo Bay in 1960, the Yoyogi Olympic stadium of 1964, and the exhibition area of the Expo '70 in Osaka. Yet the overwhelming consensus among architectural historians is that his proposal for Hiroshima was the starting point of his professional carrier.[1] Moreover, TANGE is credited as spiritus rector of the well-known and influential 1960s architectural movement Metabolism (*metaborizumu* メタボリズム), some of the members of which were also involved in the Hiroshima plan.[2] According to Jörg H. GLEITER (2017: 243) it was the first architectural avant-garde movement in Japan and marked a turning point in Japanese urban design. For his part, David B. STEWART (1987: 181) claimed that this "Tange-cum-Metabolism ideology" deliberately sought to reorganize the city "in terms of giant, multifunctional architectural units."[3]

However, the designs that came from the Metabolist movement would have been unthinkable in the absence of both their pre-war and their war-time predecessors. This paper aims to outline the reliance of postwar Japanese architects on the experiences of those periods by delineating a certain continuity between the two, both in terms of designs and of the individuals involved. Firstly, I analyze urban planning in Manchuria, specifically the new capital Shinkyō, as a city that was representative of the past. This analysis serves to reveal how architects and city planners seized the chance to make a politically modernist statement to the tune of made in Japan. Though their visions were not conclusively brought to fruition, Manchuria can nonetheless be thought of as a playground in which planners gained experiences that would come into use following the country's ruinous defeat in the war. The second part of the paper will then outline the reconstruction of the shattered cities in Japan, itself a sort of rebirth of the nation as a democratic state supervised by the Allied Powers. The third and last part will focus on TANGE Kenzō's Hiroshima Peace Park as representative for this new era in Japanese history, an era in which the city of the future was destined to rely on the past experiences of the planner. The debate around Japanese tradition and the appropriate inclusion of Western notions is seemingly timeless. Yet it should not be forgotten that Japanese architects always participated in the international discourse of urban planning and were heavily influenced by Western concepts and ideas on the functions of a city vis-à-vis its inhabitants. As such, the analysis of TANGE's plan for Hiroshima reveals two dimensions of its conception: a combination of Japanese architecture with Western thoughts, and a connection between past and future.

2 CITY OF THE PAST: URBAN PLANNING IN MANCHURIA

The Japanese Empire actively sought to establish a new Lebensraum (living sphere) in its colonies, effectively constructing the Greater East Asia Co-prosperity Sphere (Daitōa kyōeiken 大東亜共栄圏) in Asian countries, most notably in China and Manchuria.[4] Japan's first colonies—Taiwan (1895–1945) and Korea (1910–45)—were already endowed with significant urban infrastructure and Japanese construction was thus limited to the addition of office buildings and museums as well as infrastructural improvements to water and sewerage systems, for example. There was little space for city planning on a large scale. In China and in Manchuria, by contrast, vast expanses could be allocated to urban development.[5] SHIINA Etsusaburō (1976: 106) goes so far as to speak of a "great experimental ground" (*daijikkenjō* 大実験場), which he claims presented itself in Manchuria after the 1931 Mukden Incident (Manshū jihen 満州事変) and the foundation of Manchuria (Manshūkoku 満州国)—subsequently the Empire of Manchuria (Manshū teikoku 満州帝国)—as a nominally independent state in 1932.[6] For its part, the ambitious development program of the South Manchurian Railway (SMR, Minami Manshū tetsudō 南満洲鉄道) constituted a concerted effort to precipitate industrialization. Modelled after the workings of the East Indian Company of British India as a representative of the state in the colonies, it included the construction of a railway network along with communication infrastructure throughout the territory. In addition, various agricultural plans were set in motion in 1933 with a view to attracting up to 1,000,000 Japanese farmers to Manchuria. The intention was for these workers to be self-supporting, relieving the home country of poor farmers while forming an enclave that would serve as a protective shield against Russia.[7] In 1933, Japanese architects proposed 50 new villages in Northern Manchuria with a view to accommodating more than 100,000 farmers; the communities would be built as small cities, bringing urban comfort into proximity with agricultural land. The planners—notably UCHIDA Yoshikazu 内田祥三 (1885–1972)—combined two distinct design concepts in their proposals. David TUCKER (2005: 58) summarizes these two design concepts as comprising: "[T]he modern, industrial city that had given rise both to modern planning and architectural modernism, and the almost premodern village housing of rural Japan."[8] Though TUCKER draws a connection with Japanese architecture, it is clear that UCHIDA was equally following a worldwide trend toward the functional city as proposed in the 1933 Charter of Athens. To wit, the city aimed to support collective life by embracing the countryside as well as more strictly urban design concepts (GOLD 1998: 230). With significantly less farmers and families immigrating from Japan, the country's attempt to colonize Manchuria—the supposed clean sheet ripe to be filled with utopian fantasies—was ultimately a failure. Speaking to this dynamic, John R. STEWART (1939: 40)—of the National Credit Office and the Institute of Pacific Relations—has observed that "Japanese colonization appears as an artificial and forced movement that must be heavily subsidized and fostered by patriotic appeals."

However, Japan's lack of success in Manchuria did not prevent the Kwantung Army (Kantōgun 関東軍) from planning a new capital for the State of Manchuria in place of the existing city of Changchun 長春: Shinkyō 新京. The Kwantung Army—alongside the SMR—was an active authority of the Japanese Empire in the colony. Changchun was chosen over the traditional capital of Manchuria—Mukden (Hōten 奉天, today Shenyang)—as the rather small town lacked both traditional Chinese urban culture or any form of formal landownership. Speaking to the unique possibilities for the Japanese conquerors, Anke SCHERER (2012: 47) remarks: "For architects and city planners, the various infrastructural projects and especially the rare chance to construct a national capital from scratch offered the unique possibility to tackle projects on a scope that crowded urban space in Japan did not provide." As chief advisor for the construction project, the Kwantung Army appointed prominent architect and head of the architecture department at the Imperial University Tokyo, SANO Toshikata 佐野利器 (1880–1956).[9] SANO had already been involved in the reconstruction of Tokyo after the Great Kanto Earthquake of 1923, assisting acting home minister GOTŌ Shinpei 後藤新平 (1857–1929).[10] Planners intended on making Shinkyō a modern city, and this would include: open spaces and parks which divided residential, commercial, and industrial zones; cultural institutions such as museums, libraries or public halls that would be situated along wide boulevards; infrastructure including highways, railway stations, telephone and telegraph cables throughout the city, and even an airport, all of which served to represent modern technology.[11] While the Russian government had originally designed Changchun as a railway town, the Japanese architects had something more ambitious in mind, namely to have Shinkyō city physically embody the values of the new state. Qinghua GUO (2004: 104) confirms that the plan was supposed to "reflect the Japanese vision of an ideal city and their new role in East Asia." Construction on the city of Shinkyō began in spring 1933, yet the Japanese architects' plans for the city were never completed. The design envisages two overlapping patterns: First, there had been a Beaux-Arts plan, as GUO (2004: 106) calls it, for which grand representative buildings with roofs designed in the mandatory Imperial Crown style (*teikan yōshiki* 帝冠様式) would be displayed, as well as large boulevards, parks, and axial approaches connecting the old and the new railway station. Second, the architects had outlined a Chinese grid plan with a north-south axis. Areas for different purposes were designed to keep administrative, commercial, living, and industrial zones separate yet linked by a modern transportation system. Integrating the Imperial Palace into the existing plans proved extremely difficult. In point of fact, construction never got off the ground, leaving Puyi 溥儀 (jap. Fugi, 1906–67)—last emperor of China and subsequently emperor of Manchuria—to content himself with provisional living arrangements. As GUO (2004: 110) points out, the ensuing controversy in the dispute between the Kwantung Army and Puyi undermined the "intrinsic purpose of the city".

Many Japanese architects were invited to contribute to the new city, including SAKAKURA Junzō 坂倉準三 (1901–69), who was working in the office of LE CORBUSIER in Paris at the time. SAKAKURA designed an area with offices and apartment buildings integrating the surrounding landscape and thus combining the natural and urban environments. He relied on concepts originating with LE CORBUSIER, who wrote as early as 1925 that modern cities needed to incorporate more green space in urban planning so as to guarantee a livable environment for their citizens (LE CORBUSIER 2015: 139).[12] As per the Charter of Athens—composed after the CIAM IV in 1933—inhabiting was recognized as the most important function of a city; the other three key factors—work, leisure and circulation—were, however, also of great significance for a functional city according to the CIAM.[13] SAKAKURA was not the only follower of LE CORBUSIER's working at the site; MAEKAWA Kunio, who was also working on housing problems in Manchuria, brought his employee TANGE Kenzō, a man whose later postwar activities would rely heavily on his experiences in Manchuria.[14]

Shinkyō combined Chinese, Japanese, and modern Western standards in urban planning, thereby falling under what Carola HEIN (2016b: 464) terms transnational urbanism. Though the Japanese architects used their knowledge of Western urban and architectural designs to improve the cities in the colonies, they aimed to create some form of modernity that was made in Japan. According to Cherie WENDELKEN (2000: 821) the architect Itō Chūta 伊東忠太 (1867–1954) had already called for a "new national style for Japan that would reflect its broader cultural origins in Asia" at the beginning of the 20th century. The Imperial Crown style, which was mandatory for official buildings, became a symbol of the Japanese state (ISOZAKI 2011: 8–13). However, Western influence in various fields was a frequent topic of discussion in Japan. The combination of Western and Japanese thoughts and aesthetics, in particular, had been an ongoing concern in every period since the Meiji Restauration beginning in 1868. As the architect ISOZAKI Arata 磯崎新 (1931–2022) stated in a series of essays that were published in an English translation in 2006, it was always a question of defining "Japan-ness". In July 1942, a group of intellectuals gathered in Kyoto to debate the issue of "overcoming modernity" (*kindai no chōkoku* 近代の超克), yet failed to establish a working definition of modernity itself.[15] H. D. HAROOTUNIAN (1989: 68) points out that: "For the most part, 'modern' meant the West, its science, and the devastating effects it had inflicted on the face of traditional social life." The controversy concerning the traditional versus the modern was of serious concern to architects throughout Japan and the colonies, though no architects are recorded as having participated actively in the "overcoming modernity" debate. As such, Arata ISOZAKI (2011: 20–21) comes to the following conclusion:

> The 'Overcoming Modernity' debate remained essentially sterile because participants simply either praised or rejected the modern vis-à-vis a Japanese aesthetic or ethos. In contrast, architects at least came to see modernity and tradition as two sides of a single issue, articulating a stance by means of which to critique both at the same time.

The debate continued in different forums after the conclusion of the war, with architects equally involved. However, the experimental ground in Manchuria was an important experience for Japanese architects and engineers. It allowed them to incorporate Western and Japanese design ideas in urban planning, and therefore to lay the ground for their urban designs in the postwar period. The city of the past—namely Shinkyō—became a model for the city of the future, as architects and engineers found themselves back on the mainland confronted by the clean sheet left in the wake of destroyed cities requiring total reconstruction.

3   ORDER OUT OF CHAOS: REBUILDING THE NATION

Following Japan's surrender, around 40 % of its urban areas had been destroyed, leaving around 30 % of the population homeless (DOWER 1999: 45). Major losses were sustained in heavily bombed main cities such as Tokyo and Osaka, not to mention the degree of destruction the atomic bombs had left in Hiroshima and Nagasaki. Yet, minor towns throughout the country also suffered extensive damage, such as was the case for Wakayama or indeed Toyama, which was practically razed to the ground.[16] Despite the tragic loss of life and widespread destruction, this was also a period of creation for architects who could dream of all the ways they might fill the newly emerged expanses. The War Reconstruction Agency (Sensai fukkōin 戦災復興院)—established by the postwar government—, sent architects and city planners throughout the country to assess the extent of the damage, and propose plans for the reconstruction and development of urban and rural areas. That architects and city planners were presented with a virtual tabula rasa in many parts of Japan marked the beginning of the city of future, or as ISOZAKI (2011: 23) put it: "The future of the city lies in ruins."

However, the most pressing problem was homelessness in the destroyed cities. As HEIN (2003: 3) has pointed out: "Rebuilding in Japan thus concentrated on survival and the satisfaction of urgent needs through pragmatic measures." To this end, both MAEKAWA Kunio and SAKAKURA Junzō experimented with prefabricated houses using techniques and even equipment from their time in Manchuria to provide much needed housing units.[17] In addition to the need for housing, factories and infrastructure, however, there was also an urgent need for cultural institutions, as Jonathan M. REYNOLDS (2012: 326–27) has emphasized. Indeed, the demand for such can clearly be seen in the construction of numerous municipal halls and museums in the 1950s. However, the first permanent new building in Tokyo was Antonin RAYMOND's (1888–1976) Reader's Digest Building (Rīdāzu Daijesuto Tōkyō shisha リーダーズ・ダイジェスト東京支社) built opposite the Imperial Palace in Takebashi in 1949 (demolished in 1963).[18] The modern two-story building with underground ducts for electricity and telephone cables had a double cantilevered front supported by columns featuring large glass windows. The surrounding garden was designed by the Japanese-American sculptor Isamu NOGUCHI (1904–88). David B. STEWART (1987: 165) states: "It was the first large building in which Raymond had succeeded in applying the Japanese-style principles he had made use of so freely in much of his residential work."

Antonin RAYMOND's example shows that private investment from foreigners was quickly approved by the Allied Powers, yet an overall plan for the reconstruction of Japanese cities was still very much lacking. Indeed, Yorifusa ISHIDA (2003: 19) goes so far as to say that the Allied Powers were markedly unimpressed by the reconstruction ideas proposed by Japanese architects and "criticized Japanese plans as being inappropriate for a defeated nation". One way or another, reconstruction was absolutely necessary and the Japanese government enacted several provisional laws for land readjustment to increase public space through the creation of wider streets, for example. Under pressure from the General Headquarters of the Allied Powers (GHQ) the common practice of land readjustment (kukaku seiri 区画整理) without compensation of the land holder was abandoned. However, land readjustment remained the most powerful tool for urban planning.[19]

Whereas the Japanese government and the GHQ took a purely practical approach to reconstruction, more ambitious architects and planners joined forces and founded the New Architects' Union of Japan (Shin Nihon kenchikuka shūdan 新日本建築家集団, NAU) in 1947 to channel their visions for modern rural architecture. As chairman, they first appointed KOIZUMI Kashirō 小泉嘉四郎 (1911–95) who was subsequently followed by TAKAYAMA Eika 高山英華 (1910–99).[20] In the mission statement, the group emphasized the need for communication between architects and the public in order to create a popular culture of interest in architecture that would be independent of political ideologies such that architectural engineers "might contribute to the construction of a democratic Japan by the Japanese people" (IZUMI 2016: 104). Group members such as TANGE Kenzō or NISHIYAMA Uzō 西山夘三 (1911–94) were interested in creating a city whose main function would be to serve the primary needs of its inhabitants, while at the same time availing of modern technology and materials.[21] Disagreement arose around the interpretation of functionalism, a somewhat polemic issue at the time that centered specifically on the four functions of a city as discussed in the CIAM and the aforementioned Charter of Athens in particular. Though the group aimed to find new visions for the modern city, its members were still processing the concepts which had been globally dominating urban planning for years. Architects hoped that Japanese cities would be renewed according to modern standards after the war. This can clearly be seen in ISHIKAWA Hideaki's 石川栄耀 (1893–1955) visionary plan for the reconstruction of Tokyo, for instance, which followed a strict zoning concept that focused on the function of the city.[22]

In addition to large cities such as Tokyo or Osaka, Hiroshima also came to be of great interest for architects and urban planners, as 90 % of the city area had been destroyed. The first city in history to have been attacked with an atomic bomb, Hiroshima urgently needed new housing as well as a reinvention of its image, from a war-time military and industrial city to one which would above all propagate peace. Despite prevailing fears in terms of radiation, planners and citizens alike saw a chance of rebuilding the city anew. Norioki ISHIMARU (2003: 91) summarizes Ishikawa HIDEAKI's ideas, stating that for the latter "it was very important to decide the characteristics of Hiroshima, [Ishikawa] commented on how to design shopping streets, and proposed that Hiroshima make the best use of its waterways by creating landscaped river parks." Politicians such as KŌRA Tomi 高良とみ (1896–1993) for example, proposed relocating the entire city, whereas others hoped to elevate the ground level given that the city had often suffered from floods in the past.[23] Discussion of the city's future equally included its size and its population. It had a population of 380,000 inhabitants during the war and most commentators expected Hiroshima to return to its status as a rather small town with a new population between 100,000 and 400,000.[24] Today, however, there are more than 1,190,000 inhabitants.

TANGE Kenzō and his research team were sent by the War Reconstruction Agency to Hiroshima in 1946 to assess the damage and propose a plan for reconstruction. For TANGE, it might have been a personal shock, as he had been a high school student in Hiroshima in 1930 and thus had a personal connection to the city.[25] The team was quickly confronted with very different ideas and expectations, and was forced to learn that their visionary plans for reconstruction would have to be in line with political and economic realities.[26]

4 CITY OF THE FUTURE: TANGE'S HIROSHIMA PLAN

The reality of Hiroshima as a shattered city would have come as a serious shock to visiting planners. Yet TANGE (1997: 63) equally describes the excitement of being able to produce "a master plan for a new Hiroshima"—an endeavor that would dominate his professional life for more than fifteen years. The impact the ruins had on TANGE would become visible some years later in his essay "Boundless Energy: Concrete" (*Mugen no enerugī: Konkurīto* 無限のエネルギー：コンクリート), published in the journal "Culture of Architecture" (*Kenchiku bunka* 建築文化) in 1958:

> People say that there is nothing more beautiful than a ruin. Certainly, there is a beauty of transience in the sight of potential that stood up against gravity returning back to zero and in the sight of work built by humans through their struggle against nature succumbing to its power and returning back to nature. Personally, however, I see more vital beauty in the sight of humans getting back onto their feet no matter how much they are battered or beaten in their struggle against nature.[27]

Like many other planners, TANGE took a practical approach to rebuilding Hiroshima by making his plans in accordance with the Charter of Athens, dividing the city into commercial zones in the city center surrounded by residential ones and finally industrial zones further out.[28] When the city called for entries to the Hiroshima Peace Park competition in 1949, TANGE submitted a plan that relied on his earlier land-use plan.[29] His winning entry encompassed a far greater area than the proposed Nakajima District, with the architect—according to HEIN (2016c: 204)—arguing that "housing for citizens and the creation of an international memorial complex had to go hand in hand". He therefore planned facilities for children's activities, museums, and art galleries as well as recreational facilities in the expanded area of the Peace Park. Completed in 1952 (demolished in 1978), the Hiroshima Children's Library (Hiroshima-shi jidō toshokan 広島市児童図書館) was ultimately the only building erected. It was situated in the zone intended for children's activities at the northern end of the Peace Park next to the Central Park. The two-story structure for the library consisted of a "shell roof formed by an inverted cone connected on top of a cylindrical shell column via a toroidal shell" (KAWAGUCHI 2011: 259). Often mistakenly interpreted as representing a mushroom shape, TANGE had intended Japan's first concrete-shell structure to represent a tree.[30] The original plan had offered a "sense of the full intended scale and character of the Peace Park complex as an urban intervention" (KUAN / LIPPIT 2012: 29).

The core of the Peace Park—including the memorial—was based on an earlier plan TANGE's, the 1942 Commemorative Building Project for the Construction of Greater East Asia (*Daitōa kensetsu kinen zōei keikaku* 大東亜建設記念造営

計画). The plan consisted of an expressway linking the Imperial Palace in Tokyo with Mount Fuji as well as of a shrine and a plaza at the foot of the sacred mountain. The shrine was modeled after the Grand Ise Shrine (Ise jingū 伊勢神宮), with an extraordinary stature of 60 meters—a representation of state power—while the plaza foregrounded the memorial as a participatory space open to broader public activity.[31] Whereas Western monumental structures would frequently consist of high-rise structures, TANGE drew on "the Japanese traditions of natural objects and horizontal development" (HEIN 2002: 71) that would expand laterally into the distance. Projecting the proposal for the Greater East Asia Memorial on Hiroshima, Mount Fuji equals with the Atomic Bomb Dome, which is separated by a plaza from the cenotaph and the Peace Memorial Museum. TANGE himself describes the Peace Park in the following terms:

> The place that is called the Peace Park occupies a part of Hiroshima's ground zero. It is shaped like an elongated triangle that stretches in the north-south direction over an area of roughly 30,000 tsubo [approx. 99,000 square meters], and rivers flow along both of its legs. The base of the triangle abuts a tree-lined boulevard of 100 meters in width that runs across Hiroshima in the east-west direction. The construction of the Peace Park was planned on this location to memorialize the hypocenter. In my plan, I placed the cenotaph at the center of this park and positioned the Peace Hall in a linear fashion to follow the boulevard running along the base of the site. The Peace Halls comprise three buildings. The building at the center is a museum for displaying materials related to the atomic bombing; the building on the east side contains exhibition rooms, small and medium-sized assembly rooms, a library, and an office; and the building on the west side holds a large assembly hall. The area between the museum and cenotaph is a plaza designed to hold several tens of thousands of people. This is why the ground floor of the museum is made as a space that consists only of columns. The building doubles as a gate for entering the plaza from the boulevard, and it also offers a clear view toward the cenotaph beyond it. (TANGE 2015b: 50)[32]

TANGE's construction constitutes a forum akin to Michelangelo's Capitoline in Rome, which invites the public through the museum building onto a plaza facing the Atomic Bomb Dome (Genbaku dōmu 原爆ドーム) on the other side. Saikaku TOYOKAWA (2021: 34) has exaggerated the symbolic meaning of this construction insinuating Buddhist connotations: "people who had gathered to renew their hopes for peace on one riverbank (*shigan*, or this world) would face the Atomic Bomb Dome on the other (*higan*, or otherworld)."

The Peace Memorial Museum—a rectilinear building atop a 6.5-meter piloti topped by a flat roof—was a fundamental element in the overall design concept of the Peace Park. Zhongjie LIN (2010: 51) has observed that: "the building's horizontality and the lattice pattern on the façade captured the proportions of Japanese traditional architecture." Due to the piloti, the museum building is often associated with the Grand Ise Shrine and the Katsura Detached Palace (Katsura rikyū 桂離宮) in Kyoto. However, in his essay "The Japanese Character of Tange Kenzō" (*Tange Kenzō no Nihonteki seikaku* 丹下健三の日本的性格), published 1955 in "New Architecture" (*Shin kenchiku* 新建築), the art critic KAWAZOE Noboru claims that TANGE himself saw an analogy with the *azekura* 校倉 construction of the Shōsōin 正倉院 in Nara. As KAWAZOE (1976: 8) explains: "He [Tange] immediately associated the function of the proposed memorial—to preserve and display artifacts, memoirs, and personal effects related to the atomic bombing of Hiroshima—with that of the ancient treasure house, which could be considered Japan's oldest 'museum'". However, in light of TANGE's admiration of LE CORBUSIER, it is worth noting the striking resemblance to the 1947 Unité d'Habitation in Marseille. For his part, Antonin RAYMOND rather saw similarities to his own work—the Reader's Digest Building in particular—even alluding to supposed imitation (STEWARD 1987: 168). Clearly, TANGE pursued and combined two distinct directions in his work here: Japanese traditional architecture on the one hand, and Western modern architecture on the other.

Another central element of the Peace Park is the cenotaph, which is located on the plaza in front of the museum building, representing the pivotal central piece of the memorial to the dead of Hiroshima. TANGE Kenzō had originally commissioned Isamu NOGUCHI with the design of the cenotaph in 1951. This was the same NOGUCHI who had designed the garden of the Reader's Digest Building by Antonin RAYMOND two years earlier. NOGUCHI's idea was to create an underground space for the commemorative chamber, which would serve as a counterpart to the arch arising from below the surface above ground. While the underground part was built to represent a womb for the future generations that would replace the dead, the arch above ground—which was topped with a protective roof of a *haniwa* 埴輪—was to serve as a symbol to commemorate the dead.[33] Bert WINTHER (1994: 25) has given particular attention to the use of *haniwa*, drawing attention to their metaphysical significance. *Haniwa*, of course, could take whatever shape the deceased in the afterlife required, whether that be a house, a horse, or a warrior for protection: "[T]he ancient Japanese allegorical architecture of immortality is invoked as a way to reconstrue the dead of Hiroshima into a symbol of hope."

The design of the Japanese-American NOGUCHI was rejected by the Hiroshima Peace Memorial Park Special Committee (Hiroshima heiwa kinen kōen kensetsu senmon iinkai 広島平和記念公園建設専門委員会) in March 1952 because of "[o]pposition to appointing a citizen of the country that dropped the atomic bomb as the designer of the memorial" (OKAZAKI 2014: 305). Subsequently, TANGE took it upon himself to design the cenotaph, clearly inspired by NOGUCHI's concept. He created—to put in WINTHER's terms (1994: 27)—"a similar though much less convincing monument" shaped like the roof of a *haniwa* as an offering to the dead.[34]

TANGE's plan consisted of two central buildings, namely the museum building and the cenotaph. The elements of traditional Japanese architecture they featured had an enormous impact on the "tradition debate" (*dentō ronsō* 伝統論争) of the mid-1950s regarding the nature and origin of traditional Japanese architecture. At the center of this debate was the dichotomy between the Jōmon period (11,000 to 300 BCE) on the one hand, and the Yayoi period (300 BCE to 200 EC) on the other. Of these two prehistoric periods in Japanese history: "Yayoiesque Japonism was deemed traditional, or elitist, while Jōmonesque nativism was seen as populist", as ISOZAKI (2006: 39) summarizes. In his 1955 article "How to Understand Modern Architecture in Japan Today: For the Creation of Tradition" (*Genzai Nihon ni oite kindai kenchiku o ika ni rikai suru ka: Dentō no sōzō no tame ni* 現在日本において近代建築をいかに理解するか：伝統の創造のために), published in "New Architecture", TANGE Kenzō classifies Jōmon as Dionysian and Yayoi as representative of Apollo, in accordance with Friedrich NIETZSCHE's (1844–1900) dichotomy in *The Birth of Tragedy*, published in 1872. According to Kenji KAJIYA (2015: 11), TANGE "regarded the Jōmon or Dionysian as a negative agency to be sublimated for the Yayoi or Apollonian glory". In the early stage of his development, TANGE's architecture "remained Yayoi or Apollonian from the prewar period onwards, as is seen from his plan for the Greater East Asia Memorial Building to Hiroshima Peace Center [...]" (KAJIYA 2015: 11).[35] Later in his career, however, TANGE grew very fond of the Katsura Detached Palace as a prototype for Japanese traditional architecture, referring to it as Jōmonesque. With the completion of the Hiroshima Peace Memorial, TANGE was able to leave behind his war-time artistic style and experiences, moving into a new period of international renown. As far as TANGE is concerned, Jonathan M. REYNOLDS (2012: 327) was thus not wrong to state that: "The construction of the Peace Park in Hiroshima was a significant milestone in the healing process." We would do well to bear in mind, however, that the domestic Japanese debate concerning tradition and origin should be understood in an international context given that Japanese architects were always influenced by Western debates and ideas. It is, thus, hardly surprising that TANGE's plan for the Hiroshima Peace Park received significant international attention in 1951 when he was invited to Hoddesdon near London to speak at the CIAM 8, the theme of which was "The Heart of the City". Given the decimation suffered by so many cities during the war, the event focused on the zoning method in urban planning as well as the division of cities into four functions as laid out in the Charter of Athens of 1933.

The growth of modern cities following the war was unregulated and lacking direction. The very title of the conference—"The Heart of the City"—was a subject for some debate given the diversity of interpretations that emerged, particularly around what precisely the heart represented. As Leonardo Zuccaro MARCHI

(2016: 136) points out, there is a particular semantic charge to the image of the heart: "As an organic metaphor, the urban structure mirrors the presumed physical properties of the organ [...]. As a symbol the heart involved a more 'Abstract Idea' [...], with a social and humanist aim." The symbolic dimension of the Hiroshima plan—representing peace and hope through urban planning—received a great deal of praise. Indeed, as MARCHI (2016: 138) confirms, it represented "the synergy between symbol, monument and heart in relation to the destruction of the War".

5 Conclusion

Destruction of any sort can leave a clean sheet in its wake and this, in turn, can be filled with visions. Japanese architects saw the possibilities presented to them during different time periods and filled these clean sheets with visionary urban planning. Yet whether in Manchuria or in postwar Japan, the page was not entirely blank. Rather, it was framed by political, economic and social demands, each having a particular constellation in accordance with the historical circumstances. As such, architectural visions had to be adapted, and some were never executed to their full potential.

Looking at urban planning in Manchuria and at TANGE's plan for Hiroshima, certain parallels emerge. The function of the city was a subject of prolonged debate, ever embedded in an international context of the influence of the CIAM. The four functions of the city proclaimed in the Charter of Athens—dwelling, work, leisure, circulation—were important features in the planning of Shinkyō and equally led to the zoning plan that was laid out for Hiroshima. The difference between the two instances is clearly the importance placed on the inhabitants of the cities. In contrast to the Manchurian plans, for which consideration of the city-dwellers themselves played a relatively minor role, the creation of participatory space—as discussed with a broader public—played an important role in Hiroshima. In both cases, architecture would serve to achieve political objectives, rendering tangible to the public the ideology of the time. In Shinkyō, as in the entire Japanese empire, official buildings had to be topped with roofs in the Imperial Crown style, that is, a Japanese style of roofing placed on top of a neoclassical building, thus blending Japanese and Western architectural elements in a political statement of superiority. Postwar architecture, as exemplified by the Peace Park in particular, incorporated subtle elements of classical architecture. This was equally the case for buildings such as the Grand Ise Shrine or the Shōsōin, and yet a political message was equally always present. Speaking to the symbolic meaning of the Peace Park, Vinayak BHARNE (2010: 39) states that it was "not only a spatial emblem of peace, but also of Japan's new democracy". The intellectual debate concerning Japanese authenticity had a particular moment in every major period since the Meiji Restoration. During the war, the issue was one of "overcoming modernity" while after the war it focused on "tradition". In both cases, the Japan-ness was at stake and the extent to which Western design elements could reasonably be included. Discussions in the immediate postwar period, however, placed greater, more urgent emphasis on the origins of Japanese tradition, a focus that could well have been informed by the quest for a new national identity in the wake of Japan's defeat.

Postwar architecture had a significant connection with the personal experiences resulting from the time planners and engineers had spent in the colonies during the war. The continuity in architectural concepts and realization was primarily based on the personnel involved architects such as TANGE Kenzō who had been professionally trained during the war. ISOZAKI Arata draws a rather negative conclusion when he states in an interview with Rem KOOLHAAS and Hans Ulrich OBRIST (2011: 29) that: "Basically, I thought nothing had changed. Those 20 years—the decade before the war and the decade after—were unbroken progressions. There was no discontinuity. The end of the war came right in the middle. The ideological disruption was only superficial." As I see it, however, the completion of the Peace Park marked the end of the war in terms of architectural development. With the "tradition debate" and the subsequent emergence of the Metabolist movement, architecture and urban planning took a new turn in Japan.

One avenue for further research—particularly as an instance of creative possibilities following the appearance of clean sheets—would be to look at the Tōhoku region and the aftermath of the 2011 earthquake. Would one equally find, in this case, a similar debate regarding architectural Japan-ness against the backdrop of international developments in the discipline? The question of urban planning in rural areas like Tōhoku as well as the political turmoil surrounding the tragedy and the ensuing recovery are all in need of further discussion. The architects who were drawn to the destroyed region were primarily concerned with the need for housing, as was also the case for MAEKAWA Kunio and SAKAKURA Junzō immediately following the war. Among others, the well-known architect Itō Toyō 伊東豊雄 initiated the shelter project 'Home for all' (*Minna no ie* みんなの家), which worked with local residents to plan and build houses in the affected area. In terms of his reasons for working in Tōhoku, one need to look no further than a statement made by the Tokyo-based, international, Pritzker Architecture Prize laureate after his first visit to the area: "It was as if I was standing at the ground zero of Hiroshima" (FUJI 2018).

Notes

1 In the introduction to their extensive overview of TANGE Kenzō's work, Seng KUAN and Yukio LIPPIT (2012: 9) state that "the opening of the Hiroshima Peace Memorial Park announced the arrival of Tange Kenzō as a defining figure of postwar architecture." The architect and researcher in the history of architecture, Saikaku TOYOKAWA (2016: 10)—who published two volumes containing the collected essays of TANGE in 2021—begins his book "Tange Kenzō. Planner of Post-war Japan" (*Tange Kenzō: Sengo Nihon no kōsōsha* 丹下健三：戦後日本の構想者) with the plan for Hiroshima as a starting point for TANGE's career. Even David B. STEWARD (1987: 169)—in his standard work on Japanese architecture—confirms that the Hiroshima Peace Park was "Tange's first major work".

2 The group of Metabolists including the architects KIKUTAKE Kiyonori 菊竹清訓 (1928–2011), KUROKAWA Kishō 黒川紀章 (1934–2007), MAKI Fumihiko 槇文彦 (*1928), ŌTAKA Masato 大高正人 (1923–2010), and critic KAWAZOE Noboru 川添登 (1926–2015) was formed in 1960 following the World Design Conference in Tokyo. It aimed to incorporate design and technology to create new urban landscapes that could adapt to the people living in the structures. Zhongjie LIN (2010) gives a thorough overview and analysis of the Metabolist movement. Rem KOOLHAAS and Hans Ulrich OBRIST (2011) trace the development of the Metabolist movement in numerous interviews with its members as well as people in their personal surrounding.

3 Zhongjie LIN (2010: 56) demonstrates that they went even further: "Like many of their contemporaries, Tange and the Metabolists believed that thoughtful architectural forms embodied cultural meaning and national spirit and that they afforded the surest defense against the erosion of time."

4 The German term Lebensraum, a key concept of the Nazi ideology, was introduced to Japan in the mid-1920s and was gradually changed by the government through the 1930s and 1940s from "living space" (*seikatsu kūkan* 生活空間) to "living sphere" (*seikatsuken* 生活圏), justifying the concept of the Greater East Asia Co-prosperity Sphere as a territorially unbound sphere of political, cultural, and economic influence (FURUHATA 2019: 226–28).

5 Noriko ASO (2014: 95–126) points out that by the mid-1930s the Japanese government had established national museums throughout the colonies—especially in Taiwan, Korea, Sakhalin, and Manchuria—to propagate the vision of Japanese imperial identity, placing the colonial subject in different positions within the Japanese social hierarchy. Museums were used "to celebrate heritage and consolidate colonial power" (ASO 2014: 95).

6 SHIINA Etsusaburō 椎名悦三郎 (1898–1979), nephew of GOTŌ Shinpei 後藤新平 (1857–1929), held various positions in the administration during the interwar, war and postwar periods. At the request of KISHI Nobusuke 岸信介 (1896–1987), SHIINA joined the bureaucrats in Manchuria to develop the industrial sector of the new state and was appointed chief of the planning section in Manshūkoku's Industrial Department in 1933. See MIMURA 2011: 72–75.

7 J. R. STEWART (1939: 43) describes the Japanese government's planning for the colonization of Manchuria in detail and from a contemporaneous perspective. In his conclusion, he asks whether Japan would be able to continue its efforts and "find the men and the money for colonization in Manchuria?"

8 For a detailed description of the Agricultural Immigrant Plan of 1933, see TUCKER 2005: 58–73. The same UCHIDA Yoshikazu also drew up a plan for the extension of the city of Daitō 大同 (Datong), which relied heavily on the neighborhood concept of the earlier plan for rural areas. See HEIN (2016a: 174–77 and 2017: 7–9) and NAKAJIMA (2023).

9 In 1932, Japanese bureaucrats arrived in Manchuria and effectively took over the strategic planning and development of the new state from the Kwantung Army. Under the leadership of KISHI Nobusuke a tight "system of technocratic coordination and control" was established (MIMURA 2011: 70).

10 GOTŌ Shinpei's role for urban planning in Japan and in the colonies can hardly be underestimated. After the First Sino-Japanese War of 1894/95 he was head of the civil administration of Taiwan from 1898 to 1906, then becoming the president of the South Manchurian Railway (SMR) in 1906. Later, in 1920, he became mayor of Tokyo and, as home minister, he was in charge of the reconstruction of the capital after the Great Kanto Earthquake. For GOTŌ's role during the reconstruction of Tokyo, see TUCKER 2003: 163 and SEIDENSTICKER 2019: 298. On GOTŌ's significance in urban planning see KOSHIZAWA 1988: 9–29. A historical overview of urban planning from the Edo period (1603–1868) to the 21st century can be found in SORENSEN (2002); he briefly comments on the implementation of the legal planning system in the colonies (2002: 142–44).

11 For detailed description of the development and building of Shinkyō as capital of Manchuria see GUO 2004: 103–16, DENISON/GUANGYU 2016: 105–25 and KOSHIZAWA 1988: 110–49.

12 LE CORBUSIER wrote his manifesto *Urbanism*, translated into various languages, in 1925. Consequently, LE CORBUSIER was an internationally well-respected architect for urban planning. He was also one of the main authors of the Charter of Athens in 1933.

13 The Congrès Internationaux d'Architecture Moderne (CIAM) was founded in 1928 to address city planning issues, mostly in Europe. Urban problems like housing and land use were the focus of the conferences, leading to the Charter of Athens concerning the "Functional City". See GOLD 1998: 240–43.

14 DENISON/GUANGYU (2016: 124) have indicated that the architectural experience of Manchuria made a "significant impact on a generation of young Japanese architects" whose heyday would arrive with the postwar period.

15 Prominent intellectuals like KOBAYASHI Hideo 小林秀雄 (1902–83), NISHITANI Keiji 西谷啓治 (1900–90), KAMEI Katsuichirō 亀井勝一郎 (1907–66) and KAWAKAMI Tetsutarō 河上徹太郎 (1902–80) to name a few, were invited by the Literary Society (Bungakkai 文学会). In the September and October 1942 issues of the journal "Literary World" (*Bungakkai* 文学界), essays of the participants and the record of the

roundtable were published. For further information, see HAROOTUNIAN 1989: 67–78; FUJITA 2010: 75–94; TIPTON 2007: 201–02; and also TAKEUCHI 2005: 103–47.

16 For a map of Japan, which shows its heavily destroyed cites as well as the estimated rate of deconstruction, see KOOLHAAS / OBRIST 2011: 78–79.

17 MAEKAWA Kunio called his concept of prefabricated houses PREMOS; he came up with the idea after the mass-production of automobiles, which he applied to prefabrication of houses assembled on site. See REYNOLDS 2001: 142–49.

18 The Czech-American architect Antonin RAYMOND (1888–1976) came to Japan in 1919, assisting Frank Lloyd WRIGHT (1867–1959) in constructing the Imperial Hotel in Tokyo. Afterwards he stayed in Japan working, amongst others, with MAEKAWA Kunio and YOSHIMURA Junzō 吉村順三 (1908–97) before leaving again in 1937. RAYMOND had been granted permission to enter Japan after the war by General Douglas MACARTHUR (1880–1964), Supreme Commander for the Allied Powers (SCAP), in order to help with the reconstruction of the country. For a description and images of the Reader's Digest Building, see STEWARD 1987: 165–68.

19 For a detailed description of the different postwar legal acts concerning reconstruction, urban planning, and building standards, see ISHIDA 2003: 25–40.

20 KOIZUMI Kashirō was an engineer, who was employed by the City of Osaka during the war, becoming known for theater architecture thereafter. TAKAYAMA Eika had been involved in plans for Daitō in Manchuria with his teacher UCHIDA Yoshikazu. After the war, he became one of the most influential figures in urban planning in Japan. Like TANGE Kenzō, he held a position at the University of Tokyo and had worked on the plans surrounding the Olympic Games in Tokyo in 1964 and the Expo '70 in Osaka. However, he is much less known than TANGE. For a discussion of the NAU and their goals, see IZUMI (2016).

21 The urban planner and theorist NISHIYAMA Uzō was professor at Kyoto University. He proposed the "model core of a future city" concept for the Expo '70 in Osaka, which was not realized.

22 ISHIKAWA Hideaki, who had worked on plans in Shanghai, did not agree with the land adjustment concept of the government, emphasizing instead the cultural aspects of cities. See HEIN 2017: 9–11 and HEIN 2003: 322–27.

23 KŌRA Tomi was professor of psychology at Japan Women's University (Nihon joshi daigaku 日本女子大学) and a feminist activist throughout her life. In 1947, she was elected as a member of the Diet and active in the postwar pacifist movement (TAKATSUNA 2020: 4–5). An overview of different proposals can be found on the research web pages of Hiroshima Prefecture: "Hiroshima for Global Peace: II Hiroshima Reconstruction Planning".

24 A list of the various prognoses can be found in *Hiroshima shinshi* 広島新史, in the volume *Toshi bunka-hen* 都市文化編 1984: 33.

25 TANGE Kenzō had worked with MAEKAWA Kunio until 1942, before returning to university for further studies and to become an assistant professor at the Imperial University Tokyo, which is where he established his research group, the Tange Lab (Tange kenkyūshitsu 丹下研究室). After the war, he would be joined by ŌTANI Sachio 大谷幸夫 (1924–2013), ASADA Takashi 浅田孝 (1921–90) and others who, themselves,

would later form the Metabolists group. Tange Lab was the base for many of the large-scale projects TANGE realized in the postwar period. See KOOLHAAS / OBRIST 2011: 106; and TANGE 1997: 61.

26 At least two occasions are documented when TANGE and members of his team participated in meetings with the city council to discuss their reconstruction plans. See *Hiroshima shinshi* 広島新史 in the volume *Shiryō-hen II: Fukkō-hen* 資料編II・復興編 1982: 75; 107.

27 English translation by Saikaku TOYOSAWA (TANGE 2015a: 197). For a reprint of the article see TANGE 2021a: 101–16.

28 For a map of the different zones according to the plan of TANGE, see HEIN (2016c: 203).

29 ISHIMARU (2018: 390) reconstructs the planning process and mentions the fact that "Tange's original concept changed from Peace Memorial Park to Peace Memorial City Plan or Peace City Plan" combining "local planning, existing before Tange's arrival, and novel ideas by Tange and his group."

30 The Hiroshima Children's Library marks the beginning of a long cooperation effort between TANGE and the structural engineer TSUBOI Yoshikatsu 坪井善勝 (1907–90), which cumulated famously in the construction of the Yoyogi National Gymnasium for the Olympic Games in Tokyo in 1964 with its high-tension cable suspended roof. COALDRAKE (1996: 260–61) points out that many other architects like LE CORBUSIER had experimented with the technology of suspension-roof structures, but it was TANGE and TSUBOI who could put the technology on a monumental scale. For a detailed description and analysis of the design development see TOYOKAWA 2021: 82–127.

31 KUAN / LIPPIT (2012: 16–17) give a detailed overview on the implied political and symbolic meaning of the memorial. TANGE had already written about the concept of a plaza or a forum in the essay "An Ode to Michelangelo: As Prologue to the Study of Le Corbusier" (*Michelangelo shō: Le Corbusier ron e no josetsu to shite* Michelangelo頌：Le Corbusier論への序説として), published in the journal "Modern Architecture" in 1939 (reprint in TANGE 2021b: 9–30).

32 The Japanese original essay was published in the journal "New Trends in Art" (*Geijutsu shinchō* 芸術新潮; January 1956: 76–80) under the title "Gomannin no hiroba: Hiroshima Pīsu Sentā kansei made 五万人の広場：広島ピース・センター完成まで". The citation used here is a translation of an excerpt by Saikaku TOYOKAWA.

33 *Haniwa* are clay models found in the tombs from the prehistoric Kofun period (fourth century to sixth century). No more than a few photographs of NOGUCHI's design have survived. However, OKAZAKI (2014: 306) states that a dismantled model of the monument was discovered in the Museum of Modern Art, Kamakura, which led to its reconstruction. Furthermore, NOGUCHI himself reconstructed the part of the cenotaph that sits above ground, although it differed from the original design, according to OKAZAKI (2014: 317).

34 Though Isamu NOGUCHI's design of the cenotaph was rejected, his plan for the railings of the bridges over the river Ōta were accepted: The bridge Tsukuru 創る (to

create) had a sun-shaped railing, while the second bridge, Yuku 行く (to depart) was modeled after an old Japanese boat. See TANGE 2015b: 51.
35 ISOZAKI (2006: 39–42) claims that TANGE's logic in shifting to Jōmonesque was flawed, something which he ascribes to the "immediate political climate".

REFERENCES

ASO, Noriko (2014): *Public Properties. Museums in Imperial Japan*. Durham / London: Duke University Press.

BHARNE, Vinayak (2010): "Manifesting Democracy. Public Space and the Search for Identity in Post-War Japan". In: *Journal of Architectural Education* 63 (2): 38–50.

COALDRAKE, William H. (1996): *Architecture and Authority in Japan*. London / New York: Routledge.

DENISON, Edward / REN, Guangyu (2016): *Ultra-modernism: architecture and modernity in Manchuria*. Hong Kong: Hong Kong University Press.

DOWER, John W. (1999): *Embracing Defeat. Japan in the Aftermath of World War II*. London: Penguin.

FUJI, Mihoyo (2018): "Toyo Ito and 'Home for all'—Can people shape architecture?"; https://www.interactiongreen.com/toyo-ito-home-for-all-minna-no-ie/ (last access 2023 / 02 / 14).

FUJITA, Masakatsu 藤田正勝 (2010): "Zadankai 'Kindai no chōkoku' no shisō sōshitsu: Kindai to sono chōkoku o meguru tairitsu 座談会「近代の超克」の思想喪失：近代とその超克をめぐる対立". In: SAKAI, Naoki 酒井直樹 / ISOMAE, Jun'ichi 磯前順一 (eds.): *'Kindai no chōkoku' to Kyōto gakuha: Kindaisei, teikoku, fuhensei* 「近代の超克」と京都学派：近代性・帝国・普遍性. Tōkyō: Ibunsha, 75–94.

FURUHATA, Yuriko (2019): "Tange Lab and Biopolitics. From the Geopolitics of the Living Sphere to the Nervous System of the Nation". In: INOUE, Mayumo / CHOE, Steve (eds.): *Beyond Imperial Aesthetics. Theories of Art and Politics in East Asia*. Hong Kong: Hong Kong University Press, 219–42.

GLEITER, Jörg H. (2017): "The traumata of modernization: Architecture in Japan after 1945". In: KOHTE, Susanne / ADAM, Hubertus / HUBERT, Daniel (eds.): *Encounters and Positions. Architecture in Japan*. Basel: Birkhäuser, 238–49.

GOLD, John R. (1998): "Creating the Charter of Athens. CIAM and the functional city, 1933–42". In: *The Town Planning Review* 69 (3): 225–47.

GUO, Qinghua (2004): "Changchun: unfinished capital planning of Manzhouguo, 1932–42". In: *Urban History* 31 (1): 100–17.

HAROOTUNIAN, H. D. (1989): "Visible Discourses / Invisible Ideologies". In: MIYOSHI, Masao et al. (eds.): *Postmodernism and Japan*. New York: Duke University Press, 63–92.

HEIN, Carola (2002): "Hiroshima. The atomic bomb and Kenzo Tange's Hiroshima Peace Center". In: OCKMAN, Joan (ed.): *Out of Ground Zero: case studies in urban reinvention*. München: Prestel, 62–83.

HEIN, Carola (2003): "Rebuilding Japanese Cities After 1945". In: HEIN, Carola / DIEFENDORF, Jeffry M. / ISHIDA, Yorifusa (eds.): *Rebuilding Urban Japan After 1945*. New York: Palgrave Macmillan, 1–16.

HEIN, Carola (2016a): "Imperial Expansion and City Planning: Visions for Datong in the 1930s". In: WIGEN, Kären / SUGIMOTO, Fumiko / KARACAS, Cary (eds.): *Cartographic Japan. A History in Maps*. Chicago / London: The University of Chicago Press, 174–77.

HEIN, Carola (2016b): "Japanese Cities in Global Context". In: *Journal of Urban History* 42 (3): 463–76.

HEIN, Carola (2016c): "Tange Kenzō's Proposal for Rebuilding Hiroshima". In: WIGEN, Kären / SUGIMOTO, Fumiko / KARACAS, Cary (eds.): *Cartographic Japan. A History in Maps*. Chicago / London: The University of Chicago Press, 203–06.

HEIN, Carola (2017): "The Urban Core in Japanese Planning (1930s–1950s): Evolving Perceptions on the Spatial and Social Form of the Metropolitan Center on the Mainland and in the Colonies". In: *Histories of Postwar Architecture* 1 (1): 1–14.

HIROSHIMA-SHI 広島市 (1982–86): *Hiroshima shinshi* 広島新史, 13 vols. Hiroshima: Hiroshima-shi.

ISHIDA, Yorifusa (2003): "Japanese Cities and Planning in the Reconstruction Period: 1945–55". In: HEIN, Carola / DIEFENDORF, Jeffry M. / ISHIDA, Yorifusa (eds.): *Rebuilding Urban Japan After 1945*. New York: Palgrave Macmillan, 17–49.

ISHIMARU, Norioki (2003): "Reconstructing Hiroshima and Preserving the Reconstructed City". In: HEIN, Carola / DIEFENDORF, Jeffry M. / ISHIDA, Yorifusa (eds.): *Rebuilding Urban Japan After 1945*. New York: Palgrave Macmillan, 87–107.

ISHIMARU, Norioki (2018): "Studies on the Relation to plan-making of Conception of Hiroshima Peace City Construction Plan after winning of Hiroshima Peace Memorial Park Competition by Kenzo Tange". In: *International Planning History Society Proceedings* 18 (1): 389–400.

ISOZAKI, Arata (2006): *Japan-ness in Architecture*. Cambridge / London: The MIT Press.

ISOZAKI, Arata (2011): *Welten und Gegenwelten* (Architektur Denken 3). Bielefeld: transcript.

IZUMI, Kuroishi (2016): "Rethinking the Social Role of Architecture in the Ideas and Work of the Japanese Architectural Group NAU". In: *Review of Japanese Culture and Society* 28: 99–117.

KAJIYA, Kenji (2015): "Posthistorical traditions in art, design, and architecture in 1950s Japan". In: *World Art* 5 (1): 21–38.

KAWAGUCHI, Mamoru (2011): "Yoshikatsu Tsuboi, Distinguished Researchers, Warmhearted Teacher and Talented Structural Designer". In: *International Journal of Space Structures* 26 (3): 257–69.

KAWAZOE, Noboru 川添登 (1976): "Tange Kenzō no Nihonteki seikaku 丹下健三の日本的性格 (1955)". In: *Kawazoe Noboru hyōron shū* 川添登評論集 1. Tōkyō: Sangyō nōritsu tanki daigaku shuppanbu, 3–36.

KOOLHAAS, Rem / OBRIST, Hans Ulrich (2011): *Project Japan. Metabolism Talks*. Köln: Taschen.

KOSHIZAWA, Akira 越沢明 (1988): *Manshūkoku no shuto keikaku* 満州国の首都計画. Tōkyō: Nihon keizai hyōronsha.

KUAN, Seng / LIPPIT, Yukio (2012): "Tange Kenzō and Postwar Japanese Architecture: An Expanded View". In: KUAN, Seng / LIPPIT, Yukio (eds.): *Kenzō Tange. Architecture for the World*. Zürich: Lars Müller, 9–14.

LE CORBUSIER (2015): *Städtebau*. Translated and edited by Hans HILDEBRANDT. München: Deutsche Verlags-Anstalt (Original: *Urbanism*, 1925).

LIN, Zhongjie (2010): *Kenzo Tange and the Metabolist Movement. Urban Utopias of Modern Japan*. London / New York: Routledge.

MARCHI, Leonardo Zuccaro (2016): "CIAM 8—The Heart of the City as the Symbolical Resilience of the City". In: *International Planning History Society Proceedings: History Urbanism Resilience. The Urban Fabric* 2: 135–44.

MIMURA, Janis (2011): *Planning for Empire. Reform Bureaucrats and the Japanese Wartime State*. Ithaca / London: Cornell University Press.

NAKAJIMA, Naoto (2023): "The Datong City Plan (1938): the three week-process of organizing planning ideas and techniques towards the construction of a new urban area under Japanese occupation". In: *Planning Perspectives* 38 (1): 99–125.

OKAZAKI, Kenjirō (2014): "A Place to Bury Names, or Resurrection (Circulation and Continuity of Energy) as a Dissolution of Identity: Isamu Noguchi's 'Memorial to the Dead of Hiroshima' and Shirai Sei'ichi's 'Temple Atomic Catastrophes'". In: *Review of Japanese Culture and Society* 26: 304–17.

REYNOLDS, Jonathan M. (2012): "Can Architecture Be Both Modern and 'Japanese'? The Expression of Japanese Cultural Identity through Architectural Practice from 1850 to the Present". In: RIMER, J. Thomas (ed.):

Since Meiji. Perspectives on the Japanese Visual Arts, 1868–2000. Honolulu: University of Hawaiʻi Press, 315–39.

REYNOLDS, Jonathan M. (2001): *Maekawa Kunio and the Emergence of Japanese Modernist Architecture*. Berkeley: University of California Press.

SCHERER, Anke (2012): "The colonial appropriation of public space. Architecture and city planning in Japanese-dominated Manchuria". In: BRUMAN, Christoph / SCHULZ, Evelyn (eds.): *Urban Spaces in Japan. Cultural and social perspectives*. London / New York: Routledge, 37–52.

SEIDENSTICKER, Edward (2019): *A History of Tokyo 1867–1989. From Edo to Showa: The Emergence of the World's Greatest City*. Tokyo / Rutland / Singapore: Tuttle Publishing.

SHIINA, Etsusaburō 椎名悦三郎 (1976): "Nihon sangyō no daijikkenjō, Manshū 日本産業の大実験場・満州". In: *Bungei shunjū* 文藝春秋 54 (2): 106–14.

SORENSEN, André (2002): *The Making of Urban Japan. Cities and planning from Edo to the twenty-first century*. London / New York: Routledge.

STEWART, David B. (1987): *The Making of a Modern Japanese Architecture. 1868 to the Present*. Tokyo / New York: Kodansha International.

STEWART, John R. (1939): "Japan's Strategic Settlements in Manchoukuo". In: *Far Eastern Survey* 8 (4): 37–43.

TAKATSUNA, Miki (2020): "The First Generation of Japanese Women Psychologists". In: *Genealogy* 4 (2): 1–11.

TAKEUCHI, Yoshimi (2005): *What Is Modernity? Writings of Takeuchi Yoshimi*. Translated by Richard CALICHMAN. New York: Columbia University Press.

TANGE, Kenzō 丹下健三 (1997): *Ippon no enpitsu kara* 一本の鉛筆から. Tōkyō: Nihon tosho sentā.

TANGE, Kenzō (2015a): "Boundless Energy: Concrete". In: TOYOKAWA, Saikaku (ed.): *Tange by Tange 1949–1959. Kenzo Tange as seen through the Eyes of Kenzo Tange*. Tokyo: Toto, 197.

TANGE, Kenzō (2015b): "A Plaza for Fifty Thousand People. Leading up to the Completion of the Hiroshima Peace Center". In: TOYOKAWA, Saikaku (ed.): *Tange by Tange 1949–1959. Kenzo Tange as seen through the Eyes of Kenzo Tange*. Tokyo: Toto, 50–51.

TANGE, Kenzō 丹下健三 (2021a): "Mugen no enerugī: Konkurīto 無限のエネルギー：コンクリート". In: TOYOKAWA, Saikaku 豊川斎赫 (ed.): *Tange Kenzō kenchiku ronshū* 丹下健三建築論集 (Iwanami bunko 岩波文庫 33–585–1). Tōkyō: Iwanami shoten, 101–16.

TANGE, Kenzō 丹下健三 (2021b): "Michelangelo shō: Le Corbusier ron e no josetsu to shite Michelangelo頌：Le Corbusier論への序説として". In: TOYOKAWA, Saikaku 豊川斎赫 (ed.): *Tange Kenzō kenchiku ronshū* 丹下健三建築論集 (Iwanami bunko 岩波文庫 33–585–1). Tōkyō: Iwanami shoten, 9–30.

TIPTON, Elise K. (2007): "Intellectual Life, Culture, and the Challenge of Modernity". In: TSUTSUI, William M. (ed.): *A Companion to Japanese History*. Chichester: Wiley-Blackwell, 189–206.

TOYOKAWA, Saikaku 豊川斎赫 (2016): *Tange Kenzō: Sengo Nihon no kōsōsha* 丹下健三：戦後日本の構想者 (Iwanami shinsho 岩波新書 1603). Tōkyō: Iwanami shoten.

TOYOKAWA, Saikaku (2021): *Yoyogi National Gymnasium and Kenzo Tange*. Tokyo: Toto.

TUCKER, David (2003): "Learning from Dairen, Learning from Shinkyō: Japanese Colonial City Planning and Postwar Reconstruction". In: HEIN, Carola / DIEFENDORF, Jeffry M. / ISHIDA, Yorifusa (eds.): *Rebuilding Urban Japan After 1945*. New York: Palgrave Macmillan, 156–87.

TUCKER, David (2005): "City Planning without Cities. Order and Chaos in Utopian Manchukuo". In: ASANO TAMANOI, Mariko (ed.): *Crossed Histories. Manchuria in the Age of Empire*. Honolulu: University of Hawai'i Press, 53–81.

WENDELKEN, Cherie (2000): "Pan-Asianism and the Pure Japanese Thing: Japanese Identity and Architecture in the Late 1930s". In: *Positions* 8 (3): 819–28.

WINTHER, Bert (1994): "The Rejection of Isamu Noguchi's Hiroshima Cenotaph: A Japanese American Artist in Occupied Japan". In: *Art-Journal* 53 (4): 23–27.

Internet Sources:

HIROSHIMA PREFECTURE: "Hiroshima for Global Peace: II Hiroshima Reconstruction Planning"; https://hiroshimaforpeace.com/fukkoheiwakenkyu/vol1/1-19/ (last access 2023 / 02 / 12).

The Atomic Bomb Victims in Nagasaki and Hiroshima on Display: Differing Exhibition Strategies and International Trends in Musealization

André Hertrich[1]

1 INTRODUCTION

The Nagasaki Atomic Bomb Museum (NABM) is devoted to the events and aftermath of the atomic bomb dropped on the city in western Kyūshū on August 9, 1945, at 11:02 AM. The main part of the exhibition shows the effects of the nuclear explosion in terms of heat, radiation, and the force of the blast. The heat wave is represented by various objects, including a stained, tattered blouse with singed edges (see Figure 1). The object description reads as follows: "Children's outer clothing. This clothing was worn by an eight-month old infant killed in the atomic bombing [...]. The back side is scorched and torn apart. The infant's mother, who survived, kept this as a memento of her child for 25 years." Neither the name of the child nor the photo reveal the child's gender. Nothing is said about the exact circumstances or about the further life of the surviving relatives.

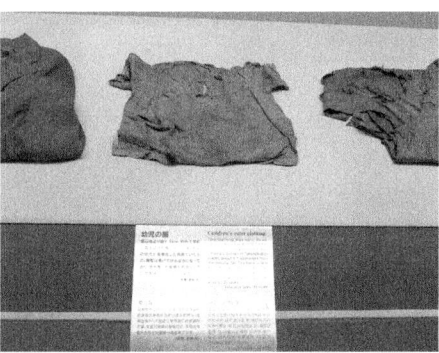

Fig. 1: Three burnt items of clothing, the one in the middle was worn by a little girl when she was killed by the atomic bomb. (Photo by André Hertrich, 2020)

The U. S. also dropped an atomic bomb on Hiroshima three days earlier, on August 6, 1945, at 8:15 AM. The victims are commemorated in the far more prominent Hiroshima Peace Memorial Museum (HPMM). Clothing worn by atomic bomb victims at the time of the explosion is also exhibited there, such as the pants of a small child, brown-spotted with traces of soot. Unlike in Nagasaki, however, visitors in Hiroshima not only learn that the child's name was Taoda Hiroo[2], and that he was two years old, but also under what circumstances Hiroo died and what the pants stand for: "[He] was wearing these pants at the time of the bombing. Carried on his mother's back, he was exposed to the heat from behind and was burned badly." There is also a photo of the boy above the object in a glass case, showing a laughing baby (see Figure 2). His last words, "Hot, I'm hot!" (*atsui, atsui* あつい、あつい) are used as a caption for the whole ensemble of object, text and photo. Finally, Hiroo's older sister adds her statement on the emptiness and pain Hiroo's death left behind: "My little brother died before my mother's eyes. He was at an adorable age, just learning to talk." Hiroo finally died crying for water which his mother refused to give him because she thought water would kill him. This would become her greatest regret (HIROSHIMA PEACE MEMORIAL MUSEUM 2020: 66).

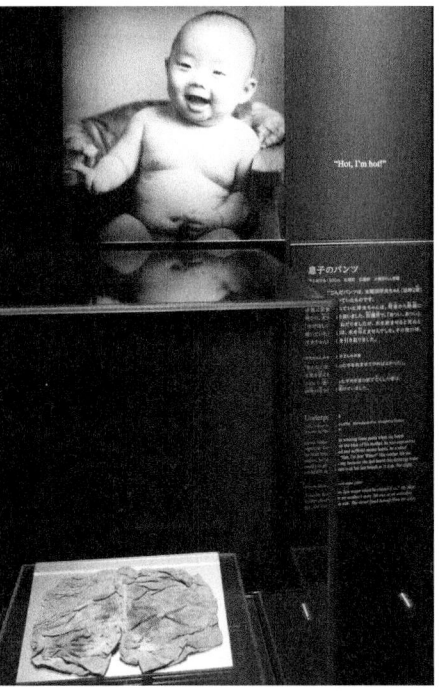

Fig. 2: The typical structure of remains on display at HPMM: A display showing Taoda Hiroo's scorched pants together with a picture of the boy, a quote, and an explanatory text next in Japanese and English. (Photo by André Hertrich, 2020)

While this example from Hiroshima is one of many artifacts exhibited there in similar ways, only very few objects in Nagasaki are connected to the stories of individual victims. The exhibition in Nagasaki is presented much more neutrally and matter-of-factly. The exhibition in Hiroshima, on the other hand, is designed to achieve an emotional impact through personal stories. This is also a current development in exhibition design which can be observed internationally in all kinds of memorial museums, i. e. museums commemorating atrocities or human rights violations (WILLIAMS 2007: 8), where the victims, their stories, their personal belongings and their individual photos receive more attention. Personalizing the victims in this way annuls their status as anonymous faces in the mass of victims, rendering them instead as individual human beings who had lives, families, hopes and dreams. This global trend emerged predominantly in Holocaust memorial museums. The United States Holocaust Memorial Museum (USHMM) in Washington DC and Jerusalem's Yad Vashem, in particular, make extensive usage of private photos taken before the persecution in order to commemorate the Jews who were killed (KÖHR 2012; RADONIĆ 2018: 486). Meanwhile, this musealization trend can be found in countless memorial museums worldwide with private photos of victims of the Rwandan genocide, of 9/11, of the Chilean military dictatorship or of kamikaze pilots in the case of Japan.

Private photos of atomic bomb victims can be found all over the exhibition in Hiroshima, but hardly any in Nagasaki. By examining the exhibitions at NABM and HPMM, I not only want to point out the differences in museum design but also show the differences in the emotional approaches to the representation of victims. I will also examine the assemblage of objects in the exhibition to show how different emotionalizing layers were added. One prominent example is "Shin'ichi's tricycle", which has been on display for many years. On the basis of the different representations of the tricycle—and the boy who owned it—before and after the renovation in 2019, I explore the impact of global trends in the practice of musealization in memorial museums. In particular, I focus on the internationally widespread modes of exhibiting individual victims with the help of their names, private photographs, objects, or short biographies. I thus argue that the newly redesigned Hiroshima exhibition is also an expression of this international trend and stands in contrast to the older 1996 Nagasaki exhibit.

The methodological approach of this paper is based on source-critical museum analysis, which analyzes museums and exhibitions along the lines of politics and aesthetics (THIEMEYER 2010: 83–84). The term politics describes the fact that the museum requires a process of creation, a construction phase, and preliminary conceptual considerations within an institutional framework of donors, curators, and local or national authorities or governments. The term aesthetics refers to the representation of content, the selection of objects, the sensory experience, and the use of textual and graphic elements; in short, the means of representation. In this approach, museums and exhibitions can be examined and analyzed with the methodological toolbox of historical source studies, which understands the museum as a historical source "created with the intention(s) of conveying selected knowledge, i.e., for instance, a certain version of history" (THIEMEYER 2010: 84). With this premise, questions of historiographical source analysis—Who are the authors of the source? What positions do they represent? What circumstances influenced the creation of the source?—can be posed to the museum as if it were a historical text. But museums are also spaces of material culture and it is, thus, vital to understand the roles of objects in the exhibitions: "The museum collects relics, things of the past, in order to turn them into things for us, into information carriers" (KORFF / EBERSPÄCHER / BAUMUNK 2002: 141). These objects possess a "memory inducing power" (KORFF / EBERSPÄCHER / BAUMUNK 2002: 143), upon which the emotive effects of the museum rest. But as things, they are, first and foremost, mute. Only the re-contextualization by the exhibition makers allows the objects to tell a story. This is achieved by means of text documents, pictures, or photographs that illuminate the background. Often, however, exhibition objects serve to substantiate a statement, an interpretation made by the museum. "Here, [...] a 'first person' (the exhibitor) intends to tell a 'second person' (the visitor) something about a 'third person' (the exhibited object)" (SCHOLZE 2010: 131).[3] As such, it is, in many cases, not the text that explains the object, but rather the object that serves as evidence of the veracity of the museum's message. At the same time, in showing the object and its interpretation, the truthfulness of the statement is asserted.

Based on this museum analytical and object related premise, the research project on "Globalized Memorial Museums"[4] examined a lengthy list of questions on all aspects of the museum, including the national / local politics of memory, the proponents of the museum's founding, spatial aspects of the museum site, the selection and arrangement of the objects, the usage of accompanying media, and so on. I visited the museums in Hiroshima and Nagasaki several times during my research trips to Japan in Autumn 2020 and 2022 respectively. During this time, I spoke with representatives of both museums and documented the entirety of the exhibitions including photographic documentation of every object on display.

There is a long list of differences and even animosity between the cities of Hiroshima and Nagasaki with regards to memorializing the atomic bomb. This does not come as a surprise when one considers the fact that "the only major historical similarity between the cities is the atomic bombings" (DIEHL 2018: 172). Hiroshima secured, monopolized, and capitalized on the title of City of Peace in the aftermath of the war. Nagasaki, by contrast, was left with the significantly less catchy title of City of International Culture and sought to include its international heritage as an open port for the Dutch and home for many Catholics into its city marketing (DIEHL 2014; MCCLELLAND 2020).[5] In terms of nuclear remembrance, Nagasaki is the "Forgotten City" (SOUTHARD 2015b). While the focus in politics, arts or academia is very often on Hiroshima as the solemn example of an A-bombed city, Nagasaki is often only mentioned dutifully and merely for the sake of completeness. But whereas the commemoration of the explosion in Hiroshima dwells on the city's victimhood and, thus, ignores the context of the Asia-Pacific War, Nagasaki's remembrance does not ignore Japan's war responsibility (DIEHL 2018: 169). Contextualizing the atomic bombing within the events of the war constitutes a major difference in the memorialization and musealization of the bombing in Hiroshima (YONEYAMA 1999; ZWIGENBERG 2015) and Nagasaki (KAMATA 1996; YOSHIDA 2014: 164–72). As I aim to demonstrate over the course of the following pages, we should expect neither the same nor even similar results when it comes to representing the victims of the atomic bomb at museums in both cities.

2 Representation of Victims in Nagasaki

The city-funded Nagasaki Atomic Bomb Museum (Nagasaki genbaku shiryōkan 長崎原爆資料館) opened its doors in 1996, replacing the International Culture Hall (Kokusai bunka kaikan 国際文化会館), established in 1955 and located in the same Urakami area. The museum is situated above the Hypocenter Park and close to the Peace Park (Heiwa kōen 平和公園), partially dug into the hill on which it was built. Observing the museum building from the outside, there is not much in the way of remarkable architectural features. The only exception is a glass dome—under which the exhibition rooms are located—on an otherwise grass-covered hill.

Over the following pages, I introduce the Nagasaki permanent exhibition and demonstrate its reliance on graphic images of anonymous victims. Attributing objects to individuals by name is rather the exception in the exhibition itself and occurs in the audio-guide on a less prominent level of attention.

The first object on display is a damaged clock, which stopped at 11:02 AM. While visitors listen to an ominous ticking sound, they pass photos with scenes of daily life in Nagasaki and cheerful inhabitants prior to the bombing. The section ends with the screening of footage filmed from one of the departing B-29 bombers showing the rising mushroom cloud. The visitors then enter the next exhibition hall, a scenographic representation of Nagasaki immediately after the bombing. It is a dark hall, filled with debris and rubble, the ruins of Urakami Cathedral, situated on the opposite side of the hall, take up most of the visitor's attention. The ruins are accompanied by statues of holy figures, burnt and scorched, their eyes turned pleadingly upwards.

A photo of Nagasaki in ruins takes up one side of the wall, while in front, the hall is filled with bricks and rubble along with bent and twisted steel beams. Video screens placed amid the debris show photos from the ruins and wastelands as well as shocking images of charcoaled corpses, black and with unrecognizable faces. Graphic images of the dead along with gruesomely mutilated bodies are often exhibited in this way in an attempt to emotionally overwhelm the visitor (HEYL 2018: 50–51), a strategy also known as the "pedagogy of horror" (SÁNCHEZ-BIOSCA 2009: 113).

This form of display of violence was often deployed, amongst others, by Holocaust memorial museums and concentration camp memorials, but many of them changed their approach from the 1990s onward. The focus of the exhibitions then shifted from the suffering of anonymous masses to the loss of individual lives, not of heroes or martyrs, but of "ordinary" victims. Within the Holocaust memorialization "there was also a growing realization that empathy with the murdered could not be awakened by photographs of mounds of corpses, but through individual testimonies" (RADONIĆ 2016a: 233). As a result of this changing understanding of putting victims on display, Holocaust memorial museums started giving visibility to the individual, for example, by presenting a personal object, a letter or a photo depicting the individual life from before being deported. The 1996 Nagasaki exhibition could, on the one hand, thus be considered outdated/antiquated in this regard. Yet, on the other hand, the trend to move away from (enlarged) graphic images or have them displayed in a rather hidden manner[6] is by no means universal, and remains common in China for example (BAJGEROVÁ VERLY 2022: 298).

After passing through this dark hall, visitors enter a brightly lit exhibition hall where the focus is on the three-fold destructive power of the atomic bomb: heat, radiation, and the force of the blast. The scorched, melted, bent, or broken objects serve as evidence of the intensity of the heat and blast, including tiles with discolorations and heat induced blisters, scorched bamboo rods, molten glass bottles, fused coins, and bent and broken steel girders. Moreover, these objects testify to the "fractures created by the trauma of the bombing, and the state of the aftermath [...] they subtly point to the terror of the bombing, but the story told is not of what the objects are, but more often what they represent—the non-embodied; and the no longer visible—that was destroyed, pulverized, atomized on 9 August 1945" (MCCLELLAND 2022: 126–27).

The items at NABM are lined up in the show case in front of black and white panoramic photos of the ruins of Urakami (see Figure 3). The photos in this display case all show collapsed buildings and empty ruins. The overall atmosphere is sober, neat, and less emotionally charged, since this part of the exhibition appears to be strangely unenlivened with almost no photos of victims, whether as corpses or survivors. It is not until halfway through the exhibition hall that the first photos of human victims appear, consisting of the aforementioned photos of burnt corpses as well as others of injured survivors next to dead bodies. Beneath these photos are items of clothing with scorched edges that survived the fire and heat. Here we also find the blouse of the eight-month old girl, mentioned earlier,

Fig. 3: Here is an example of an unenlivened glass case with photos of ruins. In front of them, objects melted or scorched by heat and fire are displayed. (Photo by André Hertrich, 2022)

whose name is not given. Next to the blouse is a jacket that shows discoloration due to the heat, and a shirt which a Ms. Ryu Suiso was wearing when she was exposed to the explosion. The caption reads as following: "Ms. Ryu Suiso was wearing this jacket when exposed to the atomic bombing in Motohara-machi, only one day after being evacuated there for safety. (Donated by Ryu Suiso)". There is no further information as to what happened to her afterwards, but the donor's name indicates that she survived the bombing.

A large color photo of a shirtless middle-aged man showing his scar-blotched face and upper body is centrally located next to these items of clothing. The photo must have been taken in the 1980s or 90s. The caption reads as follows: "Scars from flash burns suffered in the atomic bombing (Senji Yamaguchi)". No further information is given in the exhibition but we learn from the English version of the audio guide that:

> He was 14 years old [...]. He was bombed while he was digging an air-raid shelter [...]. He became unconscious at the moment of the flash, and when he regained consciousness, he found himself surrounded by a number of corpses and many people running back and forth around him. He suffered severe burns on his face and body, and was hospitalized [...]. He suffered from high fever and diarrhea along with pain from the burns. He was hospitalized for eight months.
> He was embarrassed when infants carried on their mother's back started to cry as soon as they saw him in the town. He could not find a job upon taking physical examinations, even if he passed the paper screening.
> He participated in No. 2 Special Session of the UN General Assembly on Disarmament held in 1982 as a civilian representative of NGO, and talked about his own atomic bombing experience while showing his photograph, and appealed for the elimination of nuclear weapons.[7]

I chose to quote this lengthy description to show that the audio guide often adds another level of explanation unseen in the exhibition. Above Yamaguchi's photo is another one from the immediate postwar era, depicting "a boy who suffered severe burns all over his back (Sumiteru Taniguchi)".[8] Here too, the audio guide gives far more information than the caption. Both Yamaguchi and Taniguchi were active in the anti-nuclear movement and gave speeches all over the world (SOUTHARD 2015a). Thus, the depicted victims and their testimonies represent the voice of activists, not those of ordinary *hibakusha* 被爆者, that is, persons affected by the bomb. Furthermore, Taniguchi's and Yamaguchi's experience allows visitors to feel compassion and empathy for them considering their injuries and suffering as well as their experience of social stigmatization. But their testimonies hardly address the deaths of loved ones and therefore do not primarily focus on the feeling of loss and grief (see Figure 4).

This approach contrasts quite markedly with what I explained regarding the pants belonging to Hiroo exhibited in Hiroshima, which are displayed in a much more personal context insofar as they are displayed next to a photo, together with his last words and his sister's testimony. In Nagasaki there are only two items which come close to such a dense representation. But still, even these two examples show that adding a photo and a background story does not produce the same emotional connectedness at the exhibition in Nagasaki as it does in Hiroshima. The first object ascribed to a named individual is a lunch or bento box, which still holds charcoaled rice as evidence of the tremendous heat caused by the raging fires that followed the explosion. This bento box belonged to a girl called Tsutsumi Satoko, who is shown on a group photo together with other female students (see Figure 5).

The audio-guide states: "Schoolgirl's Lunchbox. This burned lunchbox belonged to Miss Satoko Tsutsumi. She was 14 years old then, and was exposed to the atomic bomb in Iwakawa-machi, about 700 meters from the hypocenter. The rice in the lunchbox was charred by the subsequent fire after the atomic bombing. Her name and classroom number are written on the back of the smaller lunchbox on the right." We also learn that due to an earlier air raid alarm, she had stayed home with her grandparents and was killed alongside them by the bomb. Her father found the lunch box when he was looking for her corpse.[9] The second object is a pair of trousers, which is on display in the section of the museum focusing on the blast. The trousers are blood-stained and torn, and the caption reads: "Trousers torn in countless places and stained with blood. Dr. Susumu Tsuno'o, dean of Nagasaki Medical College, was wearing these trousers when exposed to the atomic bombing during a lecture." The trousers are accompanied by a photo of an earnest looking, middle-aged, bespectacled man with lightly graying hair. He is wearing a black suit with a necktie.[10]

Fig. 4: The photos of Taniguchi Sumiteru (on top) and Yamaguchi Senji (below) next to two worker's uniforms. (Photo by André Hertrich, 2020)

Fig. 5: Tsutsumi Satoko's lunch box filled with charcoaled rice. Behind the lunch box is a small group photo depicting Tsutsumi Satoko and her classmates. (Photo by André Hertrich, 2020)

These are the only objects at NABM that form an ensemble of object and photo. Both descriptions are rather sober and brief, and neither focus on the grief of the surviving family members. While Hiroo's pants add more emotionality to the display, the examples from Nagasaki, by contrast, seem rather distanced and unattached. This is also true for the survivors' testimonies in the next room, none of which are accompanied by photos. There are eight testimonies altogether consisting of six from Japanese victims, one from a drafted Korean laborer, and one from a Dutch prisoner of war. This part of the exhibition addresses the issue of Korean forced laborers in Japan as well as the fact that captured foreign soldiers were also forced to work, thus implicating Japan in its responsibility for the war (Yoshida 2014: 164–72).

The testimonies are printed on text boards that explain how survivors experienced the bombing, the fires and the encounter with dead bodies scattered all over the city. As there are no eye-catching elements attached to this part, however, and due to the awkward placing of the tables above eye-level, they are very likely to go unnoticed by the visitors. Thus, when it comes to emotionalizing musealization techniques, the Nagasaki exhibition prefers graphic depictions of burned corpses, but does not highlight the sorrow and loss experienced by the individual survivors. This is even more interesting when considering the fact that some of the testimonies explicitly speak about emotions, such as the embarrassment experienced by Yamaguchi when seen by a child. And yet, emotionality generally ranks rather low in the exhibition's 'hierarchy of visibility'. I argue that the exhibition in Nagasaki displays most objects primarily as evidence for the enormous and devastating powers of the atomic bomb, the blast, the heat, and the radiation, whereby representation rather focusses on those who vanished. Only in the case of a small number of objects did the curators connect the immediate impact of the blast or heat with that of the injuries or death of the former owners of the items. And only two of the items consist of photos depicting the victim (the photo depicting Satoko is rather small and shows her as one person among many). A few more captions mention the names of the former owners of objects on display. Even the story of a deceased baby girl is represented in a matter-of-fact and unemotional fashion as evidence for the impact of the heat. In stark contrast to the depiction of such tragedy in Hiroshima—as I demonstrate in the next section of the article—the heartbreaking loss of a young child's life is only hinted at in the Nagasaki exhibition.

3 Representations of victims in Hiroshima

In terms of war memorialization, the Hiroshima Peace Memorial Museum (Hiroshima heiwa kinen shiryōkan 広島平和記念資料館) is probably the most famous museum in Japan. The city-funded museum opened its doors in 1955 and has since undergone several renewals (SCHÄFER 2018; ZWIGENBERG 2015), with the latest taking place in 2019 (HIGASHI 2018; SHIGA 2020). The museum made significant changes to large parts of its exhibition in accordance with the prevailing trend to focus on the individual victim and the void of a loss of life. As the planning committee states: "In order to convey the tragedy of the A-bomb from a human (hibakusha) point of view, we should articulate the reality of the bombing in a straightforward manner, through the exhibition of actual materials [from the time] such as the ihin 遺品 (objects left behind by those who died) of the survivors, pictures of the victims, A-bomb paintings by city residents, and recorded hibakusha testimonies". Accordingly "the victims' ihin and pictures" are supposed to "psychologically impact and [...] emotionally grip the visitors to the museum" (ZWIGENBERG 2021: 55, citing HIROSHIMA-SHI 2019).

In order to explain what exactly has changed, a brief look back at the former exhibition is necessary. I shall focus, in particular, on how one specific object was previously displayed given that it has now become one of the central items in today's exhibition. The two approaches to representation tell us a lot about the change in curatorial methods. This central object in question is Shin'ichi's tricycle. In the previous exhibition, the tricycle did not feature very prominently. It was presented in a glass case, which was situated amongst bricks and other rubble, in front of a large wall of black and white photos depicting scenes of Hiroshima in ruins. There were no photos of any person accompanying the tricycle. Thus, the entire setting surrounding the tricycle was one of death and destruction. The accompanying text explained Shin'ichi's death and did not, in and of itself, greatly differ from the content of the current text. Yet, in stark contrast to the emotional focus of the recently opened 2019 exhibition, the overall aim of this part of the former version was to convey the enormity of the destructive power of a bomb that left Hiroshima in ruins and many people dead. This current exhibition chooses to emotionally engage visitors in a different way, which follows several international musealization trends stemming, first of all, from Holocaust memorial museums. To wit, they create a gloomy, dark atmosphere and focus on personal stories such that visitors feel the vastness of the event and the void it left in its wake. While the first aspect is also true for Nagasaki, the latter is not.

The new exhibition begins with photos from the pre-bomb-era in which citizens of Hiroshima can be seen living their daily life. Visitors enter a large, almost empty hall where they first pass a photo of a clock showing 8:15 AM, the time of the explosion on August 6. The walls are covered with a huge panoramic photo of Hiroshima as a wasteland. In the middle of the hall, a computer-animated floor screen shows the outlines of Hiroshima within which we see the B-29 bomber, Enola Gay, releasing the bomb over the city. As with Nagasaki, this exhibition begins with the disruption of daily life and goes on to lead visitors toward the ensuing destruction, pain, and death. In the Hiroshima exhibition, this happens by way of a long, dark corridor (connecting the East Wing with the entrance and the main building) that leads visitors to a photo of a young girl with wounds on her face and a bandaged arm. Covered in blood stains, she stands in front of the ruins staring into the camera.[11] This is the first photo to depict one of the many victims of the atomic bomb. It is not a grotesquely crooked charcoaled corpse, as in Nagasaki, but a girl whose wounds appear to be relatively minor and whose eyes meet the camera—and thus the museum visitor—directly as the latter emerges from the corridor.

In point of fact, the exhibition in Hiroshima barely depicts any dead bodies at all (RADONIĆ 2023: 74–75). There are photos of victims severely wounded, many with deformed facial features or burned beyond recognition. With only a few exceptions, however, these victims were all alive, at least when the photos were taken. The images of individuals receiving care, as such, are less evocative of the "pedagogy of horror" and appear, rather, to denote a focus on evoking empathy. Although many of these individuals would not survive the following days or weeks, they are depicted as human beings receiving medical attention, not as dehumanized symbols of horror. The darkness of the corridor extends to the entire section of the exhibition, named "Reality of the Atomic Bombing". The walls and ceilings are painted with dark colors, while the photos and paintings on the walls are illuminated exclusively by spotlights. One's attention is, thus, decisively drawn to the respective items on display. As compared with the bright lighting in the Nagasaki exhibition, the lighting here adds more gravitas and dignity to the atmosphere in the exhibition hall. This also creates a sacral ambience, which demands silence and respect from the visitors. The floor is covered with carpet to create an even more silent atmosphere (SHIGA 2020: 54), which I also witnessed during my numerous visits in 2020 and 2022. These dark exhibition rooms are often used as an aesthetic feature in Holocaust memorial museums, such as the USHMM in Washington DC or Yad Vashem in Jerusalem, as well as other Holocaust-related museums worldwide. Such museums often serve as role models for the design of museums that are not related to the Holocaust (RADONIĆ 2021: 44–50; ZWIGENBERG 2021: 53–54). This is also true for the HPMM, especially when we consider another trend that emerged in the 1990s and was equally inspired by Holocaust museums, namely the focus on personal stories and individual photos.

After entering this section, visitors soon see a room filled with rubble, twisted and bent iron girders, fused metals, tiles with blisters, and a part of a brick wall. There are bent pots, melted glasses, scorched gravestones. While Nagasaki exhibits each of these items separately, the exhibition in Hiroshima seems to imply a dense and concise display method, involving a sort of "best-of" collection of objects exposed to the heat and blast force of the A-Bomb. In so doing, the exhibition in Hiroshima places less emphasis on the material evidence of the impact of the atomic explosion. Instead, it shifts one's focus onto the victims. This becomes clear when looking at the center of the exhibition room, consisting of a large glass case (approx. 2m x 3m), which the visitors can walk around. The text-board reads: "Children Killed in the Bombing". On display are photos of 19 children, all around 12–14 years old. These are not victim photos but portrait shots of boys and girls, many of them in school uniform. In the glass case there are burnt and scorched items of clothing, ripped and torn trousers and shirts with scorch marks and blood stains, fire-protection hoods, little bags, personal belongings such as buttons, a purse with bank notes and stamps, a drinking bottle and shoes. This display of objects slightly resembles the exhibition at the Auschwitz Concentration Camp Memorial where there is an overwhelming number of personal items such as shoes in one show case, glasses in another, and suitcases in yet another. Especially the shoes, or sometimes a single shoe, became a signifier of the life lost in the gas chambers (UNITED STATES HOLOCAUST MEMORIAL MUSEUM 2017).[12] In contrast to the shoes and clothing of Holocaust victims, the items left behind by atomic bomb victims testify to the violence experienced by those bodies, as revealed by burn marks, blood stains, or damage caused by shards. Thus, they serve a slightly different function, but, as I argue, their inclusion in the 2019 exhibition clearly demonstrates recent international influences.

These items, the *ihin*, possess a strong aura of authenticity. As Ran ZWIGENBERG (2021: 44) puts it: "ihin 'have a power that only actual belongings from the bombing possess.' [They put the] 'focus on reality,' suggesting that 'real life artifacts [...] connected to memories of individual victims,' more adeptly convey the reality of the bombing [...]".

Personal *ihin* also serve as evidentiary objects for the ferocious power of the bomb, but first and foremost as empathy-evoking objects that personalize the story. The next section of the exhibition, "Cries of the Soul" makes the importance of those emotionalizing personal objects even clearer—the *ihin* exhibited there do not necessarily attest to the heat or other effects of the atomic bomb, instead primarily commemorating the loss of life and the suffering and grief of those who were left behind. These objects are coupled with photos of their deceased owners, a statement, last words, or the testimony of family members or friends. One wall is dedicated to adult victims, another entirely to children. The example I gave in the introduction—concerning two-year-old Taoda Hiroo's story—is also taken from this section. There are two walls covered with photo/object/testimony assemblages. The introductory text to "Cries of the Soul" quotes some of the statements which are used as captions for the individual photos: "'It hurts, hurts!'; 'Hot! Water!'; 'Help! Mother!'; 'I don't want to die…'; [...] 'My son, where are you?'; 'I am sorry I cannot save you'". It is important to stress that these quotes are both brimming with emotion and, at the same time, are contextualized in detail with the help of the private objects, photographs, testimonies, and stories, rather than with empty phrases about human suffering in general. They, thus, attest to the new curatorial strategy of evoking empathy through personal stories of ordinary citizens.

The space in the middle of the room is filled with three glass cases—from its set-up, this serves as the center, the heart of the exhibition, and puts the personal objects, private photographs, and individual stories center stage. One of the show cases contains a rather odd piece: a mannequin made of braided bamboo rods clothed with the remains of clothes belonging to three different boys, which, taken together, would make up one complete schoolkid's uniform. Another showcase holds a lunch box and a water bottle, which belonged to a boy called Shigeru Orimen and whose photo is next to the glass case. The text reads:

> Lunch box and water bottle. [...]
> Shigeru Orimen (13), a first-year student at Second Hiroshima Prefectural Junior High School, was exposed to the A-bomb and died at his building demolition worksite. This scorched lunch box and water bottle were found beneath his burned, skeletal remains by his mother. In the lunch box were a mixture of rice and barley, soy beans, and sautéed shredded potato. Shigeru left home looking forward to this lunch, but never got to eat it.

As demonstrated above, the museum in Nagasaki also displays Tsutsumi Satoko's lunch box with charred rice, where it primarily serves as an evidentiary object for the extreme temperatures caused by the subsequent fires. While the inclusion of the name and photograph next to the object may evoke some empathy for museumgoers in Nagasaki, those who visit the exhibition in Hiroshima learn not only of the gruesome detail whereby Shigeru's mother found his lunch box beneath his skeletal remains, but also that he was looking forward to his lunch, consisting both of essential plain rice as well as additional vegetables lovingly prepared by his caring mother (KODAMA 1995: 12) (see Figure 6).

Fig. 6: This glass case holds Orimen Shigeru's water bottle and lunch box, still filled with charcoaled rice. His photo and the explanatory text are alongside on a glass panel. (Photo by André Hertrich, 2020)

This and other examples show how the museum in Hiroshima attempts to create a whole range of emotional connections with the victims and survivors on a far larger scale than in Nagasaki: huge photos, captivating statements, moving testimonies of tragic stories regarding how people died and their bodies found by others. These efforts are further amplified by the darkness of the hall and dim lighting of the objects with single spotlights. This approach endows the whole setting with an emotional depth, a feeling of emptiness, a sacred sadness and grief over the loss of these persons. As the authors of the concept paper cited above would have it, the space aims to: "psychologically impact and [...] emotionally grip the visitors" (HIROSHIMA-SHI 2019). While many museums have decided to leave behind the "pedagogy of horror" which works through an overwhelming display of graphic images showing anonymous (piles of) bodies, this newer trend in memorial museums seeks to grip visitors by playing on their heart strings and "forcing" them to feel sad (as I experienced it during my visits). It should, of course, also be critically re-examined.

The object with the greatest emotional impact in this exhibition hall is almost certainly the item in the third glass case—Shin'ichi's tricycle and helmet.[13] It is no longer presented in the context of ruins, rubble, and bricks, as it was in the exhibition prior to the renovation in 2019. Instead, there are no additional elements except for a large, almost life-sized illuminated black and white photo of Shin'ichi wearing a rather plain woolen pullover and something akin to a bib. His sister, dressed in a festive kimono, stands alongside him. In front of this photo, we see a glass case with the rusty helmet and the torn tricycle from which the saddle is missing and one wheel is broken. The caption reads as follows:

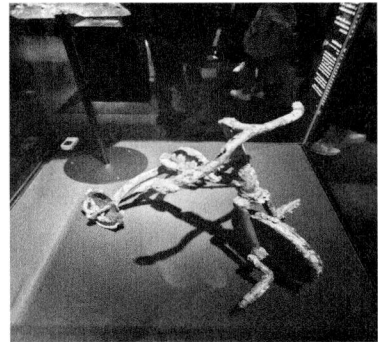

Shinichi Tetsutani (2 years, 11 months) was riding this tricycle when the A-bomb exploded. Suffering serious injuries and severe burns all over his body, he died that night groaning, 'Water, water...' His father Nobuo put this metal helmet on Shinichi's body in the back yard so he could ride it even after his death. Forty years later, Nobuo decided to place Shinichi's remains in the family grave. When Nobuo dug them up, he found Shinichi's skull intact inside the helmet. Nobuo also lost his daughters, Michiko (7) and Yoko (1), who were trapped and burned under their collapsed house.

By removing rubble and bricks as well as the background photo of the ruins, the tricycle does not stand for the destruction of buildings by the bomb anymore. It is now a symbol for the loss of young Shin'ichi's life and the void

Fig. 7: Tetsutani Shin'ichi's tricycle in a glass case is located in front of a large black-and-white photo of Shin'ichi and his sister Michiko. Due to copyright issues the photo of the children needs to be omitted. (Photo by Ljiljana Radonić, 2023)

left behind by his death. The tricycle is one of the most representative objects of the Hiroshima Peace Memorial Museum. It is, for example, the first thing visitors see on the website.[14] It is probably the most emotionally charged object in this museum and is situated in what is arguably the very center of the exhibition (see Figure 7).

This section aimed to show that in contrast to Nagasaki's very sober and matter-of-fact representation of the atomic bombing with many objects serving as evidence for the destructive power of the new weapon, Hiroshima adds a personal emotional layer to the objects and includes individual victim-photos and statements, thus putting the painful experience of individual death, loss, and emptiness in the foreground. Nagasaki also chooses an emotionalizing approach, but one that very much relies on a "pedagogy of horror" by showing photos of charcoaled corpses on the one side and material evidence of the power of destruction on the other side. Hiroshima follows newer international trends of commemorating individual victims in memorial museums. The next section discusses the roots of these trends.

4 Exhibiting Individual Victims as an International Trend

Some of the characteristics of the representation of victims, especially those in the new Hiroshima exhibition, can be found in other memorial museums around the globe. These are specific architectural and aesthetical features like the use of dark and immersive spaces and the evocation of feelings of abandonment and isolation as well as a focus on the individual victim (RADONIĆ 2021: 120–21). By combining *ihin* and photos of the victims, the HPMM creates emotionalizing assemblages and thus follows a global trend in memorialization. As ZWIGENBERG writes: "The move toward the personal [...] is [...] a process one can see in Holocaust commemoration. These changes reflect a move in global memory culture from a focus on communal and national narratives, to the experience of the individual victims, the survivors themselves and their testimonies. With the passage of time, and especially the dwindling of survivors' numbers, objects supply us with the most immediate and concrete reality of the Holocaust and the A-bomb" (ZWIGENBERG 2021: 57).

As for Holocaust memorialization, the USHMM in Washington DC, and Yad Vashem in Jerusalem serve as role models, not only for Holocaust museums but also for a whole range of different memorial museums. With the focus on private photos of victims in the "Tower of Faces" at USHMM (KÖHR 2012: 121–28) and the "Hall of Names" at Yad Vashem (KÖHR 2012: 128–29), both museums shifted their approach away from the "pedagogy of horror" that was characteristic for older exhibitions and toward individualization.

Instead of exhibiting photos of anonymous masses of deported Jews, piles of corpses, humiliating images taken by perpetrators or emaciated concentration camp inmates, the two museums installed private photos of individuals prior to their persecution by German and Austrian Nazis and their local collaborators.

Thus, the victims were de-anonymized and regained their individuality (KÖHR 2012: 120). Before the 1990s, there was hardly any individualization of "ordinary" people in memorial museums, only famous heroes, martyrs or combatants. From then on, a new trend began to spread whereby people were no longer depicted primarily as anonymous victims or dead bodies, but as living individuals. Their lives—soon to be extinguished—would be shown in humane detail, including depictions of them together with friends and families, sometimes smiling, sometimes in their professional attire as parts of a community. Thus, each photo serves as evidence, as private testimony of a life lost, and—in contrast to dehumanizing dead or "living skeletons"—can evoke empathy. For example, the 1,032 photos at USHMM's Tower of Faces depict Jewish children, women, men and families, all from the Lithuanian town Eišiškės, where in 1941 almost all Jewish inhabitants were killed by the SS. Ultimately, the Tower of Faces—as well as the Hall of Names at Yad Vashem from 2005—became a template for memorial museums worldwide (see Figure 8).[15]

Thus, photo walls covered with private photos became a common feature in memorial museums globally. To name just a few: the National September 11 Memorial and Museum in New York City; the Museum of Memory and Human Rights in Santiago de Chile (Museo de la Memoria y los Derechos Humanos) commemorating the victims of the Pinochet regime; and the Kigali Genocide Memorial in Rwanda, which memorializes the victims of the Genocide against the Tutsi in 1994. Also, the Nanjing Massacre Memorial Museum displays photos of victims of the Japanese massacre in 1937 on large-scale photo walls.

Fig. 8: The "Tower of Faces", at the USHMM in Washington DC, is covered with more than 1,000 photos of inhabitants of Eišiškės, depicted before most of them were killed by SS troops. Photo: View of a section of the Tower of Faces (the Yaffa Eliach Shtetl Collection) in the permanent exhibition at the U.S. Holocaust Memorial Museum. United States Holocaust Memorial Museum Photo Archives #N04219. Courtesy of Bruce Katz. Copyright of United States Holocaust Memorial Museum.

In Japan, this exhibition feature is also used, for example by the Yūshūkan 遊就館, the museum attached to the Yasukuni Shrine 靖國神社 in Tokyo, which honors men (and a few women) who lost their life in service for the tenno. The men in these photographs are not depicted as victims, but rather, on the one hand, as heroic martyrs and noble spirits (*eirei* 英霊) while, on the other, as "ordinary human beings with family, friends and personal feelings" (SAKAMOTO 2015: 168). Another example is the Women's Active Museum on War and Peace (WAM) in Tokyo, a place commemorating the plight of the so called "comfort women".[16] Here, the persons on the photo wall at WAM are all elderly women who had courage enough to come out as former "comfort women", thus showing the empowerment of a shared experience (WATANABE 2015: 241–52). These examples demonstrate that a mere aesthetical resemblance should always make us look closer at how the installation is embedded in the respective national narrative.

I argue that the 2019 exhibition at the Hiroshima Museum clearly follows this international trend of focusing on private photographs. The photos are combined with "auratic" objects and do not, as such, follow the photo wall aesthetic. As is the case for many other memorial museums, however, the 2019 exhibition opts for installations that combine personal objects, photographs, and short biographies.[17]

It comes as no surprise that Holocaust memorial museums initiated this trend. Memorial Museums—as a type of museums—are relatively new. In his *Memorial Museums: The Global Rush to Commemorate Atrocities*, Paul WILLIAMS defines them as a "specific kind of museum dedicated to a historic event commemorating mass suffering of some kind" (WILLIAMS 2007: 8). By exhibiting past atrocities and traumas, memorial museums are very much influenced by the growing importance of the Holocaust, which from the 1990s onward came to constitute the global symbol for the "most fundamental evil" (ALEXANDER 2002: 44). As such, the commemoration of the Holocaust was not only about remembrance of the victims but called for prevention of discrimination, hate, and repression everywhere. This created "a universal imperative, making the issue of universal human rights politically relevant to all who share this new form of memory" (LEVY / SZNAIDER 2006: 132). This universal imperative of Holocaust memory would then inspire memorial museums (also non-Holocaust related) to "focus on what is most painful in the past, [and] [...] to critically engage with past violence to build a more tolerant, democratic culture through the promotion of human rights and an ethic of 'never again'" (SODARO 2017: 5). At the same time, the universalization of the Holocaust since the end of the Cold War made it possible to produce shared memories, which would also allow all kinds of victim groups to identify their suffering with that of Holocaust victims (LEVY / SZNAIDER 2002: 100–01).

This means that there are two major ways in which museums can reference the Holocaust: either by arguing that "we" also suffered a Holocaust[18] or by incorporating musealization trends stemming from Holocaust museums, such as the aforementioned dark and immersive architectural features and the display of personal items and private photos. This form of Holocaust musealization originated at the USHMM and Yad Vashem and proved to be a powerful tool in evoking emotions and empathy for the victims. It is an approach that has been used to a significant extent in Hiroshima, but much less so in Nagasaki.

5 CONCLUSION

As Paul WILLIAMS has noted, there is a crucial tension in memorial museums "between the authenticity and evidence, on the one hand, and the desire to create an emotive, dramatic visitor experience on the other" (WILLIAMS 2007: 21). I have sought to show that the Nagasaki Atomic Bomb Museum rather tends to lean toward authenticity and evidence, while the recently opened Hiroshima Peace Memorial Museum stresses the emotive, dramatic visitor experience. Nagasaki seems to avoid emotionalizing assemblages and offers only a very limited number of individual photos of victims and their personal belongings. The main focus is on the destructive powers of the atomic bomb, namely the effects of the heat, blast force and radiation on the city, on buildings and on the human body. Thus, the older Nagasaki permanent exhibition does not really follow current international trends of memorializing atrocities but, rather, offers a sober, matter-of-fact, and scientific presentation.

On the other hand, by going beyond the mere physical effects of the bomb and demonstrating that the destroyed human body was once a person with family, friends, hopes and dreams, the exhibition in Hiroshima relies heavily on an emotive and dramatic visitor experience. Moreover, the exhibition at HPMM offers an emotional layer of loss and grief by attaching private photographs of the victims, personal items, and heart-wrenching testimony of family members, and by shrouding the exhibition space in silent and dignified darkness.

The most interesting cases for both the diachronic and synchronic comparison of the exhibitions are the objects used in Nagasaki and Hiroshima. The bento box, for example, demonstrates the different meanings attributed to it at the two sites. In Nagasaki, the charcoaled rice serves as evidence for the effects of the heat of the subsequent fires after the bombing, while in Hiroshima the charcoaled rice represents the vanished boy, who will never be able to enjoy the delicious meal prepared by his loving mother. The different settings in which the tricycle is displayed in the previous and contemporary exhibition at HPMM equally demonstrate the changed approaches to personalization and individualization of objects. Placed amongst rubble and bricks, the tricycle represented the destructive power of the atomic bomb in the earlier exhibition. Now, by contrast, with the rubble gone and a private photo added, it stands for the void as well as for the grief of a father whose beloved children were killed by the bomb.

In this paper, I have shown that the forms of representation of the victims differ profoundly in both museums, despite the significant comparability of their focus: the atomic bombing of two Japanese cities. But how do we explain the phenomena I have delineated? The most obvious factor, perhaps, is the fact that NABM was inaugurated in the 1990s and has since undergone only minor changes. HPMM, on the other hand, changed its exhibitions in the 1970s, the 1990s and, finally, again in 2019. It was not until the last renovation that the exhibition makers at HPMM picked up on musealization trends from Holocaust museums. This might have been facilitated by the second factor: the usage of a musealization strategy that puts so much focus on the victim's perspective was possibly promoted by the prevalent understanding of victimization in the city of Hiroshima. The city of Nagasaki, on the other side, at least tries to frame the atomic bombing within the context of Japan's attack on Pearl Harbor and other Japanese aggressions during the war (DIEHL 2018: 169).

M. G. SHEFTALL's observations regarding differences in topography and everyday life in Hiroshima and Nagasaki may well constitute a relevant third factor (SHEFTALL 2019). While the bomb in Hiroshima killed many who commuted to the city and mobilized school children who had to clear firebreaks along the main roads, the bomb in Nagasaki killed people in the residential area of Urakami, among them a larger percentage of entire households and families from which the only surviving members were those who commuted every day. In Hiroshima by contrast, a large number of "bereaved-but-intact" households constituted a "robust 'memory agent' community" (SHEFTALL 2019: 7) in districts that suffered less extreme damage. This community later donated all the personal items, photos, and letters—often from mobilized school children from suburban areas who were killed in downtown Hiroshima, to the HPMM, which displays them in great numbers in the new exhibition.

With the last survivors passing away, it becomes more and more important to represent their experiences—including the death and loss of their loved ones—in order to convey the "reality of the bombing", as HPMM puts it. Though the NABM has not paid much attention to this personal aspect, I recently learned—during my last visit to the museum in autumn 2022—that it has begun an outreach project asking remaining survivors for their atomic-bomb related stories, objects, and photographs. It will be interesting to see if the museum then adopts a form of representation more akin to that pioneered at the Holocaust museums as described above and, indeed, what a display of the victims specific to Nagasaki will look like in the years to come.

NOTES

1. The research project "Globalized Memorial Museum. Exhibiting Atrocities in the Era of Claims for Moral Universals" is funded by the European Research Council (ERC) under the European Union's Horizon 2020 research and innovation program (GMM – grant agreement No 816784).
2. Taoda Hiroo and his pants were no longer part of the exhibition in November 2022. These and other objects were annually replaced for preservational purposes by other items from the museum's rich collection, without changing the overall character of this part of the exhibition (SHIGA 2020: 47).
3. I have translated all quotes from THIEMEYER, KORFF / EBERSPÄCHER / BAUMUNK, SCHOLZE, as well as RADONIĆ 2016a and RADONIĆ 2021 from their original German into English.
4. The research project "Globalized Memorial Museum. Exhibiting Atrocities in the Era of Claims for Moral Universals" is based at the Austrian Academy of Sciences in Vienna. It comprises of a team of international museum experts and conducts research on memorial museums in Europe, in the U.S., in Japan, China, but also Rwanda or Bosnia; see https://www.oeaw.ac.at/projects/gmm/ (last access 2023 / 03 / 28).
5. In order to secure financial aids from the Japanese government for reconstruction measures in the early postwar years, Hiroshima concentrated its reconstruction efforts around the idea of becoming a City of Peace, a concept Hiroshima would vigorously defend, especially vis-à-vis Nagasaki. With the title City of Peace gone, Nagasaki chose instead to stress its long history as the only open port for European merchants during the Tokugawa period (1603–1868, port since 1640) and its Christian heritage (since the 16th century) and thus became the International Culture City.
6. At the Museum of the History of Polish Jews (Polin) in Warsaw, for example, humiliating photos of half-naked women—taken shortly before their execution by Nazi murderers—are hidden behind a symbolic forest such that visitors need to make an effort to see them, RADONIĆ (2016b: 188–89).
7. The texts of the audio guide in English can be found online: *Nagasaki Atomic Bomb Museum Audio-Guide* (n. d.), https://qrkaisetsu.nagasakipeace.jp/index.php?lid=2&pid=19 (last access 2023 / 11 / 08).
8. Caption of the photo. The text from the English-language audio guide reads as follows: "This is Mr. Sumiteru Taniguchi. He was 16 years old when he was bombed [...]. His skin was burned and bleeding [...]. He moved from one hospital to another and spent about 3 years and 7 months in hospital while mostly lying on his stomach. He often begged to be killed due to the agony of the pain. In spite of the recurrence of his skin cancer, he has talked about his atomic bombing experience not only in Japan but in foreign countries as well and continues to call for the elimination of nuclear weapons"; see: *Nagasaki Atomic Bomb Museum Audio-Guide* (n. d.), https://qrkaisetsu.nagasakipeace.jp/index.php?lid=2&pid=18 (last access 2023 / 11 / 08).
9. *Nagasaki Atomic Bomb Museum Audio-Guide* (n. d.), https://qrkaisetsu.nagasakipeace.jp/index.php?lid=2&pid=14 (last access 2023 / 11 / 08).

10 According to the audio guide, Dr. Tsuno'o witnessed the situation in Hiroshima the day after the bombing. Back in Nagasaki he was wounded by shards from a window. He finally died from the effects of radiation There are differences in the description on the audio guide and the caption to the objects; see: *Nagasaki Atomic Bomb Museum Audio-Guide* (n. d.), https://qrkaisetsu.nagasakipeace.jp/index.php?lid=2&pid=24 (last access 2023/11/08).
11 After visiting this part of the exhibition, the museumgoers would once again pass by this photo and learn that the girl on the photo, Fujii Yukiko (then ten years old), ultimately died of cancer in 1977. Due to copyright issues I cannot show the photo here, but you will find the photo and more information here: THE MAINICHI (2020).
12 See, for example, the poem "We are the Shoes" by Moshe SZULSZTEIN: "We are the shoes, we are the last witnesses. / We are shoes from grandchildren and grandfathers / From Prague, Paris and Amsterdam, / And because we are only made of fabric and leather / And not of blood and flesh, / Each one of us avoided the hellfire."
13 The fact that the museum shop sells children's picture-books by KODAMA Tatsuharu on Shin'ichi's and Shigeru's death, the mourning of their families as well as their remaining tricycle / bento box show the centrality of these two objects for HPMM's exhibition, see KODAMA (1994) and KODAMA (1995). The books are in English but with several words translated to Japanese and an inset with the Japanese transcript, thus facilitating the use of the book during English classes for Japanese high school students.
14 Also, a random search of the term "Hiroshima Museum" on Google Images results in repeated appearances of the image of the tricycle. See also this website with photos of the tricycle and other objects mentioned in my paper: THE MAINICHI (2019).
15 As part of our research in the research project on "Globalized Memorial Museum", we examined more than 50 memorial museums worldwide to determine if and how USHMM and Yad Vashem had served as role models for all kinds of memorial museums. For further information, see https://www.oeaw.ac.at/projects/gmm/ (last access 2023/11/14).
16 The term "comfort women" (*ianfu* 慰安婦) is a euphemistic term for Japan's military sexual slavery during the Asia-Pacific War. An estimated 200,000 to 400,000 women and girls, mostly from Japanese occupied territories, were forced into serving Japanese soldiers sexually.
17 In Hiroshima, there is a "photo wall" at the Hiroshima National Peace Memorial Hall for the Atomic Bomb Victims, directly opposite the HPMM. The "photo wall" is a giant screen with changing photos on display.
18 This again raises the problem of victim competition, as the "double genocide" theory in Central Eastern Europe shows, see HIMKA / MICHLIC 2013: 17–18.

REFERENCES

ALEXANDER, Jeffrey C. (2002): "On the Social Construction of Moral Universals". In: *European Journal of Social Theory* 5 (1): 5–85.

BAJGEROVÁ VERLY, Markéta (2022): "Survivors, victims and soldiers as figures of nationalism: Representations of women in the War of Resistance against Japan museums in mainland China". In: *East Asian Journal of Popular Culture* 8 (2): 291–309.

DIEHL, Chad (2014): "Envisioning Nagasaki: from 'atomic wasteland' to 'international cultural city', 1945–1950". In: *Urban History* 41 (3): 497–516.

DIEHL, Chad (2018): *Resurrecting Nagasaki: Reconstruction and the Formation of Atomic Narratives*. Ithaca: Cornell University Press.

HEYL, Matthias (2018): "Mit Überwältigendem überwältigen? Emotionalität und Kontroversität in der historisch-politischen Bildung: Ein Plädoyer für die Schärfung des Profils historischer Bildung". In: SCHMIDT, Jochen / SCHOON, Steffen (eds.): *Politische Bildung auf schwierigem Terrain. Rechtsextremismus, Gedenkstättenarbeit, DDR-Aufarbeitung und der Beutelsbacher Konsens*. Schwerin: Landeszentrale für politische Bildung Mecklenburg-Vorpommern, 37–58.

HIGASHI, Julie (2018): "The Destruction and Creation of a Cityscape in the Digital Age: Hiroshima Peace Memorial Museum". In: *Museum International* 70 (1–2): 104–13.

HIMKA, John-Paul / MICHLIC, Joanna Beata (2013): "Introduction". In: HIMKA, John-Paul / MICHLIC, Joanna Beata (eds.): *Bringing the dark past to light: The reception of the Holocaust in postcommunist Europe*. Lincoln, Neb.: University of Nebraska Press, 1–24.

HIROSHIMA PEACE MEMORIAL MUSEUM (2020): *Hiroshima Peace Memorial Museum Collection Catalogue. Carrying the Legacy of Hiroshima*. Hiroshima: Hiroshima Peace Culture Foundation.

HIROSHIMA-SHI 広島市 (2019): "Hiroshima heiwa shiryōkan tenji seibi tō kihon keikaku: daisanshō 広島平和記念資料館展示整備等基本計画：第3章"; https://www.city.hiroshima.lg.jp/soshiki/48/9647.html (last access 2023 / 03 / 05).

KAMATA, Sadao 鎌田定夫 (1996): "Nagasaki genbaku shiryōkan no kagai tenji mondai 長崎原爆資料館の加害展示問題". In: *Kikan sensō sekinin kenkyū* 季刊戦争責任研究 14: 22–31.

KODAMA, Tatsuharu (1994): *Shin's tricycle*. Tōkyō: Chart Institute.

KODAMA, Tatsuharu (1995): *The lunch box*. Tōkyō: Chart Institute.

KÖHR, Katja (2012): *Die vielen Gesichter des Holocaust: Museale Repräsentationen zwischen Individualisierung, Universalisierung und Nationalisierung*. Göttingen: V & R unipress.

KORFF, Gottfried / EBERSPÄCHER, Martina / BAUMUNK, Bodo-Michael (eds.) (2002): *Museumsdinge: Deponieren–exponieren*. Köln: Böhlau.

LEVY, Daniel / SZNAIDER, Natan (2002): "Memory Unbound: The Holocaust and the Formation of Cosmopolitan Memory". In: *European Journal of Social Theory* 5 (1): 87–106.

LEVY, Daniel / SZNAIDER, Natan (2006): *The Holocaust and memory in the global age*. Philadelphia: Temple University Press.

MCCLELLAND, Gwyn (2020): *Dangerous memory in Nagasaki: Prayers, protests and Catholic survivor narratives*. Abingdon / Oxon / New York: Routledge.

MCCLELLAND, Gwyn (2022): "Digitalising Trauma's Fractures: Nagasaki Museums, Objects, Witnesses, and Virtuality". In: WALDEN, Victoria Grace (ed.): *The Memorial Museum in the Digital Age*. Sussex: University of Sussex, 121–54.

NABM AUDIO-GUIDE (n. d.): *The Eternal 11:02*; https://qrkaisetsu.nagasakipeace.jp/index.php?pid=1&lid=2 (last access 2023 / 04 / 11).

RADONIĆ, Ljiljana (2016a): "Individualisierung als Abwehr: Deutsche Erinnerungskultur versus postsozialistische Affinität zur 'Sache des Zionismus'". In: *sans phrase* 8: 233–42.

RADONIĆ, Ljiljana (2016b): "Visualizing Perpetrators and Victims in Post-Communist Memorial Museums". In: *Yad Vashem Studies* 44 (2): 173–201.

RADONIĆ, Ljiljana (2018): "Introduction: The Holocaust / Genocide Template in Eastern Europe". In: *Journal of Genocide Research* 20 (4): 483–89.

RADONIĆ, Ljiljana (2021): *Der Zweite Weltkrieg in postsozialistischen Gedenkmuseen: Geschichtspolitik zwischen der 'Anrufung Europas' und dem Fokus auf 'unser' Leid*. Berlin / Boston: De Gruyter.

RADONIĆ, Ljiljana (2023): "Displaying Violence in Memorial Museums— Reflections on the Use of Photographs". In: *Österreichische Zeitschrift für Geschichtswissenschaften* 34 (1): 59–84.

SAKAMOTO, Rumi (2015): "Mobilizing Affect for Collective War Memory: Kamikaze images in Yūshūkan". In: *Cultural Studies* 29 (2): 158–84.

SÁNCHEZ-BIOSCA, Vicente (2009): "Sombras de Guerra: Las Imágenes Cinematográficas de la Shoah". In: *Historia Social* 63: 111–32.

SCHÄFER, Stefanie (2018): *Das Atombombenmuseum Hiroshima: Erinnern jenseits der Nation (1945–1975)*. Bielefeld: transcript Verlag.

SCHOLZE, Jana (2010): "Kultursemiotik: Zeichenlesen in Ausstellungen". In: BAUR, Joachim (ed.): *Museumsanalyse: Methoden und Konturen eines neuen Forschungsfeldes*. Bielefeld: transcript, 121–48.

SHETFALL, M. G. (2019): "Hiroshima Protests, Nagasaki Prays: Divergent Narratives in Early Postwar Atomic Bombing Memorialization Discourse" (Presentation at the Asian Studies Conference, Saitama University).

SHIGA, Kenji 志賀賢治 (2020): *Hiroshima heiwa kinen shiryōkan wa toikakeru* 広島平和記念資料館は問いかける (Iwanami shinsho 岩波新書 1861). Tōkyō: Iwanami shoten.

SODARO, Amy (2017): *Exhibiting Atrocity: Memorial Museums and the Politics of Past Violence*. New Brunswick et al.: Rutgers University Press.

SOUTHARD, Susan (2015a): *Nagasaki Deluxe: Life After Nuclear War*. New York: Penguin Books.

SOUTHARD, Susan (2015b): "Nagasaki, the Forgotten City". In: *The New York Times*, 08/07; https://www.nytimes.com/2015/08/08/opinion/nagasaki-the-forgotten-city.html?_r=0 (last access 2023/03/28).

SZULSZTEIN, Moshe: "We are the Shoes"; https://www.hmd.org.uk/wp-content/uploads/2018/06/We-are-the-shoes.pdf (last access 2023/04/11).

THE MAINICHI (2019): "In Photos: Renewed Hiroshima A-bomb museum exhibit makes emotional connection"; https://mainichi.jp/english/graphs/20190426/hpe/00m/0na/001000g/20190426hpe00m0na004000q (last access 2023/11/13).

THE MAINICHI (2020): "Son who helped ID Hiroshima A-bomb girl's photo decades on gives lecture about her life"; https://mainichi.jp/english/articles/20200127/p2a/00m/0fe/027000c (last access 2023/11/13).

THIEMEYER, Thomas (2010): "Geschichtswissenschaft: Das Museum als Quelle". In: BAUR, Joachim (ed.): *Museumsanalyse: Methoden und Konturen eines neuen Forschungsfeldes*. Bielefeld: transcript, 73–94.

UNITED STATES HOLOCAUST MEMORIAL MUSEUM (2017): *The last witnesses: Artifacts from the Museum's collection*. Washington, DC: United States Holocaust Memorial Museum.

WATANABE, Mina (2015): "Passing on the history of 'comfort women': the experiences of a women's museum in Japan". In: *Journal of Peace Education* 12 (3): 236–46.

WILLIAMS, Paul Harvey (2007): *Memorial museums: The global rush to commemorate atrocities*. Oxford: Berg Publishers.

YONEYAMA, Lisa (1999): *Hiroshima traces: Time, space, and the dialectics of memory*. Berkeley: University of California Press.

YOSHIDA, Takashi (2014): *From cultures of war to cultures of peace: War and peace museums in Japan, China, and South Korea*. Portland, Maine: MerwinAsia.

ZWIGENBERG, Ran (2015): *Hiroshima: The origins of global memory culture*. Cambridge: Cambridge University Press.

ZWIGENBERG, Ran (2021): "Modern Relics: The Sanctification of A-Bomb Objects in the Hiroshima Museum". In: *Holocaust and Genocide Studies* 35 (1): 44–62.

Preserving *hibakusha* memories: Ōishi Matashichi and the Daigo Fukuryū Maru Exhibition Hall

Lauren Constance

1 INTRODUCTION

Less than a decade after the 1945 atomic bombings of Hiroshima and Nagasaki, Japan was once again subject to nuclear tragedy when the fishing vessel Daigo Fukuryū Maru (The Lucky Dragon Number 5) was contaminated by nuclear fallout from a Hydrogen bomb test. On March 1, 1954, just 160 km from the blast's hypocenter (and, importantly, outside of the warning zone established by the U. S. military) the Daigo Fukuryū Maru's 23 crew members saw a sudden flash of light to their west, followed by the enormous sound of the detonation. Several hours later, white dust began to fall from the sky and onto the ship and crew members (ROPEIK 2018). This seemingly harmless white powder later became known as the "ashes of death" (*shi no hai* 死の灰), as it turned out to be radioactive fallout from the nuclear explosion. Upon their return to shore two weeks later, all 23 crew members began displaying symptoms of acute radiation poisoning and so were transferred to hospital to undergo treatment (ŌISHI 2011). Six months after their initial exposure to the ashes of death, chief radio operator KUBOYAMA Aikichi 久保山愛吉 died, leaving behind his wife and three young daughters (DAIGO FUKURYŪ MARU EXHIBITION HALL 2015a) which led to a rise in political tension between Japan and the United States (HOMEI 2013: 216–17). Widespread news coverage of the Bikini Incident and the death of KUBOYAMA, coupled with increasing awareness of *hibakusha*'s 被爆者 (bomb-affected person) experiences of the Hiroshima and Nagasaki atomic bombings reignited Japan's anti-nuclear movement (HOMEI 2013: 214; APSEL 2016: 73; LOW 2020: 68).

This event, which has come to be known as the "Bikini Incident"[1] has been represented and memorialized in mass media, in films such as SHINDŌ Kaneto's 新藤兼人 *Daigo Fukuryū Maru* 第五福竜丸 (1959) known in English as "The Lucky Dragon" and more recently Keith REIMINK's *Day of the Western Sunrise* (2018), produced by U. S. indie film company Daliborka Films. Crew member ŌISHI Matashichi 大石又七 (2011) published several works relating to his experience of the Bikini Incident, including *The Day the Sun Rose in the West: Bikini, the Lucky Dragon, and I*, an English-language account of his experiences. Perhaps most famously, the incident inspired the creation of Godzilla (RYFLE 2017: 102), "the longest continuously running movie franchise in the world" (CHO 2019: 128). Despite the huge success of Godzilla, it has been argued that the Bikini Incident (especially in comparison to Hiroshima and Nagasaki) has been "pushed to the edge of public history and memory" (CHO 2019: 128). Indeed, the history of the boat itself reflects this. After being restored as a training vessel for Tokyo University of Fisheries until it was decommissioned in 1967, the Daigo Fukuryū Maru (at the time renamed the Hayabusa Maru はやぶさ丸) was dumped on Yumenoshima 夢の島, an island made of reclaimed land in Tokyo Bay which, at the time, was used as a rubbish heap. However, a group of Tokyo citizens learned of the original history of the boat, and set up a movement in 1969, called "The Daigo Fukuryū Maru Preservation Committee" (Daigo Fukuryū Maru hozon iinkai 第五福竜丸保存委員会) to preserve the vessel (BULLETIN OF THE JAPAN COUNCIL AGAINST A & H BOMBS 2007), and by June 1976 the Daigo Fukuryū Maru Exhibition Hall was opened by the Tokyo Metropolitan Government (DAIGO FUKURYŪ MARU EXHIBITION HALL 2015b). The Daigo Fukuryū Maru Preservation Committee later became known as "The Daigo Fukuryū Maru Foundation" (Kōeki zaidan hōjin Daigo Fukuryū Maru heiwa kyōkai 公益財団法人第五福竜丸平和協会), which carries out the day-to-day operations of the museum (HASUNUMA 2020).

Peace museums born from citizens movements or established privately by individuals in Japan are said to have brought together a wide range of materials, including survivor testimonies (APSEL 2016: 72), yet this paper demonstrates how the Daigo Fukuryū Maru Exhibition Hall does not conform to this notion. Having only 23 direct eyewitnesses to the Bikini Incident aboard the Daigo Fukuryū Maru and of whom only few were willing to speak publicly, the museum has therefore had to find alternative methods to communicate the human experience of the Bikini Incident to its visitors. While this has allowed the museum to tell a cosmopolitan narrative of the Bikini Incident (DUFFY 1997: 51), it may also result in the marginalization of the fishermen's voices within the exhibits. Museums dedicated to the atomic bombings, such as the Hiroshima Peace Memorial Museum, and the Nagasaki Atomic Bomb Museum, attract an outsized share of academic attention (CHEN 2012; YOSHIDA / BUI / LEE 2016; ZWIGENBERG 2021).[2] This paper focuses on a lesser-studied example of atomic memory, the Daigo Fukuryū Maru Exhibition Hall, to address the comparative lack of scholarship on lesser-known nuclear museums.

Relying on existing academic scholarship, interviews conducted with a representative of the Daigo Fukuryū Maru Exhibition Hall, presentations given by the curator, and observations from a site visit to the Exhibition Hall which took place in June 2022, this paper examines how the Daigo Fukuryū Maru Exhibition Hall continues to pass on the memory of the incident to current and future generations through its exhibits and activities, with a focus on eyewitness testimony. In her book *The Politics of Display: Museums, Science, Culture* Sharon MACDONALD (1998: 2) writes that "Exhibitions tend to be presented to the public [...] as unequivocal statements rather than as the outcome of particular processes and contexts", therefore it was necessary to interview the curator in order to reveal the hidden processes behind the construction of the exhibitions at the Daigo Fukuryū Maru Exhibition Hall. The interview with the Daigo Fukuryū Maru representative was initially conducted via email in June 2022. Given that the interviewee is based in Japan (a nine-hour time difference from the UK), email interviews were chosen for the ability to conduct the interview asynchronously, as well as to "allow for prolonged engagement" with the participant (HAWKINS 2018: 494–95), and indeed further email exchanges took place in November 2022 and May 2023.

The Daigo Fukuryū Maru Exhibition Hall was selected as a case study for two reasons. Firstly, it is different to the other nuclear museums (including the previously mentioned museums in Hiroshima and Nagasaki, as well as those memorializing the nuclear meltdown in Fukushima on March 11, 2011) in Japan which offer visitors to listen to in-person *kataribe* 語り部 (storytellers) narrating their experiences of the nuclear, in that, with the death of ŌISHI Matashichi in 2021, it is now "no longer possible to hear the testimonies of the Daigo Fukuryū Maru crew members directly" (Interview with Daigo Fukuryū Maru Exhibition Hall representative, June 2022). Secondly, "unlike many other facilities that were created to pass on the experiences of those involved [...] the crew of the Daigo Fukuryū Maru were not involved in any way" in the process of the establishment of the Exhibition Hall (Interview with Daigo Fukuryū Maru Exhibition Hall representative, June 2022). Therefore, this analysis of the case study of the Daigo Fukuryū Maru Exhibition Hall illustrates how eyewitness testimony is displayed in nuclear institutions where the *hibakusha* have not played an active role.[3]

2 EYEWITNESS TESTIMONY IN MEMORIAL MUSEUMS

Eyewitnesses to historical events hold significant authority in public memory and in museums (DE JONG 2018: 150). However, one of the biggest challenges facing memorial museums today concerns the question of how they can continue to provide visitors with the opportunity to listen to authentic eyewitnesses' first-hand experiences, due to the unfortunate reality that as time passes, survivors pass away (ARNOLD DE SIMINE 2013: 28). The use of eyewitness testimony in memorial museums is well-researched in Holocaust literature (HOLTSCHNEIDER 2007: 87–89), and these studies have found that eyewitness testimony is often a central feature of memorial museum exhibits, to the extent that "it seems almost surprising not to find video testimonies in memorial museums" (DE JONG 2018: 162). Eyewitness testimony is used in memorial museums because it is "strategically useful for its ability to express first-hand witnessing" (WILLIAMS 2007: 170). James YOUNG (1999: 13) describes individual experiences as a "window" to understanding historical events. It has been shown that visiting memorial museums can encourage visitors to empathize with those who have suffered (LANDSBERG 2018: 148–49), and it is argued that hearing first-hand experiences can motivate visitors to take political and social action (HIRAI 2015: 82–83).

As the only country to have suffered the impact of nuclear bombs used during wartime in Hiroshima and Nagasaki, and with the meltdown at the Fukushima Daiichi Nuclear Reactor in March 2011, Japan is home to several memorial museums which offer presentations of the nuclear. Most famously, the Hiroshima Peace Memorial Museum (Hiroshima heiwa kinen shiryōkan 広島平和記念資料館), the Nagasaki Atomic Bomb Museum (Nagasaki genbaku shiryōkan 長崎原爆資料館) (both established in 1955, ten years after the bombings) and the Great East Japan Earthquake and Nuclear Memorial Museum in Fukushima (Higashi Nihon daishinsai, genshiryoku saigai denshōkan 東日本大震災・原子力災害伝承館) (opened in 2020, nine years after the event). The Hiroshima Peace Memorial Museum, The Nagasaki Atomic Bomb Museum and the Great East Japan Earthquake and Nuclear Disaster Memorial Museum all make use of individual experiences of atomic bombing / radiation. In Hiroshima, the story of SASAKI Sadako 佐々木禎子, a nine-year-old girl at the time of the bombing who eventually died from leukemia as a result of exposure to the radiation, illustrates the suffering of children during the atomic bombing both as a monument in the Hiroshima Peace Park, as well as featuring as an exhibit along the museum route as I could confirm on my research trip in 2022. Artefacts are labelled with the names and are often accompanied by images of the people who once owned them, to add greater emotional weight to them. In Nagasaki, there is an exhibit dedicated to Dr. NAGAI Takashi 永井隆, a Catholic physician who experienced the bombing, and author of "The Bells Toll for Nagasaki" (*Nagasaki no kane* 長崎の鐘, 1949), an account of his experience (BRAW 1991: 95). In Fukushima, *kataribe* are paid by the museum to recount their experiences to visitors from across Japan and worldwide, which not only "tell us what people were thinking about at the time of the disaster", but "[i]t is also possible to get a real-time view of Fukushima Prefecture as it heads toward reconstruction" (ANDO 2021: 6). The use of individual narratives of nuclear disasters also serves to remind visitors that "it was not an unspecified number of anonymous people who experienced the event [...] at that time, but many unique people who could not be replaced" (ANDO 2021: 8).

3   EXAMINING A-BOMBED ARTEFACTS

In the context of a declining population of eyewitnesses available to speak at memorial museums, some museums prefer to employ an object-led display, such as the Imperial War Museum in London (HOLTSCHNEIDER 2007: 90; HAWKINS 2020: 212–13). Furthermore, some museums which originally foregrounded individual experiences have transitioned to rely more on objects (SHENKER 2010: 38–39). This is because original objects evoke authenticity, which is a crucial element of memorial museums—especially those which memorialize events which are denied or refused to be acknowledged, and so "the requirement to be exact is even more important" (HARTMAN 1995: 193). In fact, according to Ran ZWIGENBERG (2021: 53) the "continuing decline of *hibakusha* numbers increased the importance of A-bomb artefacts." The decline of *hibakusha* has motivated curators at the Hiroshima Peace Memorial Museum to reimagine their exhibition strategy, and it appears that more weight is being given to the original "A-bombed artefacts" (*genbaku shiryō* 原爆資料) as the only remaining authentic testifiers to the atomic bombings. In fact, during its most recent renovation in 2019, the Museum stated that their new approach is to use artefacts and present information (HIROSHIMA PEACE MEMORIAL MUSEUM 2018).

It appears that the Daigo Fukuryū Maru Exhibition Hall's approach follows in the footsteps of the precedent set by the Hiroshima Peace Memorial Museum, as its new strategy is to rely on artefacts, particularly the ship itself to communicate the history of the Bikini Incident. Ōishi Matashichi did not reveal his identity as a *hibakusha* until 1983 (DAIGO FUKURYŪ MARU EXHIBITION HALL 2021: 11). While the Daigo Fukuryū Maru Exhibition Hall believes that "it is important for the next generation to pass on what the survivors were trying to convey to the next generation" through "making the most of the testimonies recorded and collected to date [...] rather than simply relying on the testimonies of the crew members, it is also necessary to utilize a variety of materials" (Daigo Fukuryū Maru Exhibition Hall Interview, June 2022). Highlighting that the origins of the museum lie in the "public campaign" to rescue the abandoned Daigo Fukuryū Maru, the museum's curators "believe that it is necessary to emphasize the power of the message of the preserved actual vessel" (Daigo Fukuryū Maru Exhibition Hall Interview, June 2022).

Although the A-bombed artefacts on display at the museums in Hiroshima and Nagasaki, and the actual Daigo Fukuryū Maru vessel do act as a testament to the capabilities and dangers of nuclear bombs, nuclear museums which memorialize nuclear radiation cannot rely on artefacts alone to tell their stories. However, "because the damage is so great or invisible" (ANDO 2021: 7), emotion and the human experience can be difficult to convey in museum exhibits which put the nuclear on display. Yet, the curators of the Hiroshima Peace Memorial Museum realized during its 1986 renovations[4] that "it was clear that the question of the final meaning of the Hiroshima bombing could not be answered by scientific data. It had to do with the human experience and suffering, that is, emotions" (SCHÄFER 2008: 162). Similarly, director of the documentary *Day of the Western Sunrise*, Keith REIMINK (2018), stated in the film's bonus features that although they had an abundance of footage that could have been used for the film, he ultimately realized that the key aspect of the documentary was the "interviews with the fishermen" and so made this the focal point (REIMINK 2020). This paper argues that the Daigo Fukuryū Maru Exhibition Hall's approach to rely on the ship itself to tell the story of the Bikini Incident could contribute to the continuation of the historical marginalization of the fishermen's narratives, which started almost immediately after news broke of the Bikini Incident (HOMEI 2013: 219).

4 UNDERSTANDING BARRIERS TO TESTIMONY

Eyewitness testimonies are used in memorial museums to illustrate individual experiences of historical events. However, it has not always been easy for *hibakusha* to come forward and speak publicly about their experiences. The following barriers to testimony may contribute to a comparative lack of eyewitness testimony in the Daigo Fukuryū Maru Exhibition Hall, compared to that found in other Japanese memorial museums.

4.1 External Pressures

One factor which prevents individuals from speaking out publicly about their experiences is pressure from external sources. This took the form of legal censorship after the atomic bombings of Hiroshima and Nagasaki. After Japan's surrender in 1945, the Allied Forces occupied Japan until 1952. The Civil Censorship Detachment, headed by SCAP General Douglas MACARTHUR, was in charge of censorship operations. In the beginning all print publications were subject to inspection but media censorship was relaxed in 1948 (DOWER 1995: 289). Until then, it was prohibited to share information about what had taken place in Hiroshima, and so information about the bombing was extremely limited (BRAW 1991: 103). This, too, meant that there were not many "possibilities for the survivors in Hiroshima and Nagasaki to tell their version of the atomic bombings" (BRAW 1991: 142). It was only after the end of the Allied Occupation of Japan that "books, articles, poems and personal recollections of *hibakusha*" (DOWER 1995: 287) could begin to emerge. In terms of the Bikini Incident, there was unofficial self-censorship of eyewitness testimonies denouncing the use of nuclear technology, because the government had adopted a "pro-nuclear campaign" as a way of revitalizing the Japanese economy through the "peaceful use" (ICHIYO 2013: 176–77) of nuclear energy. This meant that *hibakusha* were under "pressure to remain silent" (CHO 2019: 128), and so eyewitness testimony from the event took a long time to surface (ŌISHI 2011: 87–88). Additionally, Aya HOMEI (2013: 220) argues that international politics between Japan and the United States played a role in silencing the fishermen's voices, due to contention surrounding KUBOYAMA's cause of death. HOMEI specifies that "[t]he triangular relationship among medical scientists, authorities and the media provided the foundation on which the public understanding of the fishermen's illnesses following the Lucky Dragon incident was shaped. This development sidelined the victims while granting more authority to Japanese medico-scientific researchers" (HOMEI 2013: 220). This intentional silencing was illustrated by the fishermen's transfer to a hospital in Tokyo, which was "a decisive step in a process in which patients' whereabouts were increasingly controlled by government and medical authorities and the channels through which victims could communicate their suffering in public became increasingly narrow and restricted" (HOMEI 2013: 219).

4.2 Discrimination

In addition to censorship, another barrier facing *hibakusha* who wanted to pass on their experience, was the discrimination against them. At the time of the Hiroshima bombing and for many years afterwards, it was believed that radiation sickness could be caught through proximity to those suffering with effects of radiation poisoning (NAKAZAWA 2008: 8). Those who experienced the atomic bombing often chose not to disclose their status as *hibakusha*, due to fears of social marginalization and intense discrimination (YONEYAMA 1999: 88–89). These attitudes towards nuclear radiation have far from subsided, and parallels have been drawn to the stigma faced by *hibakusha* from Hiroshima, and the ongoing discrimination faced by residents of Fukushima (WILSON / FUNAKOSHI 2017). The media portrayal of the fishermen as "objects of inquiry" (HOMEI 2013: 217) only contributed to their exclusion from society. In his autobiography, ŌISHI Matashichi recalls when he watched a television interview with a young woman from his hometown being asked whether or not she would consider marrying one of the Daigo Fukuryū Maru fishermen. "Marry one of them? Out of the question!", she replied (ŌISHI 2011: 42). After being released from hospital in 1955, ŌISHI hid in Tokyo to avoid the constant discrimination he faced in his hometown of Yaizu, both due to his *hibakusha* status, and for accepting compensation from the Japanese government as a result of his ordeal (ŌISHI 2011: 58–59).

The Daigo Fukuryū Maru Exhibition Hall was conceptualized in 1969, and opened to the public in 1976, meaning that the museum's curators and the citizens' group who preserved the Daigo Fukuryū Maru were designing the exhibitions in the museum without having been able to draw from the first-hand experiences of the Daigo Fukuryū Maru crew to create their displays. Indeed, it was not until 30 years after the Bikini Incident that ŌISHI first began to speak publicly about his experiences, giving his first talk at the Exhibition Hall in 1984.

4.3 The gap between the "experienced" and "non-experienced"

Even once *hibakusha* have decided that they will speak publicly about their experiences, they still face barriers to testimony. On a global scale, few people have experienced atomic bombings, or the effects of nuclear radiation, and so "[t]o date, there is no common understanding on how to understand such disasters" (ANDO 2021: 2). Ryoko ANDO (2021: 5) refers to this as a "groove" between those who have and have not experienced nuclear disaster, which makes it difficult for those who have experienced nuclear disaster to put it into words, and for those listening to truly understand. Writing about eyewitnesses to the Hiroshima bombing, Lisa YONEYAMA (1999: 90) writes: "At every utterance, storytellers are confronted with their language's inability to reconstruct the past as they believe they really experienced it". This is something that ŌISHI experienced firsthand when he began to give public talks about his experiences. In his 2011 autobiography, he writes "[...] those with no knowledge of the Bikini Incident have a hard time grasping it in only thirty minutes or an hour. It's already been forty-six years since it happened, and it involved complex geo-politics" (ŌISHI 2011: 3–4). In an interview published in the NHK archives, ŌISHI explains how even he himself struggled with his own identity as a *hibakusha*, and his initial reaction to hearing the news that the exhibition hall had been built was "I wanted to tell them to stop. If the boat was put there, then the fact that I am a *hibakusha* cannot be erased" (ŌISHI 2013).

The above three sections show that *hibakusha*, including the crew members of the Daigo Fukuryū Maru, encounter significant issues which may impede them from speaking publicly about their experiences immediately from the initial occurrence of the event, and this can continue many years once the event is considered to be resolved. External pressures from the Japanese government, while carefully curating its relationship with the United States government took away the crew's autonomy over their own bodies and decision-making, transforming "[...] what was initially a private and intimate corporeal experience of the victims into something discussed in public" (HOMEI 2013: 217). Once released from hospital, the crew faced social marginalization, both as a result of their illness and their acceptance of compensation (ŌISHI 2011: 58–59). Finally, even after ŌISHI revealed his *hibakusha* status, he still had to overcome the "gap" (ANDO 2021: 7) in experience between himself and his audience when telling his story in public. The next chapter looks at how these barriers to testimony have had implications on the display of eyewitnesses and their testimonies in the Daigo Fukuryū Maru Exhibition Hall.

5 The Daigo Fukuryū Maru Exhibition Hall

It is often the case in Japan that memorial museums come to fruition through the activities (e. g., donations, collection of materials, recording of testimonies) of eyewitnesses, or from groups representing those eyewitnesses, including families and supporters (APSEL 2016: 72). For example, the Hiroshima Peace Memorial Museum started life as a collection of artefacts by local geologist NAGAOKA Shōgo 長岡省吾, who went on to become the museum's first director in 1955. NAGAOKA walked through the ruins of Hiroshima each day, while suffering from radiation poisoning (HIROSHIMA PEACE MEMORIAL MUSEUM 2005). All the attendant staff at the Fukushima Denshōkan listed on the museum's website have a personal experience of the disaster. Eyewitnesses'—or groups who represent them—involvement in the establishment of these museums often results in their experiences having a prominent role within the museum's narrative and displays, including access to eyewitness testimonies (APSEL 2016: 72). It is also important to consider the museum's funding source—whereas the Hiroshima Peace Memorial Museum, the Nagasaki Atomic Bomb Museum, and the Fukushima Denshōkan receive governmental support (SCHÄFER 2008 156–57; DIEHL 2014: 515; TAKEUCHI / TAKAHASHI 2021), the Daigo Fukuryū Maru Exhibition Hall relies on a combination of funding from both the Tokyo Metropolitan Government and donations from the public (Interview with Daigo Fukuryū Maru Exhibition Hall representative, June 2022). Although the Daigo Fukuryū Maru Exhibition Hall was established by a citizen volunteer group (TANIGAWA 2015: 252), their main motivation was the preservation of the ship, rather than prioritizing the voices of its crew. This is made apparent in the organization of the museum space.

The museum is made up of six exhibits which are displayed on wall panels, following a route around the Daigo Fukuryū Maru itself. This involves climbing a set of stairs which allows visitors to see the boat's deck. The first exhibit is an introduction to the museum in the form of a video that plays at the entrance. However, at the time of visiting in June 2022 this was suspended due to the COVID-19 pandemic. The second exhibit covers the "Tragedy of the Daigo Fukuryū Maru" (*Daigo Fukuryū Maru no hibaku* 第五福竜丸の被ばく), including the effects of the nuclear test on the 23 crew members. It displays images of the crew during their hospitalization, with graphic photos of crew member MASUDA Sanjirō's 増田三次郎 facial discoloration, HANDA Shirō's 半田四郎 radiation burns on his stomach, and the backs of MISAKI Susumu's 見崎進 hands. There is significant space dedicated to KUBOYAMA, with his diaries on display, accompanied by an emotional photograph of KUBOYAMA's widow and children at his funeral service. There are also excerpts from newspapers at the time, one being a letter written by a mother expressing her child's fear of playing outside in the "radiation rain" (*hōshanōu* 放射能雨). However, the most hard-hitting exhibition is simply a list of the crew's names, separated into the living and the dead. The list of the dead gives additional information such as their age at the time, age of death, date of death and cause of death—most of which are related to cancer. On the list of the living, masking tape covers the sections which used to show the names of ŌISHI Matashichi and IKEDA Masaho, who have since been added on to the list of the dead. The third and fourth exhibits examine the impact of the Bikini Incident on others, for example "the damage of numerous fishing boats" (*takusan no hisaisen* たくさんの被災船), and "nuclear testing around the world" (*sekai no kakujikken to kakuheiki no haizetsu* 世界の核実験と核兵器の廃絶). The fifth exhibit moves away from the nuclear, and instead shows the Daigo Fukuryū Maru as a fishing boat, and finally the sixth exhibit gives an overview of the history of the Exhibition Hall.

The stated aim of the Daigo Fukuryū Maru Exhibition Hall is to call for "peace and a ban on nuclear weapons" (*gensuibaku no kinshi to heiwa o negau* 原水爆の禁止と平和を願う (DAIGO FUKURYŪ MARU EXHIBITION HALL 2015b). This is reflected in the museum's slogan, to "keep voyaging for a nuclear free future" (*Daigo Fukuryū Maru wa gensuibaku no nai mirai e to kōkai o tsuzukemasu* 第五福竜丸は原水爆のない未来へと航海をつづけます). The Exhibition Hall not only houses the Daigo Fukuryū Maru itself but also exhibits materials and educates visitors about the Bikini Incident. In addition, for 30 years the museum regularly hosted ŌISHI Matashichi, the only surviving crew member who gave public lectures about his experience of being a *hibakusha* (HASUNUMA 2020). Over the 30 years, he gave over 700 lectures about his experiences, with over 400 of those taking place at the Daigo Fukuryū Maru Exhibition Hall (Interview with Daigo

Fukuryū Maru Exhibition Hall, November 2022). In his autobiography, ŌISHI writes fondly about his lectures, but it is clear that his schedule could often be demanding:

> I go to the exhibition hall twenty or thirty times a year. I'm especially busy in the spring, the peak season for school trips. On May 14, 1998, I spoke to students from three schools in one day, and the next day I spoke about my experience at the Prince Garden Hotel in Tokyo. At sessions at the museum and elsewhere, I've met and spoken with students from all over the country. (ŌISHI 2011: 2–3)

This suggests that listening to Ōishi recount his experience of the Bikini Incident was a popular activity for those looking to learn more about it, and this was often coupled with a visit to the Daigo Fukuryū Maru Exhibition Hall. However, in an interview with a representative from the Daigo Fukuryū Maru Exhibition Hall (November 2022), it was made clear that Ōishi "[...] was not affiliated with the museum but was a private individual", reiterating that although the museum was a space where people could choose to hear Ōishi's story, it was not an experience that the museum itself provided for its visitors. This shows consistency with the museum's approach to prioritize the use of artefacts, and the Daigo Fukuryū Maru itself to illustrate the history of the Bikini Incident.

Video testimony of the fishermen was not added until the museum underwent renovations in 2018 and 2019, when testimony of then 84-year-old Ōishi Matashichi (IWAMA 2018) was introduced to the museum's upper level. The main motivation for these renovations was to conduct vital repairs to the flooring. Due to Yumenoshima being reclaimed land, the floor had begun to tilt in some places which required significant modifications to the flooring and the ceiling.[5] Additionally, "wall insulation was replaced to reduce temperature and humidity changes in the building" (Interview with Daigo Fukuryū Maru Exhibition Hall representative, June 2022). Yet it was not only the building which houses the Daigo Fukuryū Maru which underwent extensive changes during the renovations, but so did the exhibition itself, with "a third of the items from the regular display [being] switched out for new content" (IWAMA 2018).

Up until 2021, there were three surviving crew members of the Daigo Fukuryū Maru, however the other two crew members do not speak about their experiences. Unfortunately, Ōishi passed away in March of 2021 (THE JAPAN TIMES 2021), and therefore "it is no longer possible to hear the testimonies of the Daigo Fukuryū Maru crew members directly" (Interview with Daigo Fukuryū Maru Exhibition Hall representative, June 2022). The museum added the video testimony of Ōishi in 2019 "due to his advanced age" making it more difficult for him to travel to the museum and speak in person (HASUNUMA 2020). The video testimony runs for seven minutes and twelve seconds and involves Ōishi describing his experience of the event in a series of answers to questions which appear on the top left of the screen. The filming appears to have taken place in several locations around the museum, often with a backdrop of the exhibits, featuring a large-scale image of an exploded atomic bomb, or the Daigo Fukuryū Maru itself. The testimony is interspersed with images of Ōishi and written text explaining further details about his story and setting the scene and context for Ōishi's answer to the next question. This helps the video flow as one piece, and the change in scene helps the viewer to keep focus. Ōishi himself was aware of the inherent tension that he could not fully explain his story within the time allotted to him to give a testimony, yet he also knew that it would be difficult for him to keep people's attention, particularly children (Ōishi 2011: 4). The video testimony is currently only available in Japanese, and it is only accessible on-site in the museum, because the video is copyrighted by the Tokyo Metropolitan Government and cannot be shown to the public online under the museum's authority. However, the museum is planning to produce its own testimony video and make this available online in the future (Interview with Daigo Fukuryū Maru Exhibition Hall representative, June 2022).

What is most surprising about the testimony is its location within the museum route. At the time of my visit to the Daigo Fukuryū Maru Exhibition Hall in June 2022, the video was not located along the museum's main route on the ground in the section on "victims" (*norikumiin hibaku to Kuboyama-san no shi* 乗り組員被ばくと久保山さんの死), instead, it was located on the museum's upper level next to the exhibit about the Daigo Fukuryū Maru as a fishing boat. The monitor screen is relatively small, with only two seats in front of it for visitors to sit and watch the testimony. Given how popular Ōishi's talks were at the museum, it is surprising that the display of the testimony video severely limits audience numbers, in effect, leaving Ōishi's testimony at the sidelines. The museum experienced a severe drop in visitor numbers in 2020 as a result of the museum being closed for 254 days during the COVID-19 pandemic with 30,615 visitors, a 27.5 % decrease from the previous year (Daigo Fukuryū Maru heiwa kyōkai 2021). However, the testimony installation has remained unchanged since it was added to the museum in 2019, suggesting that the limited viewing spots were not a result of social distancing measures due to COVID-19 restrictions. The museum is exploring new ways of increasing visitor numbers, especially amongst school-age children and university students, including hosting concerts, film screenings, performances and workshops for schoolchildren, and volunteering programmes for university students (Hasunuma 2020).

Other new additions included a 360-degree view of the inside of the ship, which is not accessible to the public, and "new explanations of the displays in English, Chinese and Korean" (Iwama 2018). This may be part of the museum's new strategy to "expand the entrance" and have a "broader range of people meet the Daigo Fukuryū Maru" making the exhibition more "universal", and "accessible regardless of language or disability" (HASUNUMA 2020). This is perhaps related to the more cosmopolitan story that the museum wants to tell. In an interview with the museum (June 2022), it was emphasized that "[t]he crew of the Daigo Fukuryū Maru consisted of 23 people, a very small and limited number", whereas "from the perspective of the damage caused by nuclear testing, there are many different types of victims and forms of damage". For this reason, in exhibit four, "Nuclear Testing Around the World", the museum exhibits materials related to the Marshall Islands and other areas affected by the nuclear testing around the Bikini Atoll. Terence DUFFY (1997: 57) argues that through this display, the museum "makes it clear that the survivors of Bikini Atoll and so many other places have the right to be considered as Hibakusha". Yu-Fang CHO (2019: 130–31) disagrees, and instead claims that the display actually "illustrates colonial forgetting", with the photographs of Japanese victims appearing on black and white backgrounds, while "the photos of Marshallese victims are framed by the classic image of an idyllic, scenic beachfront as a site of leisure and entertainment". However, in January to March of 2023, the museum held a special exhibition called "Hibakusha around the World—Under Nuclear Development and Nuclear Testing" (TŌKYŌ SHINBUN 2023). This special exhibit introduced 16 new testimonies to the museum, with an accompanying black and white photo of the *hibakusha*, including a female eyewitness from Utah in the United States, and indigenous people of islands in the Arctic Ocean (TOKYO SHINBUN 2023). In 2020, there was also a special exhibition of 1000 letters written by members of the public to the fishermen while they were hospitalized, displayed to communicate the thoughts and feelings of the Japanese population of the time (HASUNUMA 2020). These special exhibitions are illustrative of the museum's aim to cater to a more global audience, highlight that the threat of nuclear weapons exists worldwide, and to "deal with and convey the memories" of a wide variety of people involved in the Bikini Incident, not just the crew members of the Daigo Fukuryū Maru (HASUNUMA 2020).

Although it is no longer possible to hear ŌISHI speak directly, the museum also has volunteer guides and curators "who explain the history of the Daigo Fukuryū Maru to groups visiting the exhibition hall basically in the same format" (Interview with Daigo Fukuryū Maru Exhibition Hall representative, November 2022). This talk usually lasts for around 15–20 minutes and takes place before viewing the exhibition, because the museum's curators believe that visitors get a more in-depth understanding through viewing and reading the exhibitions, encouraging them to spend longer in the museum. Upon request, visitors can ask the guides to speak for around 40–60 minutes, which is approximately the same length of time of one of ŌISHI's lectures. However, these talks are given in Japanese only, and interpreters should be arranged independently by the visitor ahead of their visit (DAIGO FUKURYŪ MARU EXHIBITION HALL 2015b). Therefore, despite the museum's attempts to universalize the museum, making it more accessible to non-Japanese speaking visitors, two of the key elements of the museum visit—the video testimony and the group guide talks—are only available in Japanese. This may mean that the museum is missing out on an opportunity to encourage non-Japanese speaking visitors to take the social action they are hoping for, i. e., the abolition of nuclear weapons and a peaceful world, through empathizing with individual eyewitnesses.

6 CONCLUSION

Although perhaps not as well-known worldwide as the nuclear museums in Hiroshima and Nagasaki, the Daigo Fukuryū Maru Exhibition Hall contributes to global discourses on abolition of nuclear weapons and efforts to promote world peace, through educating its visitors on the dangers of nuclear weapons made clear in the Bikini Incident. While some memorial museums choose to highlight the voices of individual eyewitnesses in their displays, the Daigo Fukuryū Maru Exhibition Hall is taking an alternative approach, by communicating the story of the Bikini Incident through placing more emphasis on the significance of the ship and encouraging visitors to read traditional text panels (Interview with Daigo Fukuryū Maru Exhibition Hall representative, June 2022).

The museum may take this approach because it was established without the involvement of any of the crew, and for many years did not have access to eyewitness testimony of the event, due to the "barriers to testimony" which prevented eyewitness ŌISHI Matashichi from speaking out about his experiences publicly until 30 years had passed since the Bikini Incident. Similarly to those who experienced the atomic bombings of Hiroshima and Nagasaki, ŌISHI faced social barriers, such as discrimination based on his status as a *hibakusha* and the fear of contagion, as well as the envy of those around him that he was able to collect financial compensation from the government. In addition, ŌISHI also encountered political barriers to sharing his testimony, through the suppression of information about the Bikini Incident in contemporaneous media, particularly the first-hand experiences of the fishermen, for fear of aggravating the existing political tension between Japan and the United States. Finally, even once he had revealed his *hibakusha* status and three decades had passed since the Bikini Incident, ŌISHI faced the challenge of communicating a complex situation and his personal trauma, in an attempt to bridge the gap in experience between himself as a *hibakusha* and those who have no personal memory of the dangers of nuclear weapons. It is clear to see that Ōɪsʜɪ was, and continues to be, important to the Daigo Fukuryū Maru Exhibition Hall, as his testimony was added to the permanent exhibition in 2019. However, the relationship between Ōɪsʜɪ and the museum was 'unique' (Interview with Daigo Fukuryū Maru Hall representative, May 2023) and complex, as unlike in the Fukushima Denshōkan, for example, where *kataribe* are paid to speak to museum audiences, Ōɪsʜɪ's retellings of his experiences were his own personal activities and not organized by the museum.

Where the case of the Daigo Fukuryū Maru Exhibition Hall differs to that of Hiroshima, Nagasaki and Fukushima, is the availability of eyewitness testimony. While there are still over 118,935 officially recognized *hibakusha* from the atomic bombings of Hiroshima and Nagasaki, and many more who have a personal experience of the nuclear meltdown at Fukushima, Ōɪsʜɪ has passed away, leaving no surviving crew members of the Daigo Fukuryū Maru to communicate their experiences to visitors at the museum. Furthermore, while the Hiroshima Peace and Culture Foundation has recorded over 1000 testimonies since 1986 (Hɪʀᴏsʜɪᴍᴀ Pᴇᴀᴄᴇ Mᴇᴍᴏʀɪᴀʟ Mᴜsᴇᴜᴍ 2019), the Daigo Fukuryū Maru Exhibition Hall has only one video testimony of Ōɪsʜɪ available to view on site. The lack of first-hand eyewitnesses on the Daigo Fukuryū Maru is yet another 'barrier to testimony' and has resulted in the Daigo Fukuryū Maru Exhibition Hall taking alternative approaches to displaying testimony.

For example, the museum hosts special exhibitions to highlight that *hibakusha* exist all over the world and show that there were a wide range of victims of the Bikini Incident, including other fishermen (Lᴏᴡ 2020: 99) and inhabitants of the Marshall Islands (Tᴇᴀɪᴡᴀ 2010: 18). New explanations provided in English, Korean and Mandarin have been introduced to make the exhibits more accessible, regardless of language. However, crew member Ōɪsʜɪ Matashichi's eyewitness testimony is only available to be viewed at the museum and only in Japanese. Previous research on memorial museums has shown that engaging with eyewitness testimony can aid visitors to empathize with people who have suffered disasters (Jᴏɴᴇs 2017: 147). The museum could therefore continue to expand its audience by going forward with its plans to produce its own video testimony with non-Japanese subtitles and making it available online or providing all visitors with links to where testimony can be viewed externally, such as NHK's archive.

In 2010, Ōɪsʜɪ Matashichi went to New York for a United Nations conference reviewing the Nuclear Non-Proliferation Treaty. "Around 10,000 Japanese people attended the conference", he said, "and not one of them mentioned the Bikini Incident [...]. The Bikini Incident has been lost in the depths of history" (Ōɪsʜɪ 2013). At the time of writing, the museum is currently undertaking preparations to commemorate the 70th anniversary of the Bikini Incident, which will take place in 2024. Anniversaries can be a significant moment in memorialization (NORA 1989: 12) and can encourage those involved in the management of memorial museum displays to think about ways to renew public interest and discussion around memories of the event in question (HOOD 2019: 14–15). Although it remains to be seen what impact this will have on the Daigo Fukuryū Maru Exhibition Hall, in the context of the current conflict in Russia and Ukraine, once again the fear of nuclear weapons is brought to the forefront (TOKYO SHINBUN 2023). Even now, 70 years after the Bikini Incident, the message of the Daigo Fukuryū Maru Exhibition Hall, "to keep voyaging for a nuclear-free future" is as relevant and important as ever.

NOTES

1 Although in both Japanese and English, the event is widely referred to as Daigo Fukuryū Maru jiken 第五福竜丸事件 (The Lucky Dragon Incident) derived from the Japanese name of the boat, the Daigo Fukuryū Maru Exhibition Hall prefers to use the term Bikini jiken ビキニ事件 (Bikini Incident), due to the fact that the "Daigo Fukuryū Maru was not the only vessel damaged by the hydrogen testing at Bikini Atoll" and so "the testing and its impacts must not be considered in a limited way" (DAIGO FUKURYŪ MARU EXHIBITION HALL 2015b). Therefore, the event will be referred to as the "Bikini Incident" throughout this paper.
2 CHEN (2012) analyses visitor comment books at the Hiroshima Peace Memorial Museum, YOSHIDA / BUI / LEE (2016) write about Hiroshima and Nagasaki as dark tourism destinations, and ZWIGENBERG (2021) focuses on the display of A-bombed artefacts at the Hiroshima Peace Memorial Museum.
3 It is important to acknowledge here that while ŌISHI and the other crew members did not have an active role in the establishment of the museum or the preservation of the Daigo Fukuryū Maru, ŌISHI was later made a member of the museum's board of trustees, to show that the crew were at the centre of the museum's activities. While this meant that he was therefore in a position to 'speak out on management and other matters', he did not have influence over the content of the exhibition. The secretariat of the museum is responsible for the exhibition content and the museum's policies, and the secretariat is elected by the board of directors (Interview with Daigo Fukuryū Maru Exhibition Hall representative, May 2023).
4 The Hiroshima Peace Memorial Museum has been remodelled several times since it was opened in 1955. The first renovation took place from 1972–75, the second 1989–94, and most recently from 2017–19. For details, see SCHÄFER (2008) and HIROSHIMA PEACE MEMORIAL MUSEUM (2019).
5 Due to a long-standing issue with leaks in the ceiling, the ship was getting wet and had sustained water damage.

References

ANDO, Ryoko (2021): "How to overcome the difficulty of talking about the experience of a nuclear disaster". In: *Annals of the ICRP*: 1–8.

ARNOLD DE SIMINE, Silke (2013): *Mediating Memory in the Museum. Trauma, Empathy, Nostalgia.* Basingstoke: Palgrave Macmillan.

APSEL, Joyce (2016): *Introducing Peace Museums.* London: Routledge.

BRAW, Monica (1991): *The Atomic Bomb Supressed: American Censorship in Occupied Japan.* London: M. E. Sharpe Publishers.

BULLETIN OF THE JAPAN COUNCIL AGAINST A & H BOMBS (2007): "Daigo Fukuryū Maru wa kōkai o tsuzukeru: Kakuheiki ga nakunaru hi made 第五福竜丸は航海を続ける：核兵器がなくなる日まで"; https://www.antiatom.org/Gpress/?p=137 (last access 2021 / 11 / 21).

CHEN, Chia-Li (2012): "Representing and interpreting traumatic history: a study of visitor comment books at the Hiroshima Peace Memorial Museum". In: *Museum management and curatorship* 27 (4): 375–92.

CHO, Yu-Fang (2019): "Remembering Lucky Dragon, re-membering Bikini: worlding the Anthropocene through transpacific nuclear modernity". In: *Cultural Studies* 33 (1): 122–46.

DAIGO FUKURYŪ MARU EXHIBITION HALL (2015a): "Exhibition Content"; http://d5f.org/en/en-tenji (last access 2021 / 11 / 21).

DAIGO FUKURYŪ MARU EXHIBITION HALL (2015b): "Daigo Fukuryū Maru"; http://d5f.org/en/en-about (last access 2021 / 11 / 21).

DAIGO FUKURYŪ MARU HEIWA KYŌKAI 第五福竜丸平和協会 (2021): *2020 Annual Report / 2020-nenji hōkokusho* 2020年次報告所. Tōkyō: Daigo Fukuryū Maru heiwa kyōkai.

DAIGO FUKURYŪ MARU EXHIBITION HALL (2022): Interview by the author, 06 / 26.

DAIGO FUKURYŪ MARU EXHIBITION HALL (2022): Interview by the author, 11 / 24.

DAIGO FUKURYŪ MARU EXHIBITION HALL (2023): Interview by the author, 05 / 10.

DE JONG, Steffi (2018): *The Witness as Object.* New York: Berghahn Books.

DIEHL, Chad R. (2014): "Envisioning Nagasaki: from 'atomic wasteland' to 'international cultural city', 1945–1950". In: *Urban History* 41 (3): 497–516.

DOWER, John W. (1995): "The Bombed: Hiroshimas and Nagasakis in Japanese Memory". In: *Diplomatic History* 19 (2): 275–95.

DUFFY, Terence (1997): "The peace museums of Japan". In: *Museum International* 49 (4): 49–54.

HARTMAN, Geoffrey (1995): "Learning from Survivors: The Yale Testimony Project". In: *Holocaust and Genocide Studies* 9 (2): 192–207.

HASUNUMA, Yusuke 蓮沼佑助 (2020): "Daigo Fukuryu Maru Exhibition Hall: The voyage for the nuclear free future". In: *The 10th International Conference of Museums for Peace*; https://sites.google.com/view/inmp-2020/p39-daigo-fukuryu-maru-exhibition-hall-the-voyage-for-the-nuclear-free-fu?pli=1 (last access 2023 / 01 / 13).

HAWKINS, Janice (2018): "The Practical Utility and Suitability of Email Interviews in Qualitative Research". In: *The Qualitative Report* 23 (2): 493–501.

HAWKINS, Vikki (2020): "Displaying marginalised and 'hidden' histories at the Imperial War Museum London: The Second World War gallery regeneration project". In: *War & Society* 39 (3): 210–14.

HIRAI, Kyonosuke (2015): "Storytelling as Political Practice: Habitus and Social Change in the Minamata Disease Movement". In: *Senri Ethnological Studies* 91: 81–99.

HIROSHIMA PEACE MEMORIAL MUSEUM (2005): "Let's look at the special exhibit"; https://hpmmuseum.jp/virtual/VirtualMuseum_e/exhibit_e/exh0507_e/exh050701_e.html (last access 2023 / 03 / 12).

HIROSHIMA PEACE MEMORIAL MUSEUM (2018): "We are undergoing renovations"; https://hpmmuseum.jp/modules/xelfinder/index.php/view/408/Ereneweng291212.pdf (last access 2023 / 02 / 21).

HIROSHIMA PEACE MEMORIAL MUSEUM (2019): "Preserving Testimony Through Video"; https://hpmmuseum.jp/modules/exhibition/index.php?action=DocumentView&document_id=163&l (last access 2023 / 03 / 12).

HOLTSCHNEIDER, Hannah K. (2007): "Victims, Perpetrators, Bystanders? Witnessing, Remembering and the Ethics of Representation in Museums of the Holocaust". In: *Holocaust Studies* 13 (1): 82–102.

HOMEI, Aya (2013): "The contentious death of Mr Kuboyama: science as politics in the 1954 Lucky Dragon incident". In: *Japan Forum* 25 (2): 212–32.

HOOD, C. (2019): "Developing a model to explain modifications to public transportation accident memorials". In: *Mortality* 25 (4): 449–69.

ICHIYO, Muto (2013): "The buildup of a nuclear armament capability and the postwar statehood of Japan: Fukushima and the genealogy of nuclear bombs and power plants". In: *Inter-Asia Cultural Studies* 14 (2): 171–212.

IWAMA, Riki (2018): "Exhibition hall lending out items related to H-bombed Daigo Fukuryu Maru boat for peace". In: *The Mainichi*, 07 / 03; https://mainichi.jp/english/articles/20180703/p2a/00m/0na/011000c (last access 2023 / 02 / 21).

JONES, Sara (2017): "Mediated immediacy: constructing authentic testimony in audio-visual media". In: *Rethinking History* 21 (2): 135–53.

LANDSBERG, Angela (2018): "Prosthetic memory: The ethics and politics of memory in an age of mass culture". In: GRAINGE, Paul (ed): *Memory and popular film*. Manchester / New York: Manchester University Press, 141–61.

Low, Morris (2020): *Visualizing Nuclear Power in Japan: A Trip to the Reactor*. Cham: Palgrave Macmillan.

MacDonald, Sharon (1998): "Exhibitions of power and powers of exhibition: an introduction to the politics of display". In: MacDonald, Sharon (ed.): *The Politics of Display: Museums, Science, Culture*. London: Routledge, 1–24.

Nakazawa, Keiji (2008): "Barefoot Gen, the Atomic Bomb and I: The Hiroshima Legacy". In: *The Asia-Pacific Journal* 6 (1): 1–12.

Nora, Pierre (1984): *Les Lieux de Mémoire*. Paris: Gallimard.

Ōishi, Matashichi 大石又七 (2011): *The Day the Sun Rose in the West: Bikini, the Lucky Dragon, and I*. Translated by Richard H. Minear. Honolulu: University of Hawai'i Press.

Ōishi, Matashichi 大石又七 (2013): "Hibaku no keiken o kataritsugu 被ばくの経験を語り継ぐ". In: *NHK sensō shōgen* 戦争証言; https://www2.nhk.or.jp/archives/shogenarchives/postwar/shogen/movie.cgi?das_id=D0001810403_00000 (last access 2023 / 02 / 21).

Reimink, Keith (2018): *Day of the Western Sunrise*. Daliborka Films.

Reimink, Keith (2020): *Bonus Features*; https://vimeo.com/ondemand/dayofthewesternsunrise/392262732?autoplay=1 (last access 2023 / 02 / 21).

Ropeik, David (2018): "How the unlucky Lucky Dragon birthed an era of nuclear fear". In: *Bulletin of the Atomic Scientists*; https://thebulletin.org/2018/02/how-the-unlucky-lucky-dragon-birthed-an-era-of-nuclear-fear/ (last access 2022 / 11 / 21).

Ryfle, Steve (2017): *Ishiro Honda: a life in film, from Godzilla to Kurosawa*. Middletown, Connecticut: Wesleyan University Press.

Schäfer, Stephanie (2008): "The Hiroshima Peace Memorial Museum and its Exhibition". In: Schwentker, Wolfgang / Saaler, Sven (eds.): *The Power of Memory in Modern Japan*. Folkestone: Global Oriental, 155–70.

Shenker, Noah (2010): "Embodied memory: the institutional mediation of survivor testimony in the United States holocaust memorial museum". In: Sarkar, B. / Walker, J. (eds.): *Documentary testimonies global archives of suffering*. New York: Routledge, 35–58.

Shindō, Kaneto 新藤兼人 (1959): *Daigo Fukuryū Maru* 第五福竜丸. Kindai Eiga Kyōkai.

Takeuchi, Yoshikazu / Takahashi, Ryusuke (2021): "New Fukushima museum changes 2011 nuclear disaster exhibition following criticism". In: *Mainichi Daily News*; https://mainichi.jp/english/articles/20210310/p2a/00m/0na/015000c (last access 2022 / 01 / 05).

Tanigawa, Yoshiko (2015): "The promotion of peace education through guides in peace museums. A case study of the Kyoto Museum for World Peace, Ritsumeikan University". In: *Journal of Peace Education* 12 (3): 247–62.

TEAIWA, Teresia K. (2010): "Bikinis and other s/pacific n/oceans". In: SHIGEMATSU, S./CAMACHO, K. L. (eds.): *Militarized Currents: Toward a Decolonized Future in Asia and the Pacific*. Minneapolis: University of Minnesota Press, 15–32.

THE JAPAN TIMES (2021): "Japanese fisherman exposed to 1954 U. S. nuclear test dies of pneumonia at 87"; https://www.japantimes.co.jp/news/2021/03/21/national/matashichi-oishi-nuclear-weapons-atomic-bombings-bikini-atoll/ (last access 2023/01/12).

TŌKYŌ SHINBUN 東京新聞 (2023): "Sekai no hibakusha no koe ni furete: Kao-shashin paneru mo: Yumenoshima, Daigo Fukuryū Maru tenjikan 世界の被ばく者の声に触れて：顔写真パネルも：夢の島・第五福竜丸展示館"; https://www.tokyo-np.co.jp/article/227039?fbclid=IwAR0fEQlh-qgPHQH4CiArk2YNX_ZEhgSmRIOjGayhFqn2wib2LNNmSPp6Bg8 (last access 2023/01/15).

WILLIAMS, Paul (2007): *Memorial museums: the global rush to commemorate atrocities*. Oxford: Oxford International Publishers Ltd.

WILSON, Thomas/FUNAKOSHI, Minami (2017): "Japanese school children who survived Fukushima meltdown are being subjected to 'nuclear bullying'". In: *The Independent*, 03/10; https://www.independent.co.uk/news/world/asia/japan-fukushima-meltdown-school-children-nuclear-bullying-second-world-war-hiroshima-nagasaki-a7622646.html (last access 2023/01/13).

YONEYAMA, Lisa (1999): *Hiroshima Traces: Time, Space, and the Dialectics of Memory*. Berkley/Los Angeles: University of California Press.

YOSHIDA, Kaori/BUI, Hong T./LEE, Timothy J. (2016): "Does tourism illuminate the darkness of Hiroshima and Nagasaki?". In: *Journal of destination marketing & management* 5 (4): 333–40.

YOUNG, James (1999): "The Anne Frank House: An accessible window to the Holocaust". In: *Anne Frank Magazine*, 13.

ZWIGENBERG, Ran (2021): "Modern Relics: The Sanctification of A-Bomb Objects in the Hiroshima Museum". In: *Holocaust and genocide studies* 35 (1): 44–62.

At the Crossroads to Oblivion—Ōta Yōko's Literature as a Counter-narrative to National History Writing in Japan

Stephan Köhn

1 INTRODUCTION: AN AUTHOR'S DILEMMA

For most Japanese readers nowadays, ŌTA Yōko 大田洋子 (1903–63) is a rather unknown writer whose works are hard to find in any regular Japanese bookstore. Most of her novels are out of print or reprinted in, if any, rather minor paperback series such as "Peace Library" (*Heiwa bunko* 平和文庫). Only two novels are actually published in a more renowned paperback series, namely "Kōdansha's Literary Library" (*Kōdansha bungei bunko* 講談社文藝文庫), and by that, at least accessible in quite a few well-sorted bookstores in Japan. The deliberate inclusion in or exclusion from a paperback series is vital for the accessibility and, by that, visibility of an author.

ŌTA Yōko started her literary career in 1929 with her novel "The Blessed Virgin in the Twilight" (*Seibo no iru tasogare* 聖母のいる黄昏), published in the magazine "Women's Literature" (*Nyonin geijutsu* 女人芸術). In the following years, she wrote several other romanticized novels about everyday life in Japan's new Asian colonies. These were perfectly in line with nationalist propaganda of that time. But August 6, 1945, marked a significant turning point in her literary work. ŌTA, who had experienced the atomic bombing of Hiroshima personally, published her first eyewitness account "An Unfathomable Deep Light" (*Kaitei no yō na hikari* 海底のやうな光) in the newspaper *Asahi shinbun* as early as August 30. ŌTA Yōko was one of the first professional writers in Japan who tried to represent a basically un-representable experience—an experience that became the intrinsic motivation to keep on writing about August 6 in the following years.[1] For eyewitness writers like ŌTA Yōko, writing about the bomb was not only a question of how to share a life-changing experience with the reader. First and foremost, it was a question of how to depict this experience in a way that it could pass the official censorship imposed by the Allied Powers between September 1945 and October 1949, and the unofficial self-censorship exercised by many concerned publishers. The publication of ŌTA's early masterpiece "City of Corpses" (*Shikabane no machi* 屍の街) probably demonstrates this dilemma best. In the 1948 edition, the chapter "Apathetic Faces" (*Muyoku ganbō* 無欲顔貌) was actually not cut by the Civil Censorship Detachment (CCD), as widely believed, but removed beforehand by the publisher Chūō kōronsha. Only in 1950, after the CCD was officially closed, the original edition could finally be published by Tōga shobō (NAGAOKA 1982b: 243–46). Censorship and self-restraint created a highly restrictive climate, a "sealed space for verbal utterance" (*tozasareta gengo kūkan* 閉された言語空間), as ETŌ Jun has described it in detail in his eponymous work (ETŌ 1984).

For most of her critics, ŌTA Yōko seemed to be quite a prolific writer of "atomic bomb literature" (*genbaku bungaku* 原爆文学) who tried to capitalize on her personal experiences of August 6 by publishing one account after the other. As a so-called "writer of atomic bomb literature" (*genbaku bungaku sakka* 原爆文学作家)—a label ŌTA Yōko always rejected—her work was generally considered to be literature that is not really literary. As the first out of a total of three "disputes on atomic bomb literature" (*genbaku bungaku ronsō* 原爆文学論争) that broke out in 1953 impressively showed, many writers and critics of that time seriously questioned atomic bomb literature's literariness and legitimacy in postwar society (KÖHN 2020: 110–11). For them, writers of *genbaku bungaku* were neither good authors nor reliable witnesses. As a result, writers like ŌTA Yōko were, as a general rule, excluded and shunned in the literary scene, as KURIHARA Sadako (1978: 183) has accurately pointed out.

From a modern point of view, the criticism writers like ŌTA Yōko were confronted with is quite confusing, since no one in Europe would seriously question the credibility and competency of a survivor giving testimony in the case of Holocaust literature. It somehow seems that early atomic bomb literature had become a kind of scapegoat in a wider national dispute on the questions of why, how and by whom August 6 should still be remembered in the first postwar decade. As literary critic ODAGIRI Hideo has fittingly remarked in his essay "Literature and the Problem of Nuclear Energy" (*Genshiryoku mondai to bungaku* 原子力問題と文学), published in 1955:

> [...] Hara Tamiki [1905–51] and Ōta Yōko wrote [...] their "Summer Flowers" (*Natsu no hana*) and "City of Corpses" immediately after the atomic bombing of Hiroshima. Although the publication of these works was delayed by one or two years due to the aforementioned official pressure, these two works were the first testimonies of artistic value written by Japanese authors. However, most of their readers, myself included, were not able to understand the significance of these two works until the Atomic Imperialism of the United States and, by that, the threat of a Third World War became blatantly obvious for all of us. (ODAGIRI 1955: 187)

But why was ŌTA Yōko's literature misunderstood by most of her contemporary readers? And why was her literature more often than not regarded as "non-literary" and "non-artistic" by most of her critics and colleagues?

In this paper, I will try to re-contextualize ŌTA Yōko and her work historically by paying particular attention to the different dynamics that characterized the first decade of postwar Japanese history for a deconstruction of the hegemonic discourses. For this purpose, I will focus on ŌTA Yōko's novel "City and People in the Evening Calm", which was finally published in 1955, a turning point in postwar history regarding Japan's nuclear legacy, as will be shown in my argumentation. In a first step, I will provide a close-reading of "City and People in the Evening Calm" to elucidate the narrative strategies ŌTA Yōko applied in this highly controversial text. In a second step, I will show how and why history writing—and by that history teaching—became a politically contested field for historians and pedagogues resulting in an irreversible concealment of the atomic bomb's destructive power and historic singularity. In a third step, I will focus on the processes and mechanisms of discursive marginalization of eyewitnesses and writers as unpleasant voices from the past. My analysis will show that survivors of August 6 were rather considered disruptive elements in Japan's culture of remembrance than living testimonies one should carefully listen to. Finally, I will discuss the findings in my conclusion to get an answer to the question of why and how ŌTA Yōko's literary work must be considered an important contribution to the articulations of the nuclear in postwar Japan.

2 LOOKING BEHIND THE CURTAINS OF POSTWAR HIROSHIMA

ŌTA Yōko's arguably most controversial work "City and People in the Evening Calm" (*Yūnagi no machi to hito to* 夕凪の街と人と) was published in three installments between 1954 and 1955.[2] In this novel, ŌTA Yōko's literary alter ego Oda Atsuko, first introduced to her readers in her novel "Half Human" (*Han ningen* 半人間, 1954), returns to her hometown, the anonymized city of "H", eight years after the atomic bombing for a reportage she is planning to write. Staying at her sister's house in Motomachi, a district in the center of "H" where war victims (*sensaisha* 戦災者) were officially allowed to live in small houses built close to each other in a confined area, Atsuko experiences a world that seems strangely left behind in postwar Japan. For Atsuko, "the actual situation in 1953" (1953-*nen no jittai* 1953年の実態)—this is also the subtitle of this novel—is both disillusioning and fascinating.

In this novel, ŌTA Yōko uses an authorial narrator to objectify her alter ego's personal experiences during her visit in "H".[3] Having been harshly criticized for her allegedly anachronistic I-novel-style (*shishōsetsu*) in some of her earlier works, ŌTA Yōko makes use of a large number of dramatis personae Atsuko encounters during her "field research". This is done to provide her readers with multiple viewpoints and to convey many different kinds of information. Before her readers' eyes, ŌTA Yōko unfolds a microcosm of "lost humans" (*shitsu ningen* 失人間) who are doomed to live in a kind of no-man's-land. Atsuko, feeling like a "lost human" herself, explains to the reader:

> These 'lost humans' do not belong to any social class. They do not differ from each other regarding their possessions or professions. 'Lost humans' have lost their ability to function as a human being. They have become a new social stratum that is now wriggling around here and there. (ŌTA 1982b: 86)

In Atsuko's romanticized view, these "lost humans" form a new egalitarian parallel society, living in municipal housing for rent, sending their children to school, and doing their best to make ends meet. Yet even this parallel society makes a clear distinction between the poor and the poorest, as Atsuko soon will realize. Around the confined area for municipal housing, illegally erected barracks stretch along the neighboring river banks, forming a refuge for all those who could not even afford to pay rent for the social dwellings in Motomachi.[4] Kubo Masasuke, one of the board members of the Motomachi Council of Welfare, describes the scale of this area as follows:

> So, when we add to the [3,200 houses] of Motomachi all the barracks illegally built between them, and the barracks stretching on both sides to the river banks and revetments, we will get a total of around 5,000 houses. And given that at least four persons live in each and every household, we will get a total of more than 20,000 persons living here. (ŌTA 1982b: 62)

The "lost humans" who live in this slum area are—in marked contrast to the inhabitants of Motomachi—virtually cut off from all benefits of civilization: electricity, gas, or water supply. As the poorest of the poor, these slum dwellers often earn their money as day laborers, dealers, or sex workers all over the city, and are subject to discrimination and marginalization. There is only one aspect all of Atsuko's "lost humans" have in common: the permanent fear of being evicted from their homes by the municipal government of "H", regardless if one is living legally in Motomachi or illegally in the adjacent neighborhood of Aioi.

At first glance, this novel seems not to be about "that very day" (*ano hi* あの日) in August 1945, the day that changed most of ŌTA Yōko's protagonists' lives. Atsuko deliberately avoids talking about any personal experiences of "that very day", being afraid that this would bring back memories of August 6 to her conversation partners—and by that to herself. Instead, it is her housemaid Ishida, who has accompanied Atsuko to the city of "H", who reminds the reader of the fact that most of the people in this district were actually atomic bomb victims—interestingly, the term *hibakusha* remains a "gap" throughout the entire text.[5]

For the reader, Ishida "sees" all the disfigured people Atsuko rejects to make "visible" by mentioning any disfiguration of her interlocutors. Atsuko herself is heavily traumatized by "that very day" and therefore avoids "seeing bodies of injured survivors or even talking about it" (ŌTA 1982b: 24). In the end, however, Atsuko has to face the fact that even if she wants only to write about "the actual situation in 1953", August 6 will remain the main subject behind everything: "Eventually, every attempt of mine to write about this city will end up writing about the atomic bomb. And by this, the atomic bomb, something that hitherto did not belong to the realm of literature, finally becomes a literary topic" (ŌTA 1982b: 203).

Atsuko's first point of contact for her "field research" is Shimasaki, a newspaper reporter she met before during an earlier visit to "H". It is Shimasaki who introduces Atsuko to a key person in Motomachi, Inagi. Inagi is a repatriate from Manchuria who is now running a small restaurant. For Atsuko, Inagi becomes a door-opener who introduces her to many other people, so that she quickly gets acquainted with quite a lot of individuals living in the slum area. Atsuko feels strangely drawn towards the inhabitants, spending most of her days strolling around the barracks, talking to people of formerly different social classes, professions or nationalities who live—at least in Atsuko's idealized view—harmoniously together as a community of fate. This parallel world becomes a temporary refuge for Atsuko, who makes no secret of her feelings of alienation and exclusion from postwar society.

Through the eyes of her alter ego Atsuko, ŌTA Yōko looks behind the curtains of the modern city of "H" for her readers. For the inhabitants of the slum, the whole city of "H" has turned into a phantasmagoria, a world of illusion and deception. In one of the most emblematic conversations between Atsuko and Shimasaki, ŌTA Yōko expresses the perspective of her characters as follows:

> [Atsuko:] "I have been there two or three times with my sister and her kids. But walking on Hondōri felt like strolling around in a film set. That was really awkward."
> [Shimasaki:] "Yes, it is indeed a kind of film set. Looking from the front, it looks like ordinary shops, but looking from behind you see it's all fake. [...] In this city, the [new] Hiroshima written in *katakana* is becoming more and more present. However, this should not be called a reconstruction of Hiroshima anymore."
> [Atsuko:] "Yes, you are right. Reconstruction is something different. This notation in *katakana* is really annoying." (ŌTA 1982b: 136)

This city written in *katakana* Atsuko experiences during her stay is a city that has become strangely detached from time and space. The new *katakana* notation refers to the reconstructed new postwar city of "H", a city that has been re-conceptualized as a tourist mecca of world peace. Atsuko looks at the new oversized boulevards, bridges and parks that take up more and more space in the very center of the city with a sense of alienation: "On the left side, she could see a strange looking bridge. She knew at a glance that this must be the 'Peace Bridge' Isamu Noguchi had designed [...]. However, while the bridge seemed to be completed with its unique design, the surrounding area was still in ruins" (ŌTA 1982b: 176).

Atsuko is highly skeptical of this "reconstruction" (*fukkō* 復興), since it is tantamount to an erasure of all signs of August 6, including its survivors still living there in the slums. To put it in a nutshell, the *katakana* notation Atsuko is so annoyed with brings about an irreversible de-historicization and de-politicization of "H" that conceals wartime history and the role Japan had played during this war. Needless to say, for ŌTA Yōko the city of "H" is nothing but a metonymy for Japan per se. In the society of postwar Japan, this notation change has been deliberately used to detach unpleasant memories from time and space, as it has been the case with Hiroshima, Nagasaki, and recently Fukushima, too. The German-Austrian philosopher Günther ANDERS (1902–92), who visited Hiroshima in 1958, three years after the publication of "City and People in the Evening Calm", described the "uneasiness" he felt in Hiroshima in quite a similar way in his diary "The Man on the Bridge", first published in 1959:

> No, *I* cannot see anything of what had happened here. What I do see: *new* houses concealing history in the same way newspapers and everyday conversations do. Everything seems to be "time-neutral". That is, everything looks as if it had been here ever since. The present disguises itself as "always been that way". And this disguise overgrows what really had happened. History had been forged backwards. Or more precisely, history had been forged by history itself, since this reconstruction is nothing else than history. (ANDERS 1995: 110–11; emphasis in the German original)

ŌTA Yōko's "City and People in the Evening Calm" is less about the "reconstruction" of the city of "H", as KAWAGUCHI Takayuki (2011: 94–98) has pointed out in his analysis, but about the "gap" this reconstruction inevitably has left for all the victims. It is rather a story about the fear of being encapsulated in the former *kanji* Hiroshima, and of being buried in oblivion along with it. The "evening calm" (*yūnagi* 夕凪) of the novel's title, a motif that pervades the entire text, serves as a metaphor for Atsuko's claustrophobic feeling of being locked up in a history that has come to an end with August 6, 1945. The reconstruction of "H" changes the entire topography of the city, a topography that was full of *lieux de memoires* of August 6 for its survivors. In the city of "H", Atsuko witnesses the process of local memory being erased, and replaced with new memorial sites for national commemoration and mourning.

Last but not least, "City and People in the Evening Calm" is also a self-reflecting meta-narrative. It is a novel about the "impossibility" of writing and reading about these "gaps". For us, who read about Atsuko's plan to write a reportage we are paradoxically reading right in that moment, "City and People in the Evening Calm" turns out to be an "impossible" narrative in the same way as August 6 and its aftermath remain, according to Cathy CARUTH (2016: 18–20), an "impossible" history in terms of accessibility and comprehensibility for eyewitnesses and writers like ŌTA Yōko.

3 Reconfiguring national history in the postwar period

At the time ŌTA Yōko's "City and People in the Evening Calm" was published, national history writing and teaching already stood—after a short period of critical reappraisal of recent war history at the end of the occupation period—at the crossroads to re-nationalization and re-configuration. Since school textbooks formerly used for nationwide indoctrination during the war were now banned from the Japanese education system, new textbooks that would be in line with the new democratic requirements of the time were badly needed. Hence, a new screening system for school textbooks was introduced by the Civil Information and Education Section in 1948 to foster the production of new democratic textbooks. And indeed, at least in the first years after its implementation, this screening system turned out to be a blessing for many reform-oriented writers of textbooks, before that same system became more and more instrumentalized by the Japanese government after the end of the occupation in 1952 (KÖHN 2022: 174–76).

In the early 1950s, new kinds of school textbooks were published which for the first time provided—in comparison to former textbooks—a critical view of Japan's role during the Fifteen Years War. This included the atomic bombings on August 6 and 9 respectively. OKADA Yuzuru's 岡田謙 "Democracy and Cheerful Life" (*Minshu shugi to akarui seikatsu* 民主主義と明るい生活), published in 1951, was actually the first textbook to implement a new learning unit called "World Peace" (*Sekai no heiwa* 世界の平和), with a full coverage of 65 pages. This unit did not only inform about the hitherto unknown full extent of the devastation in Hiroshima and Nagasaki, but also provided a rather different interpretation of the emperor's well-cited "wise decision" (*seidan* 聖断) to end the war and save millions of his subjects: "Japan did not accept the unconditional surrender on August 15 due to the bombings, but because all of its national resources had been exhausted during the war. Yet at least it can be said that the atomic bombs accelerated ending the war" (OKADA 1951: 191). Since information about the bombs and the aftermath were scarce even at the end of the occupation period, pieces of *genbaku bungaku* written by ŌTA Yōko and HARA Tamiki 原民喜 had to be cited as firsthand witness accounts of the nuclear devastation in this textbook. Following OKADA's critical approach, two other noteworthy textbooks were published only two years later in 1953: "Model for Social Studies in Junior High Schools" (*Mohan Chūgaku shakai* 模範中学社会) by OSADA Arata 長田新, and "Social Studies in Junior High Schools" (*Chūgaku shakai* 中学社会) by MINOBE Ryōkichi 美濃部亮吉 and KOYAMA Buntarō 小山文太郎. These two textbooks provided an even more detailed depiction of the atomic inferno by making use of many figures, graphs and pictures in this new learning unit. History now being taught at school seemed to be on its way as promulgated in the "Peace Constitution" and the "Basic Act on Education", both from 1947.[6]

However, only three years later, in 1956, things had somehow changed significantly, as a short glimpse at another textbook, "Social Studies in Junior High Schools" (*Chūgaku shakai* 中学社会) written by Konishi Shirō 小西四郎 and Ienaga Saburō 家永三郎, soon reveals. Not only did the unit "World Peace" completely vanish from the textbook, but the depiction of the bombings also was reduced to a bare minimum of information:

> On August 6, an atomic bomb was dropped on Hiroshima, on August 8, the Soviet Union declared war on Japan, and the following day, another atomic bomb was dropped on Nagasaki. Japan's defeat was now obvious. [...] The war ended with the victory of the democratic nations. (Konishi / Ienaga 1956: 196–97).

It goes without saying that Konishi and Ienaga's "Social Studies" marked a crucial turning point in Japanese textbooks, since Hiroshima and Nagasaki became more and more of a blank spot, a "gap" in the historical consciousness of younger postwar generations. Both cities lost their historical singularity in collective memory; they were reduced to mere historical "events" on a linear timeline of a war an entire nation was eager to bury in oblivion as soon as possible.

This conservative step backward in national history writing was due to a backlash in Japanese politics that had begun in the early 1950s and finally affected the education system too.[7] The report "The Problem with Alarming Textbooks" (*Ureu beki kyōkasho mondai* うれうべき教科書問題), published by the conservative Democratic Party of Japan (Nihon minshu tō 日本民主党) on August 13, 1955, is paradigmatic for the political sentiment of Japan's leading conservative parties of that time. According to this report, "it has become obvious that our education system is about to collapse due to the kind of textbooks used in school nowadays—and nobody even noticed it" (Nihon minshu tō 1955: 3). For the Democratic Party, the subversion of the education system by use of allegedly "red textbooks" (*akai kyōkasho* 赤い教科書), that is textbooks written by authors like Okada and others who were renowned members of the Teachers' Union (Nihon kyōshokuin kumiai 日本教職員組合), founded in 1947, was a serious threat that had to be eliminated immediately. Textbooks used in school were considered above all peace textbooks based on Marxism-Leninism that paved the way for a Soviet or Chinese conquest of Japan through education, as the report explicates further (Nihon minshu tō 1955: 30). Thus, a reform of the implemented screening system of textbooks and a tightening of the Guidelines for Teaching and Learning (*Gakushū shidō yōryō* 学習指導要領) were deemed necessary to save Japan from communist infiltration in the name of peace (Nihon minshu tō 1955: 42). As a result, textbooks that were not in line with the new requirements of the Ministry of Education could no longer pass the examination. Especially following a fundamental reform of these guidelines in 1958, as NAGAI Jirō (1977: 65) pointed out, the hurdles for publishers and editors were significantly higher than before; here again "censorship" and "self-restraint" created a highly restrictive climate for verbal utterances.

The modified screening system became a powerful governmental tool to silence dissonant voices and rewrite national history. The principal aim of history education after 1955 was, according to ŌSUGA Akira (1969: 99–100), to rear new generations of students with a highly biased view of recent war history. The new national narrative, now implemented in textbooks, followed a clear agenda by putting an emphasis on the emperor's role as a bringer of peace. In this new narrative, the bombings were re-contextualized in a very specific way; they were deemed necessary so that the emperor could finally end the war for the benefit of all his subjects. Hiroshima's presumed "sacrifice" became a crucial building block of Japan's "foundation myth" (*kigen no monogatari* 起源の物語), as IGARASHI Yoshikuni (2007: 31–50) stated, for creating a new identity as a democratic and peaceful nation in the new postwar world order. In other words, Hiroshima had to be sacrificed to save the nation, and to save the emperor, who successfully transformed himself from a warmonger to a bringer of peace in the public eye. It was this very transformation enabled by the atomic bombings that ultimately saved the emperor from being put on trial as a war criminal during the Tokyo War Crimes Trial between 1946 and 48.[8]

This reconfiguration of national history inevitably led to a distortion of the bombings and their aftermath in the historical consciousness of an entire nation. The real victims of the bomb were rendered invisible in virtually any form of history writing—and by that, in public discourse per se—while at the same time, all non-victims were transfigured into the new victims of August 6, 1945. This "sacrificial system" (*gisei no shisutemu* 犠牲のシステム), as TAKAHASHI Tetsuya has fittingly described it, enabled the entire Japanese nation to suppress the unpleasant question of responsibility for the Fifteen Years War by nationwide self-victimization and self-affirmation. The "sacrifice", as TAKAHASHI (2012: 27–28) puts it, "is, as a general rule, either concealed or glorified and justified as a 'precious sacrifice' for the sake of the community (state, nation, society, company etc.)."[9] Since critical voices were more often than not branded as communists, as it was the case with the editors of the "red textbooks", this revisionist view of recent war history became the hegemonic master narrative for the next decades. And needless to say, the emergence of the new *katakana* Hiroshima ŌTA Yōko was so highly critical of facilitated the implementation of this master narrative significantly.

4 MARGINALIZING UNPLEASANT VOICES FROM THE PAST

For eyewitnesses and authors like ŌTA Yōko, writing about August 1945 turned out to be quite a challenge, since the first decade after the lost war was overshadowed by fierce disputes about how postwar literature ought to be. Most renowned literati had joined the Japanese Association for the Promotion of Literature (Nihon bungaku hōkoku kai 日本文学報告会), founded in 1942, to support national policy.[10] This led to heated debates after the war regarding the question of their complicity in wartime propaganda efforts by the government. Particularly in Japan's most prominent literary magazines, "Modern Literature" (*Kindai bungaku* 近代文学) and "New Literature of Japan" (*Shin Nihon bungaku* 新日本文学), both founded in 1946, the interwovenness of "war and literature" (*sensō to bungaku* 戦争と文学) became subject of an intense controversy among Japan's intellectuals of the time (KATŌ 1997: 114–15). For many, literature per se had failed to become the badly needed voice of dissent in times of war. The predominant "I-novel-style" with its protagonists' uncritical and apolitical introspection, which became popular from the 1920s onwards, was considered the main reason for the "conversion" (*tenkō* 転向) of left-wing intellectuals into blind supporters of the system, a development many writers underwent during the war. As a consequence, quite a few writers after the war campaigned for a new postwar literature that would realistically reflect on society from a multi-perspective and highly objective point of view. This new kind of "documentary literature" (*kiroku bungaku* 記録文学), as it was called, was considered the only way to successfully overcome Japan's failed project of modernity by raising critical awareness and cultivating the subjectivity of each and every individual in society. However, this condemnation of the traditional "I-novel-style" first and foremost served to distract from the more crucial and unpleasant question of one's own "war responsibility" and "war guilt" respectively. As KURIHARA Tadaichi (1981: 69) pointed out as early as in 1946: "By becoming an ally with the nation state, Japan's artists and scientists ultimately became, just like their colleagues in Germany, an enemy of truth." Japan's literati benefited—just like the Japanese emperor did before—from the "system of irresponsibility" (*musekinin no taikei* 無責任の体系) that emerged right after the war. As a result, in public opinion, it was rather the literary system per se than the literary individual that had to be blamed for not having opposed the war. According to KATŌ Norihiro (1997: 25–73), this system of "concealing" (*inpei* 隠蔽) and "distorting" (*nejire* ねじれ) the past must be considered paradigmatic for the entire society of postwar Japan, for which the Fifteen Years War was re-conceptualized into a war with millions of victims—but without any perpetrators.

Needless to say, this identity crisis of Japan's literati after the lost war did highly affect the reception of ŌTA Yōko's literary work too. For most of her critics, ŌTA Yōko did neither write fictional works that could be called "literature", nor did she produce non-fictional works that could be called "documentaries". In this regard, ŌTA Yōko's atomic bomb literature was highly paradoxical, since it gave account of "impossible history" in the form of "impossible literature".[11] Despite the fact that they were first and foremost traumatized victims of the bomb who had decided to share their experiences, authors like ŌTA Yōko were expected to give objective accounts on August 6, with a new literary style and an innovative language that fulfilled all the prerequisites of new postwar literature as propagated by the leading literary magazines. Moreover, these eyewitness writers were generally expected to give full access to a history of August 6 and 9, which had become an essentially inaccessible part of history for most of the survivors due to their experienced trauma. The resulting "unreadability" of the written account is, as Jessica LANG (2017: 3) explains with regard to Holocaust literature, "a textual quality or condition of inaccessibility—blankness, illegibility", a kind of "textual silence" that has become an integral part of their very personal experience. In other words, the "unreadability" is an inherent quality of Holocaust literature—and not a sign of lacking literariness. However, in marked contrast to Holocaust survivors and their literary eyewitness accounts, survivors of Japan's "nuclear Holocaust" and their written accounts could not hope to find such sympathy and understanding on the side of their Japanese readers and literary critics. Quite the contrary seemed to be true; their accounts had to be both readable and accessible, without any "gap" or "silence" in the text. It is this impossibility of writing the hitherto unexperienced that probably became the greatest burden for most eyewitness writers of August 1945.

However, that seemed to be just one side of the story. Since personal accounts on August 1945 left no room for doubt about the real victims of the bomb, atomic bomb literature defied the nationwide self-victimization and re-conceptualization of history. By drawing a line between victims (*hibakusha* 被爆者) and non-victims (*hi-hibakusha* 非被爆者) of the bombings, authors like ŌTA Yōko once more addressed the issue of responsibility for the war to a society which had felt absolved from any guilt after the end of the Tokyo War Crimes Trial. Yet, as HANADA Kiyoteru strikingly pointed out in 1948: "Although even ordinary people in Germany or Japan [...] were guilty, they had not received a particular punishment for that. However, not being punished is the kind of punishment they actually had received for their guilt" (HANADA 1980: 181–82). Writers like ŌTA Yōko reminded the readers of their collectively suppressed guilt and therefore became an unpleasant voice from the past for most of their contemporaries.

In 1955, when the last installment of "City and People in the Evening Calm" had finally been published, the musealization of August 6, 1945, was almost completed. After the Hiroshima Peace Memorial City Construction Law (Hiroshima heiwa kinen toshi kensetsu hō 広島平和記念都市建設法) had come into force in 1949, the city of Hiroshima accelerated its transformation into a new mecca for world peace. According to SENBA Nozomu (2018: 133–36), this transformation actually started as early as in August 1946, when the city commemorated the first anniversary of the bombing with the Peace Reconstruction Festival (Heiwa fukkōsai 平和復興祭). But it was only with the help of the construction law that substantial national funding became available for the municipal government to put former plans for the city's fundamental reconstruction into practice (EBARA 2016: 54–79). After a systematic eviction of virtually all *hibakusha* slums in the very city center, the long-planned construction of Hiroshima's Peace Park (opened in 1954) and Peace Memorial Museum (opened in 1955) could finally start. These two highly emblematic sites became the core pillars of national commemoration and mourning, and paved the way for Hiroshima's transformation into a "holy ground" (*seichi* 聖地) of world peace.

Like many other survivors, ŌTA Yōko followed this musealization process with great skepticism. Hence, this musealization signified a "cleaning" of the city center, a cleaning of all the survivors, repatriates, and evacuees that had begun to flock into the barracks of Motomachi, Aioi etc., and a cleaning of all the other—compared to the so-called Genbaku Dome—admittedly less impressive remains of Hiroshima's devastation. The musealization of Hiroshima irreversibly encapsulates August 6, 1945, in time. However, as MATSUMOTO Hiroshi critically remarked:

> For the survivors, the atomic bombing and the aftermath are a problem they are still struggling with, even now. But for all the rest, the atomic bombing is nothing more than an event of a remote past, an event that happened more than thirty years ago at a place called Hiroshima (Nagasaki). (MATSUMOTO 1995: 83)

Without doubt, the dropping of the atomic bomb on Hiroshima has become a kind of founding trauma for Japan, in the sense Dominick LACAPRA stated; it played a crucial role in generating "the concept of a chosen people or a belief in one's privileged status as a victim" (LACAPRA 2014: 81). However, in the case of Japan, this privileged status was not given to the real victims of the bomb, but to the entire nation, a nation that believed to be the only one in the world being struck by a bomb (*yuiitsu no hibakukoku* 唯一の被爆国). Here again, the difference between the Holocaust in Europe and the "nuclear Holocaust" in Japan becomes quite obvious. In Japan, atomic bomb survivors and their eyewitness accounts were not deemed necessary for the nationwide culture of remembrance;

instead, they were marginalized as unpleasant and, above all, dissenting voices from the past.

5 Conclusion: Articulating the nuclear in the 1950s

In the mid-1950s, Ōta Yōko had to witness society's radical change in attitude toward the nuclear issue. In 1954, hundreds of thousands had participated in anti-nuclear protests after the hydrogen bomb tests on Bikini Atoll. And only one year later, in 1955, hundreds of thousands flocked to the "Exhibition on the Peaceful Use of Nuclear Energy" (*Genshiryoku heiwa riyō hakurankai* 原子力平和利用博覧会) which was—among many other cities—also held in the Hiroshima Peace Memorial Museum from May to June 1956. Shortly after, Ōta Yōko vents her frustration over a society that is enchanted by the seductive promises of the supposedly "good" and "peaceful" energy in her short novel "Half Nomad" (*Han hōrō* 半放浪): "After the hydrogen bomb test, the so-called ashes of death fell on Tōkyō. I only thought: Serves you right! All of you should be contaminated by this lethal ash and die one after another" (Ōta 1982c: 296).

Admittedly, atomic bomb literature was never a popular literary genre in Japan. But in the mid-1950s, the situation became even worse. As Nagaoka Hiroyoshi (1982: 46) pointed out, literary accounts on August 1945 were a more or less unprofitable business for many publishers, since *genbakumono* 原爆もの did not sell. So many publishers hesitated to produce new titles of *genbakumono*, and many authors hesitated to keep on writing about what happened on "that very day". Nonetheless, Ōta Yōko kept on writing about August 6 in her very unique way. In "My stance as an author" she explains this as follows:

> In my point of view, the real nature of literature lies in realism. Literature should neither be hard to read or difficult to understand, nor uselessly ornamented or notably eccentric. I want to write literature in a way that everyone can read and understand it. That is the reason why the literature I write will be in permanent conflict with [the ideals of] the literary establishment, even in the future. (Ōta 1982a: 311)

When Ōta Yōko began to write about Hiroshima's transformation into a new mecca of world peace, the memory of the former *kanji* Hiroshima already stood at the crossroads to oblivion. Ōta Yōko expresses her feeling of alienation from this new Hiroshima quite impressively, when her alter ego, Oda Atsuko, discovers the Hiroshima Children's Library in the district of Motomachi, one of Hiroshima's *genbaku* slums:

> "You mean, the idea for the library's design is taken from the shape of the mushroom cloud?"
> "Yes, that's what they say."

> Suddenly, Atsuko felt like she was going to throw up. She tried to resist this feeling of horror, but when out of a sudden she had lost her inner balance, she got tormented by this image. "Right from the beginning, I had thought 'What a weird design'... But a mushroom cloud as a model?... Why on earth did this professor Tanzan [i. e. Tange Kenzō] design a library for this city in form of a mushroom cloud, of all things?" (ŌTA 1982b: 134)

For Atsuko, this library (opened in 1949) in some way epitomizes Japan's dangerous naivety regarding its recent nuclear past. Within this musealized city center, which was also designed by TANGE Kenzō 丹下健三 (1913–2005), the atomic bomb is reduced to a mere symbolic level. By all appearances, the mushroom cloud, TANGE Kenzō's source of inspiration for the library's design, did no longer refer to the real atomic bomb that devastated the entire city of Hiroshima on August 6. Instead, it had become a signifier without an object that is signified by.

In the end, it is more than questionable if ŌTA Yōko really was blissfully unaware of the fact that August 6 had already become a "national project", and therefore kept on writing about her experiences of the bombing and the aftermath, as KUROKO Kazuo (1977: 91) stated in his analysis. Quite the contrary seems to be true. ŌTA Yōko kept on writing about her experiences because she was keenly aware of this instrumentalization and nationalization of August 6. Her literature deliberately focused on the "gaps" within the discursive articulation of Japan's nuclear legacy, and as such, her literary work became an important counter-narrative to national history writing in the 1950s.

NOTES

1 For details, see ESASHI Akiko's prize-winning biography (1981).
2 The first two installments were published 1954 in the literary magazine *Gunzō* 群像 (November and December issue), the third and last installment was published 1955 in the literary magazine *Shin Nihon bungaku* 新日本文学 (August issue).
3 The question of how and where to place the narrator in a novel about August 6 properly is a problem ŌTA Yōko had actually been struggling with in most of her works. To generalize, and by that, to objectify her personal experiences, ŌTA chose the form of a documentary to become a trustful and reliable narrator. As John W. TREAT expressed pointedly with regard to ŌTA's "City of Corpses": "A documentary will more often than not attempt to arrange its information 'objectively,' i. e., as if that information is all the more trustworthy because it is detached from an explicitly involved narrator and thus interpretation" (1995: 209). However, ŌTA Yōko's, or more precisely, Oda Atsuko's documentary about "the actual situation in 1953" misses a crucial requirement for being "objective" and "trustworthy", namely a multi-perspective view on August 6 and the aftermath. The different stories Atsuko listens to during her stay in "H" strangely complement each other—opposing views about August 6 remain a "gap" in the entire text. "City and People in the Evening Calm" is, as

NAKANO Kazunori (2005: 139) has pointed out, an experimental text; fragments of information are collected and put together into a coherent storyline by Atsuko. Yet, what NAKANO misses in his analysis, is the fact that this experimental structure of the text becomes somehow obsolete with the third installment, when Atsuko is the one who remembers August 6 and monopolizes the narration's point of view. At the end, ŌTA's objective documentary has become a personal, and by that, subjective testimony she actually had tried to avoid writing.

4   However, one should keep in mind that living standards in the social dwellings in Motomachi did only gradually differ from the illegally built barracks in Aoi; for details (including figures) see HIROSHIMA-SHI (1983: 110–48).

5   As NAONO Akiko (2015: 37–50) pointed out in her study, the term *hibakusha* became widely used in Japanese mass media from the mid-1950s onwards, after the Japanese tuna fishing boat "Lucky Dragon Number 5" (Daigo Fukuryūmaru 第五福竜丸) was contaminated by nuclear fallout from the hydrogen bomb tested by the United States in March 1954 at Bikini Atoll. In the nationwide discourse, the Fukuryūmaru became a metonymy for an entire nation that began to re-conceptualize itself as a "victim of the bomb". Interestingly, ŌTA Yōko had actually used the term *hibakusha* as early as in 1949 in her short novel "August 6, 8.15" (*Hachigatsu muika hachi-ji jūgofun* 8月6日8時15分), published in the magazine "Reconstruction" (*Kaizō* 改造), but avoided the use in her later work "City and People in the Evening Calm". It seems that ŌTA Yōko deliberately produced this "gap" in this text to draw a clear line between the "real" victims of the bomb Atsuko encounters in "H", and the "imagined" victims of the bomb after the Lucky Dragon Incident.

6   OSADA Arata's "Model for Social Studies" contains a learning unit titled "peace" (*Heiwa* 平和) with a full coverage of 56 pages. This textbook provides an even more detailed depiction of the nuclear inferno, as OSADA's portrayal is interspersed with various citations from his earlier work "Children of the Bomb" (*Genbaku no ko* 原爆の子), published two years before. It is the first collection of eyewitness accounts from Hiroshima. Although OSADA, a survivor of August 6, is highly critical of the nuclear threat the existence of atomic bombs poses for future generations, he is, however, surprisingly optimistic regarding the positive effects of nuclear energy for mankind: "Nuclear energy is not only invented for war. [...] This unmeasurable energy can become a source of power that will contribute largely to the happiness of humanity. And a peaceful usage will cost less efforts and resources than a militaristic usage" (OSADA 1953: 110). For OSADA, as he explains in the preface of his "Children of the Bomb", the "Phoenix of peace that was reborn from the ashes of death that came from the devastated city of Hiroshima [...] is full of hope for the peaceful use of nuclear energy, an energy that has the potential of bringing great benefits [to everyone]" (OSADA 1951: 72). As OSADA's case impressively shows, the discursive split between alleged good and bad nuclear energy was already apparent in the early 1950s—even among atomic bomb survivors.

7   This conservative backlash was the result of a dramatic change in the occupation policies for Japan. The so-called "reverse course" (*gyaku kōsu* 逆コース) was in actual fact a reaction to the political developments in China that finally led to the Com-

munist Revolution in 1948. The American fear of communist or socialist ideologies that could also spread out to Japan caused a withdrawal of many former reforms that were conducted for the sake of Japan's process of liberalization and democratization. Needless to say, the Allied Power's "reverse course" strengthened Japan's conservative parties, who were more interested in turning back time than reforming the country into a new democratic nation. The promulgation of the "Subversive Activities Prevention Act" (*Hakai katsudō bōshi hō* 破壊活動防止法) in 1952 is probably the most obvious attempt to silence dissonant (political) voices, as it was common practice with the "Peace Preservation Law" (Chian iji hō 治安維持法) during the war.

8   As SHINOBU Seizaburō (1992: 297–326) convincingly pointed out in his study, the emperor's "wise decision" also started Japan's war machinery and aggressive expansion policy during the Fifteen Years War. However, his last "wise decision" to end a war he had started himself 15 years ago has become the only "wise decision" remembered in national history writing.

9   Interestingly, TAKAHASHI described this "sacrificial system" with regard to the triple disaster of Fukushima and its aftermath. However, the mechanisms and dynamics depicted in this book can easily be traced back to the nationwide process of self-victimization after the bombings of Hiroshima and Nagasaki.

10 Almost four thousand literati joined the "Japanese Association for the Promotion of Literature" which was founded on the initiative of the "Imperial Rule Assistance Association" (Taisei yokusankai 大政翼賛会) and the "Cabinet Information Bureau" (Naikaku jōhōkyoku 内閣情報局). According to MAKIYAMA Raishō (1946: 38), renowned publishing houses like Bungei shunjū or Shinchōsha became pillars of this new form of literary propaganda, whose sole purpose was to justify the emperor's "holy war" (*seisen* 聖戦) for the sake of all supposedly suppressed and exploited Asian countries. For details on the various activities of this association, see SAKURAMOTO (1995).

11 The rather negative attitude that was adopted by Japan's leading literati toward *genbaku bungaku* in general, and ŌTA Yōko in particular, can be seen best in the jury-member's reactions towards ŌTA's rather surprising nomination for the prestigious Akutagawa Prize in 1954 for "Half Human". Only one jury-member, UNO Kōji 宇野浩二, provided an explanation for rejecting her work, stating that "it is indeed a remarkable fact that an author tries to write such a novel, but it is really a pity that the writing style, which is so essential to literature, is anything but good." Interestingly, 20 years later, in 1975, another atomic bomb survivor was nominated for the Akutagawa Prize, this time Nagasaki survivor HAYASHI Kyōko 林京子 (1930–2017) for her novel "Festival Ground" (*Matsuri no ba* 祭りの場). In contrast to ŌTA Yōko, HAYASHI Kyōko was awarded the prize. However, this was less—as the comments of the jury-members reveal—due to its outstanding literariness, but due to its dramatic topic. It seems that the reception of *genbaku bungaku* did not change significantly over the course of time. For the nominations and comments see the website at https://prizesworld.com/akutagawa/ichiran/ichiran161-180.htm (last access 2022/06/25).

REFERENCES

ANDERS, Günther (1995): *Hiroshima ist überall* (Beck'sche Reihe). München: Verlag C. H. Beck.

CARUTH, Cathy (2016): *Unclaimed Experience. Trauma, Narrative, and History*. Baltimore: Johns Hopkins University Press.

EBARA, Sumiko 頴原澄子 (2016): *Genbaku dōmu: Bussan chinretsukan kara Hiroshima heiwa kinenhi e* 原爆ドーム:物産陳列館から広島平和記念碑へ (Rekishi bunka raiburarī 歴史文化ライブラリー 431). Tōkyō: Yoshikawa kōbunkan.

ESASHI, Akiko 江刺昭子 (1981): *Kusazue* 草饐. Tōkyō: Ōtsuki shoten.

ETŌ, Jun 江藤淳 (1984): *Tozasareta gengo kūkan: Senryōgun no ken'etsu to sengo Nihon* 閉された言語空間:占領軍の検閲と戦後日本 (Bunshun bunko 文春文庫). Tōkyō: Bungei shunjū.

HANADA, Kiyoteru 花田清輝 (1980): "Tsumi to batsu 罪と罰". In: ITŌ, Sei 伊藤整 et al. (eds.): *Nihon gendai bungaku zenshū* 日本現代文学全集, vol. 104. Tōkyō: Kōdansha, 177–82.

HIROSHIMA-SHI 広島市 (ed.) (1983): *Hiroshima shinshi: Toshi bunka-hen* 広島新史:都市文化編. Hiroshima: Hiroshima-shi.

IGARASHI, Yoshikuni 五十嵐惠邦 (2007): *Haisen no kioku* 敗戦の記憶. Tōkyō: Chūō kōron shinsha.

KATŌ, Norihiro 加藤典洋 (1997): *Haisengo ron* 敗戦後論. Tōkyō: Kōdansha.

KAWAGUCHI, Takayuki 河口隆行 (2011): "Machi wo kiroku suru Ōta Yōko: 'Yūnagi no machi to hito to 1953-nen no jittai' ron 街を記録する大田洋子:『夕凪の街と人と:1953年の実態』論". In: *Genbaku bungaku kenkyū* 原爆文学研究 10: 83–100.

KÖHN, Stephan (2020): "Ōta Yōko's Literary Dilemma. Who Cares about the Atomic Bomb in Times of Peace". In: *Japonica Humboldtiana* 22: 83–118.

KÖHN, Stephan (2022): "How to Fill the Void in National History. Japanese Peace Education at the Crossroads in the 1970s". In: *Japanonica Humboldtiana* 24: 167–91.

KURIHARA, Sadako 栗原貞子 (1978): *Kaku, tennō, hibakusha* 核・天皇・被爆者. Tōkyō: San'ichi shobō.

KURIHARA, Tadaitsu 栗原唯一 (1981): "Heiwa to shakai shugi 平和と社会主義". In: *"Chūgoku bunka" genshi bakudan tokushūgō fukkoku narabi ni nukizuri* 「中国文化」原子爆弾特集号復刻並に抜き刷り. Hiroshima: "Chūgoku bunka" fukkoku kankō no kai, 68–69.

KUROKO, Kazuo 黒古一夫 (1977): "Sengo, aru juso to ikari no kōzō: Ōta Yōko no baai 戦後・ある呪詛と怒りの構造:大田洋子の場合". In: *Shin Nihon bungaku* 新日本文学 32 (4), 78–91.

LaCapra, Dominick (2014): *Writing History, Writing Trauma*. Revised edition. Baltimore: Johns Hopkins University Press.

Lang, Jessica (2017): *Textual Silence. Unreadability and the Holocaust*. New Brunswick et al.: Rutgers University Press.

Makiyama, Raishō 槙山雷章 (1946): "Bungakkai no sensō hanzaisha wa dare ka ... 文學会の戦争犯罪者は誰か…". In: *Genron* 言論 3: 36–39.

Matsumoto, Hiroshi 松元寛 (1995): *Hiroshima to iu shisō* ヒロシマという思想. Tōkyō: Tōkyō sōgensha.

Minobe, Ryōkichi 美濃部亮吉 / Koyama, Buntarō 小山文太郎 (1953): *Chūgaku shakai* 中学社会. Tōkyō: Shimizu shoin.

Nagai, Jirō 永井滋郎 (1977): "Nihon no kyōkasho ni okeru genbaku kijutsu no hensen 日本の教科書における原爆記述の変遷". In: NGO hibaku mondai kokusai kaigi Hiroshima senmon iinkai 被爆問題国際会議広島専門委員会 / Hiroshima heiwa kyōiku kenkyūjo 広島平和教育研究所 (eds.): *Hiroshima de oshieru: Kakujidai no heiwa kyōiku* ヒロシマで教える：核時代の平和教育. Tōkyō: Rōdō kyōiku sentā, 53–144.

Nagaoka, Hiroyoshi 長岡弘芳 (1982): *Genbaku bunken wo yomu* 原爆文献を読む. Tōkyō: San'ichi shobō.

Naono, Akiko 直野章子 (2015): *Genbaku taiken to sengo Nihon: Kioku no keisei to keishō* 原爆体験と戦後日本：記憶の形成と継承. Tōkyō: Iwanami shoten.

Nakano, Kazunori 中野和典 (2005): "Shinshō fūkei to shite no hibaku toshi: Ōta Yōko 'Yūnagi no machi to hito to: 1953-nen no jittai' ron 心象風景としての被爆都市：大田洋子『夕凪の街と人と：一九五三年の実態』論". In: *Genbaku bungaku kenkyū* 原爆文学研究 4: 130–47.

Nihon minshu tō 日本民主党 (1955): *Ureu beki kyōkasho no mondai* うれうべき教科書の問題 (Kyōkasho mondai hōkokusho 教科書問題報告書 1). Tōkyō: Nihon minshu tō.

Odagiri, Hideo 小田切秀雄 (1955): "Genshiryoku mondai to bungaku 原子力問題と文学". In: Odagiri, Hideo 小田切秀雄 (ed.): *Genshiryoku to bungaku* 原子力と文学. Tōkyō: Kōdansha, 171–94.

Okada, Yuzuru 岡田謙 (1951): *Minshu shugi to akarui seikatsu* 民主主義と明るい生活, vol. 2. Tōkyō: Chūkyō shuppan.

Osada, Arata 長田新 (1953): *Mohan: Chūgaku shakai* 模範中学社会, vol. 2. Tōkyō: Jikkyō shuppan.

Osada, Arata 長田新 (1990): *Genbaku no ko* 原爆の子, vol. 1 (Iwanami bunko 岩波文庫). Tōkyō: Iwanami bunko.

Ōsuga, Akira 大須賀明 (1969): "Kentei kijun to shite no gakushū shidō yōryō 検定基準としての学習指導要領". In: *Hōritsu jihō* 法律時報 41 (10): 95–100.

Ōta, Yōko 大田洋子 (1949): "Hachigatsu muika hachiji jūgofun 8月6日8時15分". In: *Kaizō* 改造 30 (8): 42–49.

ŌTA, Yōko 大田洋子 (1982a): "Sakka no taido 作家の態度". In: ŌTA, Yōko 大田洋子: *Ōta Yōko shū* 大田洋子集, vol. 2. Tōkyō: San'ichi shobō, 307–11.

ŌTA, Yōko 大田洋子 (1982b): "Yūnagi no machi to hito to 夕凪の街と人と". In: ŌTA, Yōko 大田洋子: *Ōta Yōko shū* 大田洋子集, vol. 3. Tōkyō: San'ichi shobō, 5–294.

ŌTA, Yōko 大田洋子 (1982c): "Han hōrō 半放浪". In: ŌTA, Yōko 大田洋子: *Ōta Yōko shū* 大田洋子集, vol. 3. Tōkyō: San'ichi shobō, 296–314.

SAKURAMOTO, Tomio 櫻本富雄 (1995): *Nihon bungaku hōkoku kai: Daitōa sensōka no bungakusha-tachi* 日本文学報告会：大東亜戦争下の文学者たち. Tōkyō: Aoki shoten.

SENBA, Nozomu 仙波希望 (2018): "<Heiwa toshi> kūkan no keifugaku <平和都市>空間の系譜学". In: HIGASHI, Takuma 東琢磨 / KAWAMOTO, Takashi 川本隆史 / SENBA, Nozomu 仙波希望 (eds.): *Bōkyaku no kioku: Hiroshima* 忘却の記憶：広島. Tōkyō: Getsuyōsha, 126–73.

SHINOBU, Seizaburō 信夫清三郎 (1992): *Seidan no rekishigaku* 聖断の歴史学. Tōkyō: Keisō shobō.

TREAT, John W. (1995): *Writing Ground Zero. Japanese Literature and the Atomic Bomb*. Chicago / London: The University of Chicago Press.

Narrating Nuclear Geographies and Victims: The Late Fiction and Essays of Tsushima Yūko

Rachel DiNitto

1 Now More Than Ever, I Oppose Nuclear Power

Contemporary writer Tsushima Yūko 津島佑子 (1947–2016) was deeply shocked and troubled by the 2011 Fukushima Daiichi Nuclear Power Plant disaster, and she turned her subsequent fictional and nonfictional writing into antinuclear activism. Within a year of the Fukushima Daiichi accident, Tsushima wrote "From 'Dream Songs'" (*'Yume no uta' kara*「夢の歌」から), a powerful essay that traced the long history of postwar nuclear harm (Tsushima 2016c). The essay is an expansive and informative exposé of the dangers of the nuclear and a censure of what Tsushima saw as the unforgivable pronuclear position of the Japanese government. Tsushima recognized the conceptual and linguistic gap between nuclear bombs (*genbaku* 原爆) and nuclear power (*genpatsu* 原発) as an intentional deception, but she was also highly attentive to the deleterious effects of the entire nuclear fuel cycle from uranium mining to waste, a cycle supported by a global network of institutions and knowledge that maintain the exercise of power in the nuclear realm. In the last five years of her life, Tsushima thematized her nuclear concerns in a number of important fictional works, including "Wildcat Dome" (*Yamaneko dōmu* ヤマネコ・ドーム, 2013), "Jacka Dofuni: A Tale of Oceanic Memory" (*Jacka Dofuni: Umi no kioku no monogatari* ジャッカ・ドフニ：海の記憶の物語, 2016b), and "Celebrating Half-Life" (*Hangenki o iwatte* 半減期を祝って, 2016a). In this paper, I read "From 'Dream Songs'" alongside these late fictional writings to demonstrate the power of narrative to critique the *genbaku*/*genpatsu* deception. I analyze Tsushima in the context of other Japanese antinuclear voices, like the late Hayashi Kyōko 林京子 (1930–2017), and demonstrate how she employs the narrative techniques of environmental and antinuclear activists to reveal multigenerational nuclear harm.

Tsushima's essay was first published in the antinuclear volume, "Now More Than Ever, I Oppose Nuclear Power" (*Ima koso watashi wa genpatsu ni hantai shimasu* いまこそ私は原発に反対します, 2012), a collective rejection of nuclear power in the form of essays from 51 authors, critics, manga writers, poets, etc.

The Japan Pen Club edited the volume and then president, ASADA Jirō, wrote the following in his preface:

> It is true that we who experienced the catastrophe of nuclear weapons brought on this "nuclear accident", as we have termed this disaster that came from the same atomic root. But before debating the consequences of our responsibility, we must turn our attention to the disgrace of the government and the betrayal of history. (ASADA 2012: 10)

ASADA's words echo those of Nobel Prize Author ŌE Kenzaburō 大江健三郎 who wrote on March 28, 2011, that the construction of nuclear power plants in Japan was "the worst possible betrayal of the memory of Hiroshima's victims" (ŌE 2011). Both statements emphasize the undeniable bond between nuclear bombs and nuclear power, a deadly linkage that both the American and Japanese governments and nuclear industries vociferously denied and successfully dismantled in their campaign to embed nuclear power in postwar Japan. This happened despite the atomic bombing of Hiroshima and Nagasaki and the widespread opposition to nuclear testing incited by Japan's third nuclear exposure, as symbolized by the Lucky Dragon incident in 1954.[1]

TSUSHIMA argues in her essay that "[t]here are those who say we must think of nuclear arms testing and nuclear power separately, but in 1974 India used plutonium made at a nuclear power plant and succeeded in creating a nuclear weapon, so the distinction is no longer recognized" (TSUSHIMA 2012: 460). TSUSHIMA exposes the fallacy of the *genbaku/genpatsu* divide, and the "myth of safety" (*anzen shinwa* 安全神話) surrounding nuclear power when she acknowledged the potential danger that radioactive materials from power plants will be used for bombs. Yet for many Japanese, this distinction of dangerous bombs vs. safe power remained powerfully embedded in their minds as they labored for the postwar economic miracle. Referring to the U.S campaign to sell nuclear as safe energy, critic YOSHIMI Shun'ya argued: "'Atoms for Peace' implied more than an affordable new form of energy to sustain affluent lifestyles. It also had a particular political connotation: the forgetting of Hiroshima and Nagasaki" (YOSHIMI 2012: 323). Arguably, this myth of safety that promoted nuclear power and cleaved it from nuclear weapons was not fully challenged until the Fukushima Daiichi disaster.

Many have spoken about the fall of the myth of safety, but some of the most moving comments come from HAYASHI Kyōko 林京子 in her semi-fictional essay "To Rui, Once Again" (*Futatabi Rui e* 再びルイへ, 2013) written in the wake of the Fukushima Daiichi accident. HAYASHI solidifies the link she made in "From Trinity to Trinity" (*Toriniti kara Toriniti e* トリニティからトリニティへ, 2000) between atomic bombs and nuclear power plants, a link engendered by the criticality accident at the Japanese nuclear fuel conversion plant in Tōkaimura, Ibaraki prefecture the night before HAYASHI went to the Trinity test site in the American desert.² The image of the atomic bombs is often confined to the spectacle of the mushroom cloud and the horrifically instantaneous incineration of its victims, at the cost of attention to the long-term, silent, deadly aspects of continued radioactive exposure, both external and internal. Rejecting the spectacle of the bombs, HAYASHI spoke for nuclear victims of both bombs and power plant accidents when she discussed her shock at the Japanese government's public recognition of "internal exposure to radiation" (*naibu hibaku* 内部被曝) for victims of the Fukushima Daiichi disaster.

> One day, I heard a government official explaining how radiation affects the human body. He used the phrase "internal exposure to radiation". Rui, I stared at the face of that official on my TV screen. So they've known about it all along—the effect of "internal exposure" on the human body.
>
> Since I was exposed to radiation from the A-bomb on August 9 1945, this is the first time I have heard a government official in a responsible position use the phrase "internal exposure." They knew about it, but they never mentioned it in public. Since August 6 and August 9, how many of my classmates have died of "A-bomb disease" brought on by years of "internal exposure" to radiation? (HAYASHI 2017a; HAYASHI 2013: 13)

A little later in the essay she echoes ASADA's themes of disgrace and betrayal:

> Hearing the words "internal exposure to radiation" coming from the mouths of people who had denied over and over again that such a thing even existed was a greater shock than exposure to the A-bomb itself. Our country has betrayed us. And now, after the nuclear disaster at Fukushima they're once again showing us how little they think of human life. (HAYASHI 2017a; HAYASHI 2013: 18)

In the wake of the Fukushima Daiichi accident, HAYASHI's voice took on even more importance in the global nuclear discourse. For TSUSHIMA, who knew HAYASHI from their days at the coterie journal "Literary Capital" (*Bungei shuto* 文芸首都) in the 1960s, HAYASHI's literary works allowed her to better understand nuclear harm and the experience of nuclear victims known in Japan as *hibakusha* (KUROKO 2018: 201).³ While not all antinuclear bomb advocates spoke out against nuclear power, HAYASHI saw beyond Japan's experience of the atomic bombs to contemplate nuclear harm in Japan and the U.S, and after the Fukushima Daiichi accident, ŌE, in the editorial mentioned above, also argued that: "I have long contemplated the idea of looking at recent Japanese history through the prism of three groups of people: those who died in the bombings of Hiroshima and Nagasaki, those who were exposed to the Bikini tests, and the victims of accidents at nuclear facilities" (ŌE 2011).⁴ TSUSHIMA expanded the scope even further, moving beyond ŌE's three groups to include those harmed by the entire nuclear fuel cycle from mining to the production and testing of weapons, to the construction of power plants and waste, as well as a wider field of fallout outside Japan including indigenous communities and peoples affected by this cycle in places including the U.S, Kazakhstan, Australia, New Zealand, Taiwan, and the South Pacific. In her fiction and non-fiction, Tsushima casts her gaze across and around the Pacific to see how this fuel cycle affects indigenous peoples, many of whom share an ocean with Japan.

2 Caught in the Nuclear Dispositive

In "From 'Dream Songs'" Tsushima admits that while she could never have supported nuclear power, she was ignorant of the nuclear build up in Japan. She had not paid attention to the construction of 54 reactors in her seismically active nation, did not know about things like the Three Electric Power Laws (Dengen sanpō 電源三法) that accelerated construction of nuclear power plants[5], the tremendous build-up of spent nuclear fuel at Japan's nuclear power plants, or that even with reprocessing the vast amount of plutonium in Japan would end up as nuclear waste. She remembers being afraid of radiation after the Lucky Dragon incident, rumors of *hibakusha* unable to marry, accidents at the Tōkaimura plant, protests against the Rokkasho Nuclear Fuel Reprocessing Facility, the Japanese Three Non-Nuclear Principles—not possessing, producing, permitting—, and the words plutonium, strontium, and Geiger counter. Her memories also reach beyond Japan to accidents at Three Mile Island and Chernobyl, protests against French nuclear tests, similar tests by China and India, and words like "nuclear deterrence" and "nuclear umbrella". But after the Fukushima Daiichi accident, Tsushima wondered how she had failed to connect all these discreet pieces of knowledge about the nuclear. How had she lived so long without seeing the interrelations? (Tsushima 2012: 466–67). She says: "Essentially, I knew nothing. My life completely dependent on electricity, I blindly assumed everything was OK and kept living in ignorance" (Tsushima 2012: 468). For Tsushima, the Fukushima Daiichi accident crystallized the falsity of the *genbaku*/*genpatsu* gap and put the two in a toxic lineage from mining to waste. Tsushima experienced a rude awakening to the dangers of the nuclear that is similar to that of residents in contaminated communities, as described by environmental scholar Lawrence Buell (1988: 647).

Rather than blaming Tsushima for her failure to piece together the evidence of nuclear harm, we can see her inability to construct a holistic vision as the result of the nuclear dispositive. There was an intentional government and industry campaign to promote nuclear power while keeping it separate from the discourse of nuclear arms. Tsushima came to recognize this as a global network. After the Fukushima Daiichi accident, she realized that the French and American politicians were only visiting Japan because the nuclear industry was a huge international business linked by these various interests. She hears in the news that "nuclear power plants are vital for Japan to maintain their 'latent nuclear deterrent power'" and realizes that nuclear power plants are tied to Japan's aspirations for nuclear weapons (TSUSHIMA 2012: 468). She further condemns the Japanese government with the following comparison: "The Japanese government's proclamations on the radioactive contamination after the Fukushima Daiichi accident were shockingly similar to the U.S response to the nuclear weapons test at Bikini Atoll" (TSUSHIMA 2012: 468). Here TSUSHIMA refers to the Japanese government and nuclear industry's attempts to downplay the radiological harm from the Fukushima Daiichi explosions and meltdowns, a situation that over ten years later is unresolved and exacerbated by the government's decision to dump radioactive water stored at the nuclear power plant into the ocean. In "From 'Dream Songs'", TSUSHIMA excoriates the Japanese government and exposes the similarities between these two supposedly discreet technologies—bombs and power—and the attitudes of the governments supporting them and denying the harm they have caused.

While TSUSHIMA discusses instances of nuclear harm in terms of specific locations (the Semipalatinsk Test Site in Kazakhstan, Three Mile Island in the U.S, and Chernobyl in Ukraine), she is not focused on pollution events that are named after the place in which they occurred and conceptualized as discrete happenings, like the Fukushima Daiichi accident. Rather, for TSUSHIMA, and Japanese writers like ARIYOSHI Sawako 有吉佐和子, these events are not confined to particular places, but are part of a global modernity of industrial pollution.[6] They are not generated by particular societies, culture, or histories, but are the product of a global nuclear colonialism. The idea of a shared globe is present in TSUSHIMA's essay when she argues that although Japan is not testing nuclear arms, they are harming their neighbors through the sale and export of nuclear power plants to places like Vietnam; their failure to join native islanders in the South Pacific, New Zealand, and Australia in renouncing the nuclear; and by allowing the Fukushima Daiichi disaster to happen and consequently spread radioactive contamination through water and wind to others who share the same ocean.

3 ANTINUCLEAR STORYTELLING

3.1 The Obligation to Narrate

TSUSHIMA's narrative style in her essays and fiction echoes the strategies and exigencies of environmental and antinuclear advocates including Rachel CARSON and HAYASHI Kyōko. In this section, I look closer at TSUSHIMA's narration with reference to these advocates and to the critical analyses in Christine MARRAN's *Ecology Without Culture* (2017), to discuss the important role of writers like TSUSHIMA in documenting harm and challenging the nuclear dispositive. Nonfiction science writer Rachel CARSON recognized the power of storytelling and used it to effect in *Silent Spring*'s opening chapter, "A Fable for Tomorrow". In this famous opening, CARSON describes a town where "all life seemed to live in harmony with its surroundings" until a "strange blight" descended on the town killing animals and people alike (CARSON 2002: 1–2). Animals and plants ceased to reproduce, birds to thrive, bees to pollinate, leaving the town silent and lifeless. CARSON uses the story of this mythical town as a warning about disasters real and immanent, but she also emphasizes the need to speak of or narrate this harm using the persuasive powers of storytelling. In *Silent Spring*, she "established rhetorical conventions that would become standard fare in the environmentalist debate" and her work is particularly relevant to the present discussion because she engaged nuclear metaphors to represent the apocalyptic threat of pollution (KILLINGSWORTH / PALMER 2012: 65).

CARSON drew parallels between radioactive fallout and pesticide pollution, a rhetorical strategy with great potency since "the image of nuclear detonation redefined popular conceptions of the end of the world [...] whilst the fear of lethal fission products such as strontium-90 undetectable to the senses provided a perfect model for the all-pervasive insinuation of pollutants such as DDT" (GARRARD 2011: 103). CARSON rallied the apocalyptic vision of life-ending nuclear war, and she also recognized harm from "radioactive waste from reactors", "fallout from nuclear explosions", and the "genetic effects of radiation", using these examples to heighten the threat of water pollution from other chemical agents (CARSON 2002: 39; 37). Additionally, she draws the readers' attention to the latent and unknowable threat that ionizing radiation could have on "comparatively innocuous chemicals", potentially rearranging the atoms and changing the chemicals in ways that are "beyond control" (CARSON 2002: 44). In linking the harm from both bombs and power plants, CARSON joins the ranks of those who refuse to propagate the *genbaku* / *genpatsu* divide. CARSON's awareness of the nuclear threat is also embedded in the title to her second chapter, "The Obligation to Endure". The chapter articulates the urgency of the public need for the facts about this chemical threat, and in doing so, she quotes from Jean ROSTAND, biologist, philosopher, and activist against nuclear bombs and nuclear power: "The obligation to endure gives us the right to know" (CARSON 2002: 13).

This quote aptly captures the state of existence of atomic bomb victims like HAYASHI Kyōko who survived the bombs by living the "obligation to endure" and insisting on the "right to know". These imperative burdens weighed upon HAYASHI on a foundational level in her essay "To Rui, Once Again". A victim of the atomic bomb at age 14, who witnessed the Fukushima Daiichi nuclear accident at age eighty, HAYASHI speaks powerfully of a lifetime of enduring the somatic harm of radiation, and of the absolute right of *hibakusha* and the Japanese public to know the truth. She remarks in "To Rui, Once Again" about the need to educate readers about radiation by explaining the concept of half-life and what happens when plutonium is inhaled. She also voices her obligation as a survivor of the atomic bombs to warn mothers in the areas affected by the Fukushima Daiichi accident to flee with their children as quickly as possible (HAYASHI 2013: 12–13; 18–19). HAYASHI speaks despite the fear that her comments will be disregarded or condemned as "damaging rumors" (*fūhyō higai* 風評被害), a term that continues to circulate post-Fukushima and has been used to silence dissent.

When speaking about the Fukushima Daiichi accident, HAYASHI unequivocally dismisses any illusions about the safety of nuclear power: the nuclear power that was developed for "peaceful use", exploded in what was the "worst possible accident". She obliterates the myth of safety when she argues: "The danger of radioactive material. The problem of 'internal radiation'. For most people it is probably the first time they have heard these words. But this was a problem of the nuclear and the human, nuclear matter and life, that atomic bomb victims have been confronting as 'internal radiation' for sixty years" (HAYASHI 2013: 18). Despite the government's attempts, HAYASHI says, they cannot hide the danger from "internal radiation" (HAYASHI 2013: 19). As a survivor, HAYASHI focuses on this long-term nuclear harm that is crucial for seeing beyond this false divide between bombs and power plants: "During the twentieth century, we lived in fear of radiation from an atomic bomb. In the twenty-first century, the threat of radiation will come from accidents at nuclear power plants" (HAYASHI 2017a; HAYASHI 2013: 21). Writing this chapter in the autumn of 2022, as the war in Ukraine drags on, there is a widespread fear that HAYASHI's words could become far more prophetic should Russia decide to use nuclear plants like Zaporizhzhia as a dirty bomb, a possibility that further erodes the myth of safety around nuclear power. This situation has been described as a "new kind of nuclear threat", a rhetoric that can only consider this threat to be new by relying upon the *genbaku/genpatsu* fallacy of dangerous bombs vs. safe energy (SMITH 2022).

In TSUSHIMA's "From 'Dream Songs'" we also see the obligation to inform, especially given her awakening to the expansive network of nuclear activities discussed above. In order to further analyze TSUSHIMA's narrative power, I adapt Christine MARRAN's concept of "obligate storytelling" that she introduces in her analysis of ISHIMURE Michiko, who advocated for victims of Minamata disease. For MARRAN, this storytelling mode "foregrounds material relations as fundamental to narrative" (MARRAN 2017: 27–28). For TSUSHIMA, the material, namely radiation (particles and waste), is also of ultimate importance. TSUSHIMA's essay and novels adhere to MARRAN's definition in that TSUSHIMA addresses the "environmentally disenfranchised on a scale that refuses an ethnic nationalist or humanist writing tradition." She "animates the bonds that obligate one life to another beyond nation and species" (MARRAN 2017: 30). TSUSHIMA is not interested in Japanese particularism or national identity. Rather she seeks to reach beyond national or ethnic boundaries to envision a world impacted by radiation from the entire nuclear fuel cycle.

The idea of refusing "an ethnic nationalist" frame to address the "environmentally disenfranchised" is clear in TSUSHIMA's post-Fukushima writing, but has roots in her earlier fiction as well. The title "From 'Dream Songs'" points to her 2010 novel, "Golden Dream Song" (Ōgon no yume no uta 黄金の夢の歌), which "relates a series of journeys (based on TSUSHIMA's own travels in the early 2000s) through central Asia: post-Soviet Kyrgyzstan, Inner Mongolia and other autonomous regions of China, spaces saturated with multiple minority languages and story-telling traditions that exist in uneasy relations with the centralizing forces of existing nation-states" (MCKNIGHT / BOURDAGHS 2018).[7] TSUSHIMA's stories of nomadism and dispossession are epic in terms of their temporal and geographic scale. And as Anne MCKNIGHT and Michael BOURDAGHS argue, her "interest in narrative forms from and about Japanese geo-peripheries, especially displaced, orphaned ones, extends to aboriginal and oral cultures around the world" (MCKNIGHT / BOURDAGHS 2018). TSUSHIMA continued to explore her interest in aboriginal cultures or indigeneity, epic storytelling, and an expansive geographical reach in the years following the Fukushima Daiichi accident when she authored a number of antinuclear essays, short stories, and novels that capture the effects of the nuclear on indigenous and marginalized communities. In her novels she reveals the deep linkage between the *genbaku* and *genpatsu* through plotlines, images, and orthographic choice. In this next section I briefly examine three of her novels: "Wildcat Dome" (Yamaneko dōmu, 2013), "Jacka Dofuni: A Tale of Oceanic Memory" (Jacka Dofuni: Umi no kioku no monogatari, 2016b), and "Celebrating Half-Life" (Hangenki o iwatte, 2016a).

3.2 Bombs, Radiation, and Buried Waste

TSUSHIMA's "Wildcat Dome" follows a wide cast of multiracial characters, illegitimate children of American soldiers and Japanese women, as they live through postwar history and traverse the globe. Their lives are haunted by the legacy of the war, the atomic bombs, and the reality of radioactive contamination from an accident at a nuclear power plant in March. The nuclear (*genbaku* and *genpatsu*) shadows her characters through Japanese and world history. Adorning the dust jacket and first page of "Wildcat Dome" is the image of the famous Runit Dome in the Marshall Islands that covers an atomic blast crater filled with radioactive soil and debris from the 1958 Cactus test in the South Pacific. In a postscript at the very end of the book, Tsushima quotes from a report by sociologist Takemine Seiichirō:

> American nuclear testing was not confined to Bikini Atoll. Between 1948 and 1958, residents of Enewetak Atoll were forced to relocate because of nuclear testing there as well. But when they were allowed to return in 1980 after American military decontamination work had finished, a number of islands had vanished. The enormous volume of radioactive material from the decontamination work was buried on Runit Island under the massive Runit Dome, a huge, concrete radioactive waste storage facility. Signs were posted in both English and the Marshall language indicating "Danger, Stay Away". Twenty-five years later, the writing is fading and is hard to read. (Tsushima 2013: no pagination)

The photograph on the cover, credited to the Defense Threat Reduction Agency, shows humans walking on top of the dome. They are presumably military personnel, but their existence in such close proximity to this place empties the danger signs of their meaning. The same image is reproduced on the inside of the book as an orange line drawing, a color that represents violence in the novel, both local and global. Orange is associated with both the death of one of the child protagonists, but also with the napalm dropped on the Vietnamese forests and Chilean death squads. The orange color of the Runit Dome image infuses the site with this violence.

Yet for all its visual prominence, neither the Runit Dome nor the American nuclear testing feature directly in the novel. However, the story opens and closes with radiation, linking *genbaku* to *genpatsu*. The Fukushima Daiichi nuclear accident enters the narrative within the first few pages of the novel. One of Tsushima's characters speculates about the effects on the natural world of the nuclear power plant accident in March: the plants are sprouting strangely large leaves, some disturbingly rust colored. The radioactive fallout contaminated the grass, trees, ground, water, and ocean. The novel ends with the fear of this spreading toxin: "The radioactive goop clung to the ground, stones, and houses, and traveled on the wind and rain. It would linger for an unfathomable length of time, and quietly bring a merciless pain to plants, birds and wildlife, insects and fish, and humans from which they could never escape" (Tsushima 2013: 329). In this passage we see Tsushima emphasize the bonds that "obligate one life to another" beyond the humans in her novel.

"Wildcat Dome" spans global nuclear history, and the dome of the cover recalls another dome: the Genbaku Dome in Hiroshima, the symbol of the atomic bombing. The mothers in Tsushima's novel are haunted by the memory of the atomic bombs, and the radioactive rain after the nuclear power plant accident recalls the Black Rain of the atomic bombs. One character, the mother of an outcast, feels history collapsing when she hears the words: plutonium, strontium, and Geiger counter. Another character, Mitch, has a son who lives in Stockholm and is threatened by the explosion at Chernobyl. In "Wildcat Dome", the Fukushima Daiichi accident is not an isolated incident, rather it is one of many nuclear betrayals in Japanese and global history alongside the years of nuclear testing symbolized by the cover image.

In an interview, TSUSHIMA discusses the contradictions inherent in these nuclear sites: "The name 'Runit Dome' refers to something inconvenient that society and people don't want to face up to, something that's been swept under the carpet." TSUSHIMA deftly links her own novel to issues that Japan has tried to sweep under the carpet—the violence of the early postwar era and the Fukushima Daiichi nuclear disaster (TSUSHIMA / KARUBE 2013: 182). The images in the book make it clear that this Runit Dome—seven meters high and 115 meters in diameter—the enormous physical remnant of nuclear testing, is hardly something that can be ignored. It also demonstrates the multigenerational lifespan of nuclear harm, something common to *genbaku* and *genpatsu*. The dome is cracking and could release radionuclides including plutonium through the permeable soil at the bottom and into the ocean—harming Pacific communities for generations to come.

3.3 Oceans, Indigeneity, and Nuclear Harm

TSUSHIMA's interest in global nuclear harm and the Pacific Ocean is woven through her sprawling novel "Jacka Dofuni: A Tale of Oceanic Memory". The novel has two alternating storylines. One details the first-person narrator's travels to Hokkaido and to a museum commemorating the indigenous peoples of northern Japan / Sakhalin in 1967, 1985, and 2011 after the Fukushima Daiichi nuclear accident. The second begins in the 1600s and follows an Ainu-Japanese character from her birth to an Ainu mother who dies in Hokkaido, through her travels with persecuted Christians to Nagasaki and Macau, and eventually her death as an old woman in Dutch Indonesia.

The two storylines intersect through indigeneity, oppression, and a contaminated, radioactive ocean. The post-Fukushima contaminated Japan casts its shadow back onto the narrator's previous visits to indigenous sites in Hokkaido, while the voyages of oceanic refuge of the Ainu-Japanese protagonist sharpen the contrast of a pre- and post-irradiated despoiled sea. Using TSUSHIMA's essay, "From 'Dream Songs'", we can analyze the untouched ocean of the 1600s as a commentary on the post-nuclear age of global environmental harm. In "From 'Dream Songs'", TSUSHIMA educates readers about nuclear harm across the Asia-Pacific including the Taiwan Power Company's nuclear waste storage facility constructed without consent on the land of the Tao people of Orchid Island, the nuclear testing on American Navajo and Inuit land, and in Kazakhstan, Polynesia, etc. TSUSHIMA has another shameful awakening to the fact that indigenous people and land have been targeted for uranium mining, nuclear testing, and final storage of nuclear waste, including a failed Japanese plan in the 1980s to dump radioactive waste in huge drum canisters into the Pacific ocean near the Mariana Islands. After 2011, TSUSHIMA learned of the link between Australian aborigines and uranium sent to Fukushima, another instance of the horrors of nuclear colonialism. She reminds the Japanese that they have forgotten the reality that they are in the same situation as all the countries on the Pacific Ocean with regard to atomic testing. Despite this, and the resolve to renounce all forms of the nuclear by native Pacific islanders, New Zealand and Australian aborigines, the Japanese, acting as if they do not face the same ocean, turned their backs and contaminated the sea with large amounts of radiation from the Fukushima Daiichi nuclear power plant accident.[8]

3.4 Radioactive Orthography

TSUSHIMA's short story "Celebrating Half-Life" is set 30 years in the future in a Japan that has suffered from another earthquake and nuclear accident, where even more dangerous radioactive particles, like plutonium, were released. The title refers to a celebration of the half-life of radioactive cesium 137 by former residents of the disaster area on the 100-year anniversary of World War II. Originally published in the literary journal *Gunzō* 群像 in 2016 in a section themed "The World Thirty Years Later: The Literary Imagination", TSUSHIMA's story moves simultaneously forwards and backwards in time. The image of a future nuclear disaster in Japan is familiar from speculative fiction like "Island of Eternal Life" (*Fushi no shima* 不死の島) by TAWADA Yōko 多和田葉子, a story that like "Celebrating Half-Life" comments on the failure of the Japanese government to competently respond to the 2011 Fukushima Daiichi nuclear accident and imagines a dark future for the nation. In TSUSHIMA's story, former residents of the disaster area use the half-life celebrations as an opportunity to visit their homes, only to find that nothing has changed, meaning the area and its residents have not recovered from the disasters (TSUSHIMA 2016a: 80–81). The title and reference to World War II also echo earlier eras of radioactive contamination, namely the atomic bombings of Hiroshima and Nagasaki. As mentioned earlier, in the wake of the Fukushima Daiichi accident, HAYASHI Kyōko felt the need to explain the term "half-life" for readers she assumed were unfamiliar with the radiological terminology that dominated the lives of atomic bomb survivors. She references cesium 137 and explains that it has a half-life of 30 years, which means it would take longer than the average life span to be freed from its harm, a concept that brings new meaning to *Gunzō*'s theme of the world 30 years later, namely a world where radiation continues to linger in human bodies (HAYASHI 2013: 18–19).

The link between the atomic bombings (*genbaku*) and nuclear accidents (*genpatsu*) goes deeper than the theme of half-life. TSUSHIMA's story is set in an era of military rule. In her fictional Japan, the declining population has led to a loss of international political power, and the nation has fallen prey to a military dictatorship and isolationism, recalling a return to the *sakoku* 鎖国, or closed-door policies of the premodern past. The military government keeps the populace in line with policies that enforce racial superiority and incite xenophobia in order to silence dissent. Those subject to discrimination include Ainu, Okinawans, and Koreans, peoples who have been the object of historical discrimination in Japan. But to this list, TSUSHIMA adds people from Tōhoku, the areas of northeastern Japan that were most heavily affected by the actual, and her fictional, nuclear disaster.

TSUSHIMA makes the shared radioactive contamination and discriminated status of these minorities visible at the word level of her story by writing the names of people and places in the phonetic *katakana* script.[9] This is a clear reference to the use of *katakana* for Hiroshima and Nagasaki, and post-2011 for Fukushima, linking the three locations despite government and industry efforts to separate the threat of nuclear weapons from the so-called safety of nuclear power. TSUSHIMA is not the only author or critic to use *katakana* in this controversial way, but she extends it beyond Fukushima to encompass Tōhoku and all of Japan. The taint of radioactive contamination that comes with this linguistic choice signals to the reader that the minorities in her story are also nuclear victims, evoking an expansive definition of the term *hibakusha* that stretches beyond the original confines of either the atomic bombs or the Fukushima Daiichi accident. This concept reverberates with her explorations and exposes of global nuclear harm in her essay "From 'Dream Songs'".

4 BY WAY OF CONCLUSION: FROM LITERATURE TO ACTIVISM

TSUSHIMA's awakening to her entrapment within the nuclear dispositive and her struggles to make her voice of protest heard pay testament to the expansive network of power surrounding the nuclear. This includes both the domestic and global institutions and networks, but also the particularities of the discursive trap in Japan. As referenced earlier, HAYASHI was also acutely aware of this when she mentioned the desire of atomic bomb victims to warn Fukushima residents not to trust the government and to flee the area, a desire muted by the fear of being dismissed as spreading "damaging rumors", or in HAYASHI's own case, of having her advanced age (over 80 years old) used against her as proof that "radiation is safe" (HAYASHI 2017a; HAYASHI 2013: 20). Amid this climate of fear and confusion, government and industry attempts to silence oppositional voices, and concern from residents of the disaster areas, TSUSHIMA's words on the nuclear move beyond their function as topics for her fiction and essays, to a call to activism.

TSUSHIMA's daughter TSUSHIMA Kai explains this in a posthumous essay titled "Along With Those Who Can Hear My Mother's Voice" (*Haha no koe ga kikoeru hitobito to tomo ni* 母の声が聞こえる人々とともに, 2016). As evident in her essays and fiction, the nuclear was always a global issue for TSUSHIMA Yūko. One week after the Fukushima Daiichi accident, she sent a report to her friends in Taiwan and India and encouraged those beyond Japan's borders to join their voices to say "No More Fukushimas!" a reference to the antinuclear slogan, "No More Hiroshimas!" Her report was published in Taiwan and soon translated into Korean; it was also commented on in India where protestors were killed by the police when demonstrating against Jaitapur Nuclear Power Project in April 2011 (SIDDIQUI 2011).[10] TSUSHIMA launched an international campaign. However, back home, she was criticized for harming the people in Fukushima with her "damaging rumors", and she stopped this campaign. She did not, however, stop writing. The essay, stories, and novels examined in this chapter demonstrate TSUSHIMA's dedication to exposing the long history of global nuclear harm at every stage of the nuclear fuel cycle. Her decision to focus on this global history serves to highlight the particular fallacy of the *genbaku / genpatsu* divide in Japan, and the way that the actions of the pro-nuclear actors and institutions in Japan continue to harm not only their own people, but their oceanic neighbors in the Pacific. Before TSUSHIMA died she went to an antinuclear protest with her daughter and resolved to find answers to the question: what was lost in the Fukushima Daiichi accident? For TSUSHIMA, those answers could only be found in the act of writing, for that is the point of literature, and she wielded the power of narrative to critique this deception and expose the multigenerational harm. TSUSHIMA Kai ends her essay with the call for readers to continue her mother's work and raise their voices to demand "No more Fukushimas!" (TSUSHIMA Kai 2016: 347).

NOTES

1 The Lucky Dragon Number 5 was a Japanese tuna trawler caught in the radioactive fallout from the U. S Castle Bravo nuclear test at Bikini Atoll in the South Pacific on March 1, 1954. The test was twice as powerful as predicted and radioactive fallout exceeded the danger zone. Residents of nearby atolls and boat crews were also ex-

posed, but this boat came to symbolize the incident for the Japanese and reinvigorated the antinuclear movement in Japan. In the wake of this incident, the antinuclear testing movement garnered 32 million signatures, "one-third of the Japanese population at the time" (YOSHIMI 2012: 324).

2   The accident happened on September 30, 1999. See HAYASHI's short story "Harvest" (*Shūkaku* 収穫, 2002) about a farmer dealing with the aftermath of the Tōkaimura accident (HAYASHI 2017b).

3   Coterie journals were published by small groups of authors (often emerging or new writers) as a means to respond to literary and social trends of the era and garner the notice of the literary establishment. The term *hibakusha* is written two ways using different Sino-Japanese characters to indicate either radiation from a bomb (被爆者) or from other exposure (被曝者), like that from an accident at a nuclear power plant.

4   Many left-wing activists, including the Communists, supported the idea of a "peaceful" nuclear in the postwar era, and in the 1980s, the organizers of antinuclear movements took the antinuclear power stance off their agenda in an effort to be as inclusive as possible (KAN 2011: 113–16; KUROKO 2013: 161).

5   "Japan's laws to promote nuclear power, including the Dengen Sampo (the Three Electric Power Laws), have accelerated nuclear reactor construction via subsidies, grants, and other incentives. These laws also have had the perverse effects of discouraging promotion of safety as the highest priority, with consequences that can be seen in the Fukushima nuclear disaster" (BEHLING et al. 2019: 411).

6   For more on ARIYOSHI's "Compound Pollution" (*Fukugō osen* 複合汚染, 1975) see MARRAN's Chapter 3, and COLLIGAN-TAYLOR's Chapter 5 (MARRAN 2017: 91–115; COLLIGAN-TAYLOR 1990: 146–90).

7   See WU for more on her work on aboriginal policies in colonial Taiwan and her depiction of the "Musha Incident", a 1930 uprising of Taiwanese aborigines against the colonial regime in her story "All Too Barbarian" (*Amari ni yaban na* あまりに野蛮な, 2008) (WU 2018).

8   After the Fukushima Daiichi accident, the Australian Mirarr people sent a letter to the UN that uranium taken from the Ranger uranium mine on their land was sent to TEPCO, and hence they have some responsibility for the accident (TSUSHIMA 2012: 454). "Yvonne Margarula, a senior Traditional Owner of the Mirarr people of Kakadu, where the Ranger uranium mine was, wrote to the UN Secretary-General on behalf of her people, expressing their grief for the people of Fukushima, and their continued opposition to the nuclear industry" (OBA 2021).

9   These proper nouns would normally be written in Sino-Japanese characters or the more standard syllabary of *hiragana*. TSUSHIMA writes the following in *katakana*: Japan (Nihon), Tōkyō, Ainu, Okinawa, Tōhoku, Yamato, and Chōsen.

10 Antinuclear protestors in a march against the Jaitapur Nuclear power project in Mumbai on October 21, 2013, held up posters saying "No More Fukushimas." See https://www.kractivist.org/jaitapur-to-witness-anti-nuclear-plant-protest-again/ (last access 2023 / 06 / 29).

References

Asada, Jirō 浅田次郎 (2012): "Hajime ni はじめに". In: Nihon Pen Kurabu (ed.): *Ima koso watashi wa genpatsu ni hantai shimasu* いまこそ私は原発に反対します. Tōkyō: Heibonsha, 8–10.

Behling, Noriko et al. (2019): "Aftermath of Fukushima: Avoiding Another Major Nuclear Disaster". In: *Energy Policy* 126 (March): 411–20; https://doi.org/10.1016/j.enpol.2018.11.038 (last access 2022 / 05 / 09).

Buell, Lawrence (1988): "Toxic Discourse". In: *Critical Inquiry* 24 (3): 639–65.

Carson, Rachel (2002): *Silent Spring*. Boston / New York: Mariner Books.

Colligan-Taylor, Karen (1990): *The Emergence of Environmental Literature in Japan*. New York / London: Garland Publishing, Inc.

Garrard, Greg (2011): *Ecocriticism*. 2nd edition. Abingdon, Oxon / New York: Routledge.

Hayashi, Kyōko 林京子 (2008): "From Trinity to Trinity". Translated by Kyoko Selden. In: *The Asia-Pacific Journal: Japan Focus* 6 (5); https://apjjf.org/-Hayashi-Kyoko/2758/article.html (last access 2022 / 05 / 09).

Hayashi, Kyōko 林京子 (2013): "Futatabi Rui e 再びルイへ". In: *Gunzō* 群像 68 (4), 7–25.

Hayashi, Kyōko 林京子 (2017a): "To Rui, Once Again". Translated by Margaret Mitsutani. In: *The Asia-Pacific Journal: Japan Focus* 15 (7.3); http://apjjf.org/2017/07/Hayashi.html (last access 2022 / 05 / 09).

Hayashi, Kyōko 林京子 (2017b): "Harvest". Translated by Margaret Mizutani. In: *The Asia-Pacific Journal: Japan Focus* 15 (10.3); https://apjjf.org/2017/10/Hayashi.html (last access 2022 / 05 / 09).

Kan, Takayuki 管孝行 (2011): "Saimu shiharai no kakugo o: Genpatsu sonkai ga kokuchi suru mono 債務支払いの覚悟を：原発損壊が告知するもの". In: Kuroko, Kazuo 黒古一夫 (ed.): *Hiroshima, Nagasaki kara Fukushima e: 'Kaku' jidai o kangaeru* ヒロシマ・ナガサキからフクシマへ：「核」時代を考える. Tōkyō: Bensei shuppan, 109–24.

Killingsworth, M. Jimmie / Palmer, Jacqueline S. (2012): *Ecospeak: Rhetoric and Environmental Politics in America*. Carbondale: Southern Illinois University Press.

Kuroko, Kazuo 黒古一夫 (2013): *Bungakusha no "kaku, Fukushima ron": Yoshimoto Takaaki, Ōe Kenzaburō, Murakami Haruki* 文学者の「核・フクシマ論」：吉本隆明・大江健三郎・村上春樹. Tōkyō: Sairyūsha.

Kuroko, Kazuo 黒古一夫 (2018): *Genpatsu bungakushi, ron: Zetsubōteki na "kaku (genpatsu)" jōkyō ni kō shite* 原発文学史・論：絶望的な「核（原発）」状況に抗して. Tōkyō: Shakai hyōronsha.

MARRAN, Christine L. (2017): *Ecology without Culture: Aesthetics for a Toxic World*. Minneapolis: University of Minnesota Press.

MCKNIGHT, Anne / BOURDAGHS, Michael (2018): "Memento Libri: New Writings and Translations from the World of Tsushima Yūko (1947~2016)". In: *The Asia-Pacific Journal: Japan Focus* 16 (12.7); https://apjjf.org/2018/12/McKnight.html (last access 2022 / 05 / 09).

OBA, Yumi (2021): "Australia is deeply 'connected' to the Fukushima nuclear accident". In: *SBS Your Language*, 03 / 11, 2021; https://www.sbs.com.au/language/english/australia-is-deeply-connected-to-the-fukushima-nuclear-accident (last access 2022 / 05 / 09).

ŌE, Kenzaburō (2011): "History Repeats". In: *The New Yorker*, 03 / 28; http://www.newyorker.com/talk/2011/03/28/110328ta_talk_oe (last access 2022 / 05 / 09).

SIDDIQUI, Danish (2011): "Protests against Jaitapur Nuclear Plant Turn Violent". In: *Reuters*, 04 / 19 (sec. Money News); https://www.reuters.com/article/idINIndia-56433720110419 (last access 2022 / 05 / 09).

SMITH, Kiona N. (2022): "Chernobyl Reminds Us What's At Stake In Russia's War On Ukraine". In: *Forbes*, 03 / 04; https://www.forbes.com/sites/kionasmith/2022/03/04/chernobyl-reminds-us-whats-at-stake-in-russias-war-on-ukraine/?sh=6087ee7a4ea2 (last access 2022 / 11 / 14).

TAWADA, Yōko 多和田葉子 (2012): "The Island of Eternal Life". Translated by Margaret MITSUTANI. In: LUKE, Elmer / KARASHIMA, David (eds.): *March Was Made of Yarn: Reflections on the Japanese Earthquake, Tsunami, and Nuclear Meltdown*. New York: Vintage, 3–11.

TSUSHIMA, Kai 津島香以 (2016): "Haha no koe ga kikoeru hitobito to tomo ni 母の声が聞こえる人々とともに". In: TSUSHIMA, Yūko 津島佑子: *Yume no uta kara* 夢の歌から. Tōkyō: Inscript Inc, 338–47.

TSUSHIMA, Yūko 津島佑子 (2012): "'Yume no uta' kara 「夢の歌」から". In: NIHON PEN KURABU (ed.): *Ima koso watashi wa genpatsu ni hantai shimasu* いまこそ私は原発に反対します. Tōkyō: Heibonsha, 454–70.

TSUSHIMA, Yūko 津島佑子 (2013): *Yamaneko dōmu* ヤマネコ・ドーム. Tōkyō: Kōdansha.

TSUSHIMA, Yūko 津島佑子 (2016a): "Hangenki o iwatte 半減期を祝って". In: TSUSHIMA, Yūko 津島佑子: *Hangenki o iwatte* 半減期を祝って. Tōkyō: Kōdansha, 73–105.

TSUSHIMA, Yūko 津島佑子 (2016b): *Jacka Dofuni: Umi no kioku no monogatari* ジャッカ・ドフニ：海の記憶の物語. Tōkyō: Shūeisha.

TSUSHIMA, Yūko 津島佑子 (2016c): *Yume no uta kara* 夢の歌から. Tōkyō: Inscript Inc.

Tsushima, Yūko 津島佑子 / Karube, Tadashi 苅部直 (2013): "Kikite Karube Tadashi 'Yamaneko Dōmu': Kakusareta 'sengo' o tadorinaosu 聞き手苅部直『ヤマネコ・ドーム』:隠された「戦後」をたどり直す". In: *Gunzō* 群像 68 (7): 180–90.

Wu, Peichen (2018): "The Remains of the Japanese Empire: Tsushima Yūko's All Too Barbarian; Reed Boat, Flying; and Wildcat Dome". Translated by Michael Bourdaghs. In: *The Asia-Pacific Journal: Japan Focus* 16 (12.2); https://apjjf.org/2018/12/Wu.html (last access 2022 / 05 / 09).

Yoshimi, Shun'ya (2012): "Radioactive Rain and the American Umbrella". Translated by Shi-Lin Loh. In: *The Journal of Asian Studies* 71 (2): 319–31.

How Shifting U. S. Perceptions of the Atomic Bombings of Hiroshima and Nagasaki Have Influenced Americans' Attitudes Toward the Nuclear Arms Race, 1945–2022

Peter Kuznick[1]

In April 2016, legendary basketball coach Bobby KNIGHT, who had been fired by Indiana University for his violent and abusive behavior 15 years earlier, returned to the Hoosier state to endorse candidate Donald TRUMP. In his typically charming and genteel fashion, KNIGHT declared:

> We gotta talk about this presidential crap just for a moment here. You older folks remember this. Your mothers, fathers have taught you this. I'll tell you who they said wasn't presidential. I don't even know what the hell 'presidential' means, but they told him he wasn't presidential. And that guy they told all these people that wanted to say, 'You're not presidential,' that guy was Harry Truman. Harry Truman, with what he did in dropping and having the guts to drop the bomb in 1944 [sic], saved, saved millions of American lives. And that's what Harry Truman did and he became one of the three great presidents of the United States. And here's a man who would do the same thing because he's going to become one of the four great presidents of the United States. (SCOTT 2016)

After the cheers died down, TRUMP, overwhelmed with emotion and gratitude, took the microphone and gushed: "Such a great guy. Wow. How do you top that? How do you top that? What a great guy. You should be very proud of him in Indiana. That is a national treasure, OK?" (WEAVER 2016).

No. Not okay. No one ever mistook either TRUMP or KNIGHT for an expert regarding nuclear history or the final victory in the Pacific phase of World War II. But TRUMP would soon win the election to the U. S. presidency and would, a mere nine months after the KNIGHT endorsement, gain control over America's immense nuclear arsenal, which would effectively give him veto power over the future existence of life on this planet. Fortunately, he never exercised that ultimate power, but he did raise the world's temperature several degrees not only with his climate change-denying lunacy but with his cavalier, ignorant, and dangerously provocative views on nuclear weapons. He asked during the campaign about the point of having nuclear weapons if the U. S. couldn't use them, but instead of opting to eliminate them due to their excessive destructiveness, as most reasonable people would, he opted to make them more usable. In office, he pushed for development of smaller tactical nuclear weapons. In July 2017, upon realizing the extent of the decline in the size of the U. S. nuclear arsenal since its mid-1980s peak, he suggested a tenfold increase, provoking Secretary of State Rex TILLERSON to call him a "fucking moron" (FILKINS 2017).

TRUMP's response to North Korean missile and nuclear bomb tests in 2017 made analysts fear that nuclear war with North Korea might be imminent. Having earlier dismissed North Korean leader KIM Jong Un as a "28 year old wack [sic] job" and a "maniac" (STONE/KUZNICK 2019: 658), TRUMP responded to KIM's threats to fire missiles at Guam by warning of "fire and fury like the world had never seen" (BIERMAN 2017). His soon-to-be national security advisor John BOLTON chimed in, "[y]ou eliminate the nuclear threat by eliminating North Korea" (STONE/KUZNICK 2019: 660). The crisis ratcheted up further following North Korea's sixth and most powerful nuclear test in September, which it claimed was a hydrogen bomb. Experts confirmed that the blast was 17 times as powerful as the Hiroshima bomb. KIM responded to new UN sanctions by threatening to "reduce the U.S. mainland to ashes and darkness" (LEE 2017). TRUMP told the UN General Assembly, "Rocket Man is on a suicide mission for himself and for his regime. The United States has great strength and patience, but if it is forced to defend itself or its allies, we will have no choice but to totally destroy North Korea." U.S. UN Ambassador Nikki HALEY warned that KIM was "begging for war" (SHELBOURNE 2017). KIM shot back that TRUMP was a "mentally deranged U.S. dotard" whom KIM would "tame [...] with fire" (CHOE 2017).

The insults and name-calling continued, backed up by massive U.S. and South Korean military exercises and U.S. deployment of three aircraft carrier strike groups to the Pacific. War talk reverberated across the U.S. and globally. Former CIA director John BRENNAN thought there was a 20- to 25-percent chance of war. Council of Foreign Relations president Richard HAASS put the odds at 50–50 as both countries began to mobilize. Republicans and Democrats sounded the alarm. Senate Foreign Relations Committee Chair Bob CORKER, a Tennessee Republican, warned that TRUMP's actions had put the U.S. "on the path to World War III" (MARTIN/LANDLER 2017). Democrats Ed MARKEY and John CONYERS introduced a bill requiring Congressional approval before an attack could be launched. The situation heated further in late November when the DPRK launched a Hwasong-15 intercontinental ballistic missile that could reach anywhere in the United States. In his nationally televised New Year's Day address, KIM announced that "the whole territory of the U.S. is within the range of our nuclear strike and a nuclear button is always on the desk of my office" (KIM 2018). TRUMP responded that his nuclear button was "much bigger & more powerful [...] and my Button works" (BAKER/TACKETT 2018).

With the insults, threats, and puerile macho posturing increasing, the experts at the *Bulletin of the Atomic Scientists* moved the hands of the Doomsday Clock to two minutes before midnight in January 2018, the closest it had been since the perilous 1950s. The TRUMP administration's *Nuclear Posture Review*, which was released the following month, elevated the role of nuclear weapons in U. S. security strategy and lowered the threshold for their use. It even called for development of a new sea-launched cruise missile and a new low-yield sea-launched ballistic missile.

TRUMP never explicitly weighed in on the morality or righteousness of the 1945 atomic bombings, but he did loudly inveigh against the idea that President OBAMA might apologize for the bombings when he visited Hiroshima in 2016. TRUMP offended many Japanese while campaigning that year by suggesting that Japan should abandon the U. S. nuclear umbrella and develop its own nuclear arsenal. As he said in a CNN town hall in March of that year, "[a]t some point, we have to say, you know what, we're better off if Japan protects itself against this maniac in North Korea" (THOMPSON 2016). Secretary of State John KERRY, who was in Japan for April's G-7 meeting, visited Hiroshima, went to the Peace Museum, and laid a wreath at the Cenotaph, finding the experience "gut-wrenching". "It tugs at all of your sensibilities as a human being" (LABOTT 2016). Yet OBAMA's State Department made clear that KERRY, the first U. S. Secretary of State to visit Hiroshima, would certainly not apologize for the atomic bombings.

KERRY's visit helped clear the path for President OBAMA's visit the following month. The only previous president to visit Hiroshima was Jimmy CARTER, but he did so long after he was out of office. The OBAMA visit was certainly controversial even though he had waited more than seven years to undertake it. OBAMA's June 2009 Prague speech ostensibly calling for the abolition of nuclear weapons had greased the way for him to win the Nobel Peace Prize that year, making him the first leader to receive the prize while conducting the invasion of two countries with a third—Libya—in the offing. OBAMA's 2010 *Nuclear Posture Review* had lowered the profile of nuclear weapons in U. S. defense strategy and he had negotiated the New Start Treaty with Russia, which further reduced the number of nuclear weapons and delivery vehicles. But while OBAMA was en route to Hiroshima, the Federation of American Scientists revealed that he had reduced the nuclear arsenal at a slower pace than his three predecessors—the two BUSHes and CLINTON—despite sharp cuts in the Russian arsenal (KRISTENSEN 2016). So there were both hopes and fears as he prepared for his visit to Hiroshima. Donald TRUMP was not the only prominent American or editorial writer to insist that OBAMA not apologize for the 1945 atomic bombings. *Newsweek* described a "wave of editorials lamenting an 'apology tour' second-guessing past decisions" (WHALEN 2020). The American Enterprise Institute's Michael MAZZA's op-ed in the *Wall Street Journal* argued, "Obama's visit will only reinforce the increasingly common but misguided view that America shouldn't have used the atomic bomb." MAZZA didn't fear the reaction in Japan or other parts of Asia;

he feared the reaction "at home". MAZZA understood that OBAMA's stated intention "to highlight his continued commitment to pursuing the peace and security of a world without nuclear weapons" would suggest that "the U.S. bombings played a role in ushering in a world that has become less peaceful and less secure" instead of the "correct lesson [...] that it is sometimes necessary to employ great violence to root out great evil." Pointing to the low level of support for the atomic bombings recently shown among 18- to 29-year-olds, he worried that such thinking "hints at a broader skepticism toward applying American power to achieve just ends" (MAZZA 2016). In fact, OBAMA felt pressure to avoid discussing the decision in its entirety. Deputy National Security Advisor Ben RHODES announced that OBAMA "will not revisit the decision to use the atomic bomb at the end of World War II" (MCCURRY/SMITH/YUHAS 2016), a decision enthusiastically endorsed by National Security Advisor Susan RICE (KUZNICK 2020: 75–76).

OBAMA stuck to that playbook and worse. First, he spent nearly twice as much time with U.S. and Japanese troops at the Iwakuni military base as he did in Hiroshima, where he spent eight minutes in the Peace Memorial Museum. After his talk at the Cenotaph in the Peace Memorial Park, he spoke with the two atomic bomb survivors (*hibakusha* 被爆者) deliberately placed in front of him who had publicly stated that they didn't want an apology, embracing Shigeaki MORI, who had spent years apologizing to the U.S. for the American POWs murdered by angry Japanese who had survived the atomic bombing. OBAMA went nowhere near the more militant and less forgiving *hibakusha* who had been carefully seated beyond his reach. But it was his speech that was most disconcerting, beginning with its opening sentence, stating: "Seventy-one years ago, on a bright cloudless morning, death fell from the sky and the world was changed." Neither here nor anywhere else in the speech did he even mention that the United States bore responsibility for death falling from the sky. Even more egregious, however, was his contention that the war "reached its brutal end in Hiroshima and Nagasaki"—the flagrant falsehood that has been used to justify the atomic bombings for 77 years (THE WHITE HOUSE 2016).

FIRST REACTIONS TO THE BOMBINGS

The belief that the atomic bombs ended World War II and forced Japan's surrender dates back to the immediate aftermath of the bombings. Although President TRUMAN's earliest statement stresses revenge for Pearl Harbor, the rationale quickly shifted to justifying the bombings as the only way to avoid a costly Allied invasion that would have resulted in the loss of "thousands" of American lives. The number of projected casualties would gradually increase and the number of potential Japanese victims would be added to the total. The *New York Times*, in its earliest reporting on August 7, stated,

> Secretary Stimson said that this new weapon 'should prove a tremendous aid in the shortening of the war against Japan,' and there were other responsible officials who privately thought that this was an extreme understatement, and that Japan might find herself unable to stay in the war under the coming rain of atom bombs. (SHALETT 1945: 1)

Another *Times* article that day described TRUMAN's notifying crew members aboard the USS Augusta of the successful deployment of the first atomic bomb. "The crew's reception of the news was uproarious", the *Times* reported. "The word heard on every hand was, 'I guess I'll get home sooner now'" (ANONYMOUS 1945b: 2). An August 7 *Times* article by Hanson BALDWIN was titled "The Atomic Weapon: End of War Against Japan Hastened But Destruction Sows Seeds of Hate." Filled with foreboding about the future and concerned that the massive scale and indiscriminate nature of U. S. bombing of civilians had made "Americans [...] a synonym for destruction", BALDWIN acknowledged that the bomb's "use will probably save American lives, may shorten the war materially, may even compel Japanese surrender" (BALDWIN 1945: 10).

The Japanese surrendered on August 14—eight days after the atomic incineration of Hiroshima and five days after Nagasaki suffered a similar fate. Americans celebrated, none more so than those troops stationed in the Pacific who were getting ready, they believed, to invade Japan. Among them was young second lieutenant Paul FUSSELL who wrote a *New Republic* piece in 1981 and a 1988 book both titled *Thank God for the Atom Bomb*. In the book, he confessed, "[F]or all the fake manliness of our facades we cried with relief and joy. We were going to live. We were going to grow up to adulthood after all" (FUSSELL 1981: 29).

The belief that the atomic bombs were responsible for the Japanese surrender and that they saved countless lives in what would have been a bloody invasion was widely held right from the start. It was commingled with deep hatred of the Japanese that resulted from stories of Japanese prewar atrocities in China and other parts of Asia, the "sneak" attack at Pearl Harbor, and the wartime propaganda about Japanese fanaticism replete with horror stories about unimaginable Japanese brutality toward U. S. and other Allied troops at Bataan and beyond, on top of some good old-fashioned American racism. Many Americans shared TRUMAN's desire to punish Japan for its transgressions, which further influenced American attitudes toward the atomic bombings as did post-New Deal faith in the government and end-of-the-war triumphalism and patriotism (GALLUP 1945: 3).

GALLUP conducted its first poll in an attempt to ascertain public attitudes about the bomb, asking between August 10 and August 15, 1945, "[d]o you approve or disapprove of using the new atomic bomb on Japanese cities." The lopsided response was not surprising. Despite the fact that the Vatican and numerous religious leaders had immediately condemned the bombings as unchristian acts and other observers like *Saturday Review* editor Norman COUSINS were appalled by what the United States had done, 85 percent of those polled supported the bombing; only ten percent opposed it. Among the college educated it was 90 percent to seven percent. George GALLUP contrasted this response to the fact that the public had opposed use of poison gas against Japan throughout the war even if it would reduce casualty among U. S. soldiers, calling the reversal an "odd twist" (GALLUP 1945; GALLUP 1972: 521–22).

Fortune Magazine released results of a poll conducted by Elmer ROPER on November 30, 1945 finding that 53.5 percent of respondents approved the bombings as conducted and another 22.7 percent were so filled with hatred of the Japanese that they wished the U. S. could have dropped "many more" bombs "before Japan had a chance to surrender." 13.8 percent preferred demonstrating the bomb "on some unpopulated region" before dropping it on a Japanese city. Only 4.5 percent were outright opposed to dropping the bombs (YAVENDITTI 1974: 225). Four years of war had numbed many Americans to the suffering of war's victims. Writer and social critic Dwight MACDONALD understood the dehumanization that had occurred. He traced the sad progression from the "unbelieving horror and indignation" people felt in 1938 when FRANCO's planes bombed Barcelona killing hundreds of civilians to the abject indifference to reports of a million casualties in the firebombing of Tokyo seven years later. "King Mithridates is said to have immunized himself against poison by taking small doses, which he increased slowly. So the gradually increasing horrors of the last decade have made each of us, to some extent, a moral Mithridates, immunized against human sympathy" (MACDONALD 1957: 97).

American hatred toward the Japanese was indeed palpable and pervasive. Pulitzer Prize-winning historian Allan NEVINS observed after the war: "Probably in all our history no foe has been so detested as were the Japanese" (NEVIN 1946: 13). *Newsweek* had voiced the same judgment in January 1945, contending, "[n]ever before has the nation fought a war in which our troops so hate the enemy and want to kill him" (ROSE 1973: 58).

When the question was phrased a little differently by the National Opinion Research Center at the University of Chicago in September of that year, the results were a little less severe. Asked if they would have dropped the bomb if they bore responsibility for the decision, 67 percent responded that they would indeed have bombed Japanese cities; 26 percent would have targeted an unpopulated location, and four percent would not have dropped them at all (SMITH 2020).

The early triumphalist euphoria occurred at a time when prospects for a unified and peaceful postwar period seemed favorable in the afterglow of victory over Germany and Japan. But, as historian Paul BOYER had shown, even that early excitement over the prospect of ending the war with a weapon that only the United States possessed was immediately soured by the realization that the future looked bleak in a nuclear-armed world. BOYER described a "primal fear of extinction" that swept the United States. NBC radio news commentator H. V. KALTENBORN warned listeners in his August 6 evening broadcast: "For all we know, we have created a Frankenstein! We must assume that with the passage of only a little time, an improved form of the new weapon we use today can be turned against us" (BOYER 1986: 5). The *St. Louis Post Dispatch* feared the bombing had "signed the mammalian world's death warrant, and deeded an earth in ruins to the ants" (ANONYMOUS 1945a: 12). NBC radio commentator Cesar SAERCHINGER noted: "[T]he atomic bomb is merely in its infancy. Indeed, mankind [...] has achieved the power to destroy himself" (GEDDES 1945: 164).

One of the starkest assessments came from Major George Fielding ELIOT, who wrote in the *New York Herald Tribune*: "Mankind stands at the crossroads of destiny." If humans fail to rise to the challenge, "this planet will vanish into darkness and roll on, a blackened cinder, through the limitless night of interstellar space." Earlier pre-atomic warnings, he explained,

> were warnings of chaos and of terror, but they were not warnings of the end of the world, only of the end of a particular phase of civilization. They were warnings of a new Dark Age, out of which man might again have arisen after a few centuries of suffering. But the forces which man has now brought into play are forces which can be utterly destructive, so that no living thing may survive their loosing—if ever they are loosed in their ultimate power. (GEDDES 1945: 167)

What BOYER refers to as the "post-Hiroshima gloom" was reflected in CBS radio correspondent Edward R. MURROW's statement that day that "[s]eldom, if ever, has a war ended leaving the victors with such a sense of uncertainty and fear, with such a realization that the future is obscure and that survival is not assured" (BOYER 1986: 7). The *New York Herald Tribune* took "no satisfaction in the thought that an American air crew has produced what must without doubt be the greatest simultaneous slaughter in the whole history of mankind, and even in its numbers matches the more methodical mass butcheries of the Nazis or of the ancients" (GEDDES 1945: 58).

Time and again in early commentary, Americans envisioned themselves as future victims of the terror they had unleashed on Japan in August 1945, but the actual human tragedy for Japanese victims was only an abstraction that could be easily ignored or justified, as the *Times* did in its August 12 commentary on the atomic bombings: "By their own cruelty and treachery, our enemies had invited the worst we could do to them" (ANONYMOUS 1945c: 8).

But some began chipping away at this narrative. On August 19, the *New York Times* carried a front-page article headlined, "Einstein Deplores Use of Atom Bomb". But the first serious dent in this triumphal narrative came with the publication of John HERSEY's extraordinary 31,000-word article "Hiroshima" in the *New Yorker* in August 1946 and soon thereafter as a book. In 1999, the experts assembled by the NYU School of Journalism named it the most important work of journalism in the 20th century. By closely recounting the often harrowing and painful experiences of six Japanese survivors, HERSEY, for the first time on a large scale in English, humanized the victims. The impact was explosive. Critic Mary McCARTHY wrote, "[u]p until August 31 of this year no one dared to think of Hiroshima—it appeared to us all as a kind of hole in human history" (McCARTHY 1968 [1946]: 367). The book did more than just fill a hole. The reviewer for *Christian Century* captured the enthusiasm with which many readers responded, writing, "[o]nce in a lifetime you read a magazine article that makes you want to bounce up out of your easychair and go running around to your neighbors, thrusting the magazine under their noses and saying: 'Read this! Read it now!'" (HUTCHINSON 1946: 1151).

And they did, including the defenders of the bombing. McGeorge BUNDY, aide to Secretary of War Henry STIMSON and son of STIMSON's special assistant Harvey BUNDY, admitted, "[w]e all exhausted ourselves" reading the article (BLUME 2020: 143). But that wasn't all. Norman COUSINS quickly published a new editorial condemning what he now called "the crime of Hiroshima and Nagasaki" and Admiral William F. "Bull" HALSEY, commander of the Third Fleet, added his voice to those of other top admirals and generals (seven of America's eight five-star officers in 1945 would go on record declaring the atomic bombs were either militarily unnecessary, morally reprehensible, or both), "[t]he first atomic bomb was an unnecessary experiment [...]. It was a mistake to ever drop it" (ALPEROVITZ 1995: 445).

Realizing that they were in jeopardy of losing the battle over the rectitude of the atomic bombings, Harvard president and top wartime science administrator James CONANT, who had chaired the National Defense Research Committee and served as deputy director of the Office of Scientific Research and Development during the war, enlisted respected senior statesman Henry STIMSON to write the "official" response. With the assistance of the BUNDYs, General Leslie GROVES, and others, STIMSON defended the decision in the February 1947 issue of *Harper's*, arguing that the bombings were the lesser of two evils and that by forcing Japan's surrender without an invasion, they avoided a million U. S. casualties (both dead and wounded) and saved a far greater number of Japanese lives. In its lead editorial, the *New York Times* proclaimed, "[t]here can be no doubt that the President and Mr. STIMSON are right when they maintain that the bomb caused the Japanese to surrender" (ANONYMOUS 1947).

That combined with the growing tension between the U. S. and the Soviet Union in 1946 and 1947 led some Americans to revamp their thinking about the atomic bombings and the wisdom of the nuclear age. In October 1947, GALLUP released results of a new survey comparing responses then with those offered in September 1945 as to whether Americans thought "it was a good thing or a bad thing that the atomic bomb was developed." Whereas 69 percent thought positively about the bomb in 1945, only 55 percent did so in 1947. The percentage of those who thought it a bad thing more than doubled from 17 to 38. The percentage of those who thought the U. S. should continue manufacturing bombs had risen from 61 to 70 over the past 18 months perhaps because the percentage of respondents who believed other countries were already producing their own bombs had jumped from 42 to 59 over that same period (GALLUP 1947; GALLUP 1972: 527, 578, 680). A ROPER poll the following year for *Fortune* found 66 percent supporting the 1945 bombings, but 14 percent thought an unpopulated area should have first been chosen and twelve percent opposed using any atomic bombs.

The parameters of the debate over the atomic bombings were established within days of the attacks as the three narratives—heroic, tragic, and apocalyptic—were clearly set before the public (KUZNICK 2007). EINSTEIN even early intuited that the real target was the Soviet Union as much as it was Japan long before British physicist P. M. S. BLACKETT argued in his 1949 book *Fear, War, and the Bomb* that the bombings were not the last acts of World War II but the first acts of the Cold War against the Soviet Union. It is stunning how closely the public debate has hewed to these formulations over the subsequent seven and a half decades although the positions embraced have not always conformed to political views or attitudes about the legitimacy of future use of nuclear weapons. In fact, the usual left-right divides didn't always apply. While many progressives became critics of the atomic bombings and the ensuing nuclear arms race and Cold War, Uday MOHAN and Leo MALEY have demonstrated that many conservatives also condemned the bombings. Medford EVANS wrote in America's leading conservative magazine, the *National Review*, in 1959 that criticism of the atomic bombings of Hiroshima and Nagasaki "was becoming part of the national conservative creed" (EVANS 1959: 525).

American nuclear fears were stoked by the Soviet Union testing its first bomb in August 1949, the U. S. decision to proceed with development of a hydrogen bomb, the Korean War, and the discovery of atomic espionage in 1950.

Americans who had been inclined to think more critically about the utility and wisdom of possessing nuclear arms were now increasingly inclined to see the bomb more as a shield and protector than a potential destroyer. Nuclear fears ran high throughout the 1950s and intensified with the Soviet launch of Sputnik in 1957 upon the heels of their development of the first ICBM.

PUBLIC OPPOSITION TO THE BOMBINGS CLIMBS IN RECENT YEARS

Among those who were watching OBAMA's visit to Hiroshima "very, very closely" was Fox News star Bill O'REILLY (O'REILLY 2016). He told viewers that he was making sure there was no "disparagement" of the atomic bombings. He had nothing to fear. But just to make certain the message was clear, O'REILLY wrote his own book later that year about the end of the war. It was part of his "killing" series co-authored with Martin DUGGARD. Dismissed as "historical fiction" by reviewers, O'REILLY's defense of the bombings ended on a very personal note. He wrote:

> Not usually introspective, my father was convinced of one certainty, which he shared with me on a few occasions—that his very existence, and therefore my life as well, was likely saved by a terrible bomb and a gut-wrenching presidential decision that is still being debated to this day. But for the young ensign and his present-day son, there really is no debate, only a stark reality. Had the A-bombs not been used, you would very likely not be reading this book. (O'REILLY / DUGARD 2016: 283)

As the 75th anniversary of the bombings approached in 2020, the debate over the 1945 decision heated up once again, but it lacked the passion of the 50th anniversary when the World War II generation was still vibrant enough to assert its views. Susan RICE, OBAMA's former National Security Advisor, kicked things off in October 2019 with a *New York Times* op-ed. "Following D-Day, my father was sent to the West Coast to prepare for deployment to the Pacific theater", she recalled, adding, "[h]e was spared combat by President Harry Truman's decision to drop atomic weapons on Hiroshima and Nagasaki, provoking the Japanese surrender" (RICE 2019). RICE was either consciously or unconsciously perpetuating the lie that had been at the heart of American triumphalism for 74 years. She should have known better, having majored in history at Stanford where the talented and prolific nuclear historian Bart BERNSTEIN had taught for decades. She later got M. Phil and D. Phil degrees in international relations at Oxford. Perhaps her judgment was clouded by the fact that she was a Harry S. TRUMAN scholar at Stanford. She had accompanied OBAMA on his 2016 visit to Hiroshima during which she joined him in visiting the Hiroshima Peace Memorial.

Rice had been involved in preparing for Obama's Hiroshima visit, even doing damage control for the notoriously cautious president. She met beforehand with a delegation of veterans at the White House and assured them that there would be no apology for the atomic bombing. She doubled down on this assurance when discussing the impending visit on CNN. Host Fareed Zakaria noted that Rice's deputy Ben Rhodes had said that Obama wouldn't be revisiting the decision to drop the bomb and commented, "I wonder why not. It seems like that is the elephant in the room and why not discuss the—you know, what the president thinks about it? [...] Surely the Japanese must be wondering." Then Zakaria asked Rice point blank whether she thought the atomic bombings were justified. She responded, "I'm not going to give you my historian's judgment on the decision." "Why? Why not?" he asked. She punted a second time: "This will be a forward-looking visit. Yes, it will happen in the context of history, but we don't think it's particularly useful to give a long discourse on the past." Zakaria pressed one more time: "And the president must have a view on whether it was the correct decision to drop the atomic bomb. He's a man who studies history deeply. Surely he has a view." "I'm not saying he doesn't", she replied testily—end of conversation (Zakaria 2016).

New York Times reporter Maria Cramer wrote a piece on July 15, 2020, exploring different aspects of the first atomic test in Alamogordo, New Mexico on July 16, 1945. She mentioned scientists' efforts to block U. S. usage of the bombs against Japan but then gave the final word to Steve Olson, who had written a book about plutonium development at Hanford. Olson averred: "It's very hard to conceive of a set of developments in 1945 that would have avoided dropping those bombs. [...] Truman wanted to end the war as quickly as possible" (Cramer 2020). Cramer didn't challenge this assertion.

Longtime Fox News stalwart Chris Wallace, who left Fox for CNN in December 2021, finally having had enough of the network's mendacity, wrote a number one *New York Times* bestselling book in 2020 titled *Countdown 1945: The Extraordinary Story of the Atomic Bomb and the 116 Days that Changed the World*, which the *Wall Street Journal* called "deservedly the nonfiction blockbuster of the season" (Hornfischer 2020). It was certainly a "blockbuster", but calling it "nonfiction" might be a bit of a stretch. Wallace wrote, "[d]espite all his misgivings, Truman knew he had to drop the atomic bomb. The Manhattan Project had given him a weapon to potentially end the war. And no matter how devastating their losses, the Japanese refused to surrender. They left him no choice" (Wallace 2021: 163).

Given the importance of the atomic bombing of Hiroshima and Nagasaki—the Newseum's[2] panel of experts named it the most important news event of the 20th century—there had been surprisingly little public opinion polling about it after those early GALLUP and ROPER polls in 1945, at least until the Cold War was in its final days. In 1965, before the anti-Vietnam War movement erupted into a broader questioning of U. S. foreign policy, 70 percent told Louis HARRIS & Associates that they supported the atomic bombings and 17 percent voiced opposition. Support went down some when HARRIS asked that question again in 1982 at a time when Ronald REAGAN's hawkish foreign policy and military buildup had rekindled widespread concerns about the dangers of nuclear war. 63 percent considered the bombings "right and proper" while the percentage of those opposed jumped a bit to 26.

GALLUP's polling showed greater opposition in the 1990s. In 1990, with tensions between the U. S. and USSR having eased markedly, largely due to the efforts of Mikhail GORBACHEV, only 53 percent supported the bombings while 41 percent expressed disapproval. The results remained the same in November 1991, a month before the collapse of the Soviet Union. The percentage supporting the bombing jumped a couple points in December 1994 and rose further in the GALLUP poll on the eve of the 50th anniversary in 1995, probably reflecting the short-lived post-Cold War triumphalism and the media's distorted reporting on the Smithsonian's Enola Gay exhibit controversy, which resulted in the Air and Space Museum canceling the planned exhibit on the atomic bombings in response to a fusillade of pressure from the Air Force Association, American Legion, and right-wing Republicans. Now 59 percent approved with disapproval dropping to 35 percent—the highest level of support registered in five GALLUP polls between 1990 and 2005 (BOWMAN 2016).

But when asked if they personally had to decide what to do, 50 percent said they would have tried some other way to force surrender and only 44 percent said they would have used atomic bombs (STOKES 2015). Oddly, the percentage who believed the bombs saved American lives dropped from 86 to 80 between 1995 and 2005. Another thing that jumped out in 2005 was the gender divide with 73 percent of men but only 42 percent of women approving of the bombings. Age was also a factor with support registering at 63 percent for those over 50 and 53 percent for those under (MOORE 2005). But of interest in 1995 was that 61 percent thought developing the bomb in the first place was a bad idea with only 36 percent saying it was a good idea, which was significantly down from the 69 percent who thought it a good thing in 1945 at a time when only 17 percent thought it a bad thing (GALLUP 2000: 234). Where GALLUP's 1991 poll showed a 53–41 margin, a *Detroit Free Press* poll that year put the margin at 63–29 (STOKES 2015). Quinnipiac University weighed in with a national poll of registered voters in 2009, after tension had begun to ramp up over U. S. efforts to fast-track Ukraine and Russia joining NATO, that showed an even greater imbalance. 61 percent said the U. S. did the "right thing" in 1945 and only 22 percent disagreed. Another 16 percent were undecided (ANONYMOUS 2009). When Pew Research Center asked the approval / disapproval question again in 2015, 56 percent judged the bombings to be justified and only 34 percent thought they were not (BOWMAN 2016). Among Japanese that year, the results were quite different. Only 14 percent told Pew that the atomic bombings were justified; 79 percent thought not (STOKES 2015).

This 2015 Pew poll may have been an outlier that did not reflect Americans' shifting views. A *CBS News* poll taken in May 2016 just prior to OBAMA's visit reported a quite different response. For the first time, a narrow plurality of respondents opposed the atomic bombings by a margin of 44 to 43. The patterns were again revealing with the highest disapproval rates among younger respondents, women, and non-whites. 58 percent of men but only 28 percent of women approved as did 49 percent of whites and only 24 percent of non-whites. 37 percent of those between the ages of 18–44 approved compared to 50 percent of those 65 and older (DUTTON et al. 2016).

The most recent poll I could find was a *YouGovAmerica* one conducted in August 2020. 39 percent approved the decision and 33 percent disapproved. A whopping 25 percent said they didn't know. Pollsters also asked participants about the U. S. apologizing to Japan with 41 percent rejecting that prospect and 34 percent supporting it. Among 18- to 24-year-olds, however, 52 percent favored an apology. Only 21 percent of those 55 and older did. Interestingly, 53 percent of those earning above $80,000 approved, but only 33 percent of those earning under $40,000 did so (JAIMUNGAL 2020).

OFFICIAL MYTHS DIE HARD, BUT THEY DO DIE

How these views have impacted Americans' attitudes on nuclear weapons is very difficult to discern. Americans generally prefer a world without nuclear weapons even if that means the U. S. forsaking its arsenal as well. *Associated Press* polls in 2005 and 2010 found that 66 and 62 percent of Americans believed that no country should have nuclear weapons. A more recent poll conducted by the Chicago Council on Public Affairs in July 2020 found 66 percent of Americans wanted to see all countries give up their nuclear weapons. Such views were held by 54 percent of Republicans, 78 percent of Democrats, and 64 percent of Independents (KAFURA 2020). It would be interesting to see how Vladimir PUTIN's nuclear saber-rattling over Ukraine has influenced these views.

But whether it's the obscene triumphalism of boorish Bobby KNIGHT and, by implication, Donald TRUMP or the more circumspect and restrained myth perpetuation by Barack OBAMA, Susan RICE, Mike WALLACE, and Bill O'REILLY, U. S.

attitudes about the 1945 atomic bombings have helped legitimize ongoing U. S. nuclear policy, which is based ultimately on the threat of ending life on this planet. In fact, TRUMAN was well aware of this threat as he stated on at least three occasions before blithely authorizing use of the bombs, despite his knowledge that the Japanese were eager to surrender, wanting only guarantees that they could keep the emperor, and that U. S. intelligence had been forecasting for months that the imminent Soviet invasion of Japan would provide the final death blow to the Japanese empire.[3] It was at Potsdam on July 25 after receiving a full report on the awesome destructive power of the July 16 Trinity test that he wrote in his diary, "[w]e have discovered the most terrible bomb in the history of the world. It may be the fire destruction prophesied in the Euphrates Valley Era, after Noah and his fabulous Ark" (FERRELL 1980: 55). BIDEN, like TRUMP before him, has doubled down on OBAMA's reckless commitment to a $1.2 trillion dollar 30-year modernization of every aspect of the U. S. nuclear arsenal—a modernization designed to make U. S. nuclear warfighting more efficient and more deadly. The current estimate is that this modernization will cost more than $1.7 trillion.

Much of the world has not been fooled by official U. S. deception surrounding the atomic bombings of 1945. In November 2019, Pope Francis made the first papal visit to Hiroshima and Nagasaki since that of Pope Paul VI in 1982. He spoke with the kind of honesty that OBAMA dared not. In Nagasaki, he deplored "attempts to build upon the fear of mutual destruction or the threat of total annihilation." He said that in this city, given what it had witnessed, "our attempts to speak out against the arms race will never be enough." He condemned those who would sanction or profit from this orgy of potential destruction: "[T]he money that is squandered and the fortunes made through the manufacture, upgrading, maintenance and sale of ever more destructive weapons, are an affront crying out to heaven" (POPE FRANCIS 2019a).

In Hiroshima, after listening to two moving *hibakusha* testimonies, the Pope reflected on what had occurred: "Here, in an incandescent burst of lightning and fire, so many men and women, so many dreams and hopes, disappeared, leaving behind only shadows and silence. In barely an instant, everything was devoured by a black hole of destruction and death. From that abyss of silence, we continue even today to hear the cries of those who are no longer." "The use of atomic energy for the purposes of war is immoral", he declared, adding, "so too the possession of nuclear weapons is immoral, as I said two years ago." He further insisted, "[w]e cannot allow present and future generations to lose the memory of what happened here" (POPE FRANCIS 2019b; O'CONNELL 2019).

The Pope's message, while poignant and stirring, was matched in raw emotional intensity by remarks made on August 6, 2009, at the Hiroshima Peace Memorial Ceremony by Nicaraguan priest Miguel D'ESCOTO BROCKMANN, who at the time was president of the United Nations General Assembly, in remembrance of what he called "one of the greatest atrocities the world has ever witnessed." "As a Roman Catholic priest, and a disciple of Jesus of Nazareth", he explained, "I want also, from the depth of my heart, to seek forgiveness from all my brothers and sisters in Japan for the fact that the captain of the fateful B-29 Enola Gay, Paul TIBBETS, now deceased, was a member of my church. I am consoled, to a certain degree, that Father George ZABELKA, the Catholic chaplain of the mission, recognized, after the event, that this was one of the worst imaginable betrayals of the teachings of Jesus." "In the name of my church", D'ESCOTO BROCKMANN implored listeners, "I ask your forgiveness" (KINGSTON et al. 2009: 4). Father ZABELKA not only apologized, he returned years later to Hiroshima, threw himself face down on the ground, wept, and begged forgiveness.

Combating these official lies—this triumphalist narrative about the necessary and justifiable atomic obliteration of Hiroshima and Nagasaki—has at times seemed like a Sisyphean effort. But dedicated historians and other public figures have not abandoned the effort, understanding the impact that the persisting myth that the 1945 atomic bombings were not only justifiable, they were humane, has had on deterring ongoing efforts to rid the world of these potentially apocalypse-inducing weapons. In 2020, Marty SHERWIN, Gar ALPEROVITZ, Kai BIRD, and I gave a series of international press briefings and webinars intended to debunk that myth before it was given new life in the anticipated spate of media coverage upon the 75th anniversary of the bombs' use. Just a few years earlier, Academy Award-winning filmmaker Oliver STONE and I had visited numerous U. S. college campuses to screen the Hiroshima / Nagasaki episode of our documentary film series *The Untold History of the United States*, which, along with the accompanying *New York Times* bestselling book by the same name, focused heavily on the insanity of the nuclear arms race from its inception up until the present. So, while our ongoing efforts and those of many of our colleagues in academia and the antinuclear movement persist, it remains an uphill struggle to break through the pervasive messaging by the myriad forces invested in perpetuating the lies surrounding this most crucial of historical issues as presented in American media and the falsehoods fed to school children via the nation's heavily and often increasingly distorted "patriotic" school curricula.

NOTES

1 Portions of this article were adapted from KUZNICK ("Hiroshima and Nagasaki—75 Years On: Some Reflections", 2020).
2 Established in Washington DC in 1997, the Newseum was a museum that was dedicated to journalism and the freedom of expression. It was funded by the Freedom

Forum, a non-profit organization founded by publisher Al Neuharth. After financial losses, the Newseum closed its DC location in 2019.

3   For more on this and for a fuller discussion of the entire controversy surrounding the atomic bombings, see Stone / Kuznick (2019), and Kuznick (2007). See also Alperovitz (1995); Hasegawa (2005); Lifton / Mitchell (1995); and Sherwin / Alperovitz (2020).

References

Alperovitz, Gar (1995): *The Decision to Use the Atomic Bomb: And the Architecture of an American Myth*. New York: Alfred A. Knopf.

Anonymous (1945a): "A Decision for Mankind". In: *St. Louis Post Dispatch*, 08 / 07: 12.

Anonymous (1945b): "News of Weapon Electrifies Truman Ship; President Makes Announcement to Crew". In: *The New York Times*, 08 / 07: 2.

Anonymous (1945c): "One Victory Not Yet Won". In: *The New York Times*, 08 / 12: 8 (Section T).

Anonymous (1947): "War and the Bomb". In: *The New York Times*, 01 / 28: 22.

Baker, Peter / Tackett, Michael (2018): "Trump Says His 'Nuclear Button' Is 'Much Bigger' than North Korea's". In: *The New York Times*, 01 / 02: 10 (Section A).

Baldwin, Hanson (1945): "The Atomic Weapon". In: *The New York Times*, 07 / 08: 10.

Bierman, Noah (2017): "Trump Warns North Korea of 'Fire and Fury'". In: *Los Angeles Times*, 08 / 08; https://www.latimes.com/la-app-north-korea-trump-nuclear-missiles-20170808-story.html (last access 2023 / 02 / 18).

Blume, Leslie M. M. (2020): *Fallout: The Hiroshima Cover-up and the Reporter Who Revealed It to the World*. New York: Simon & Schuster.

Boyer, Paul (1986): *By the Bomb's Early Light: American Thought and Culture at the Dawn of the Atomic Age*. New York: Pantheon.

Bowman, Karlyn (2016): "Historical Opinions on Hiroshima, President Obama and the Atomic Bomb". In: *Forbes*, 05 / 12; https://www.forbes.com/sites/bowmanmarsico/2016/05/12/historical-opinions-on-hiroshima-president-obama-and-the-atomic-bomb/?sh=40170bd35d61 (last access 2023 / 02 / 18).

Choe, Sang-Hun (2017): "Kim's Rejoinder to Trump's Rocket Man: 'Mentally Deranged U. S. Dotard'". In: *The New York Times*, 09 / 21; https://www.nytimes.com/2017/09/21/world/asia/kim-trump-rocketman-dotard.html (last access 2023 / 02 / 18).

CRAMER, Maria (2020): "'Now I Am Become Death': The Legacy of the First Nuclear Bomb Test". In: *The New York Times*, 07/15; https://www.nytimes.com/2020/07/15/us/trinity-test-anniversary.html (last access 2023/02/18).

EVANS, Medford (1959): "Hiroshima Saved Japan". In: *National Review*, 02/14: 525.

FERRELL, Robert H. (ed.) (1980): *Off the Record: The Private Papers of Harry S. Truman*. Columbia: University of Missouri Press.

FILKINS, Dexter (2017): "Rex Tillerson at the Breaking Point". In: *The New Yorker*, 10/16; https://www.newyorker.com/magazine/2017/10/16/rex-tillerson-at-the-breaking-point (last access 2023/02/18).

FUSSELL, Paul (1981): "'Thank God for the atom bomb'—Hiroshima: A Soldier's View". In: *The New Republic*, 185 (8/9): 26–30.

GALLUP, George (1945): "Using A-Bomb on Japs Approved by U. S. Public". In: *Los Angeles Times*, 08/26: 3.

GALLUP, George (1947): "More Now Say Atom Bomb's Invention Was a Bad Thing". In: *The Washington Post*, 10/18.

GALLUP, George (1972): *The Gallup Poll: Public Opinion 1935–1971. Volume One 1935–1948*. New York: Random House.

GALLUP, George (2000): *The Gallup Poll: Public Opinion 1999*. Wilmington: Scholarly Resources.

GEDDES, Donald Porter (ed.) (1945): *The Atomic Age Opens*. New York: Pocket.

HASEGAWA, Tsuyoshi (2005): *Racing the Enemy: Stalin, Truman, and the Surrender of Japan*. Cambridge, MA: Harvard University Press.

HORNFISCHER, James (2020): "'Countdown 1945' Review". In: *The Wall Street Journal*, 08/05; https://www.wsj.com/articles/countdown-1945-review-checkmate-in-the-pacic-11596664105 (last access 2023/02/18).

HUTCHINSON, Russell S. (1946): "Hiroshima". In: *Christian Century*, 09/25: 1151.

KINGSTON, Jeff et al. (2009): "The Nuclear Age at 64: Hiroshima, Nagasaki, and the Struggle to End Nuclear Proliferation". In: *The Asia-Pacific Journal: Japan Focus* 7/33: 1–16.

KRISTENSEN, Hans M. (2016): "U. S. Nuclear Stockpile Numbers Published Enroute [sic] to Hiroshima". In: *FAS*, 05/26; https://fas.org/blogs/security/2016/05/hiroshima-stockpile/ (last access 2023/02/18).

KUZNICK, Peter (2007): "The Decision to Risk the Future: Harry Truman, the Atomic Bomb and the Apocalyptic Narrative". In: *The Asia-Pacific Journal: Japan Focus* 5/7: 1–23.

KUZNICK, Peter (2020): "Hiroshima and Nagasaki—75 Years on: Some Reflections". In: SIMPSON, Tony (ed.): *Waging Peace: Festschrift for David Krieger* (The Spokesman 145). Nottingham: Spokesman Books, 75–82.

LABOTT, Elise (2016): "John Kerry Calls Trump Nuclear Policy 'Absurd'". In: *CNN*, 04/11; https://edition.cnn.com/2016/04/11/politics/john-kerry-hiroshima-memorial/index.html (last access 2023/02/18).

LEE, Michelle Ye Hee (2017): "North Korea's Latest Nuclear Test Was So Powerful It Reshaped the Mountain Above It". In: *The Washington Post*, 09/14; https://www.washingtonpost.com/news/worldviews/wp/2017/09/14/orth-koreas-latest-nuclear-test-was-so-powerful-it-reshaped-the-mountain-above-it/ (last access 2023/02/18).

LIFTON, Robert Jay / MITCHELL, Greg (1995): *Hiroshima in America: Fifty Years of Denial*. New York: G. P. Putnam's Sons.

MARTIN, Jonathan / LANDLER, Mark (2017): "Bob Corker Says Trump's Recklessness Threatens World War III". In: *The New York Times*, 10/09: 1 (Section A).

MAZZA, Michael (2016): "The Wrong Lesson on Hiroshima". In: *The Wall Street Journal*, 05/12; https://www.wsj.com/articles/SB12395380881288183678004582061760120232438 (last access 2023/02/18).

MCCURRY, Justin / SMITH, David / YUHAS, Alan (2016): "Obama Visit to Hiroshima Should Not Be Viewed as an Apology, White House Says". In: *The Guardian*, 05/10; https://www.theguardian.com/us-news/2016/may/10/obama-hiroshima-japan-visit-second-world-war (last access 2023/02/18).

MACDONALD, Dwight (1957): *Memoirs of a Revolutionist: Essays in Political Criticism*. New York: Farrar, Straus, and Cudahy.

MCCARTHY, Mary (1968 [1946]): "The Hiroshima New Yorker". In: MACDONALD, Dwight (ed.): *Politics*, vol. 3. New York: Greenwood Reprint Corporation, 367.

NEVIN, Allan (1946): "How We Felt About the War". In: GOODMAN, Jack (ed.): *While You Were Gone: A Report on Wartime Lie in the United States*. New York: Simon & Schuster, 13.

O'CONNELL, Gerard (2019): "Pope Francis at Nagasaki and Hiroshima Makes Impassioned Plea for Peace and Nuclear Disarmament". In: *America: The Jesuit Review*, 11/24; https://www.americamagazine.org/faith/2019/11/24/pope-francis-nagasaki-and-hiroshima-makes-impassioned-plea-peace-and-nuclear (last access 2023/02/18).

O'REILLY, Bill (2016): *The O'Reilly Factor* (Fox News Channel), 05/26.

O'REILLY, Bill / DUGARD, Martin (2016): *Killing the Rising Sun: How America Vanquished World War II Japan*. New York: Henry Holt and Company.

RICE, Susan E. (2019): "What My Father Taught Me About Race". In: *The New York Times*, 10/06: 2 (Section SR).

ROSE, Lisle Abbott (1973): *Dubious Victory: The United States and the End of World War II*. Kent, Ohio: Kent State University Press.

SCOTT, Eugen (2016): "Bobby Knight: Trump Would Drop A-Bomb Like Truman". In: *CNN*, 04 / 28; https://edition.cnn.com/2016/04/28/politics/donald-trump-bobby-knight/index.html (last access 2023 / 02 / 18).

SHALETT, Sidney (1945): "New Age Ushered". In: *The New York Times*, 07 / 08: 1.

SHELBOURNE, Mallory (2017): "Trump Calls Kim 'Rocket Man' on a 'Suicide Mission'". In: *The Hill*, 09 / 19; https://thehill.com/policy/international/untreaties/351321-trump-calls-north-korea-leader-rocket-man-at-un/ (last access 2023 / 02 / 18).

SHERWIN, Martin J. / ALPEROVITZ, Gar (2020): "U. S. leaders knew we didn't have to drop atomic bombs on Japan to win the war. We did it anyway". In: *Los Angeles Times*, 08 / 05: https://www.latimes.com/opinion/story/2020-08-05/hiroshima-anniversary-japan-atomic-bombs (last access 2023 / 02 / 18).

SMITH, David Michael (2020): "U. S. Public Opinion on the Hiroshima and Nagasaki Bombings since 1945". In: *Peace Review* 32 / 3: 342–49.

STONE, Oliver / KUZNICK, Peter (2019): *The Untold History of the United States*. New York et al.: Simon & Schuster.

THOMPSON, Mark (2016): "Trump Wants to Free the Nuclear Genie". In: *Time. com*, 03 / 30; https://time.com/4276960/trump-wants-to-free-the-nuclear-genie/ (last access 2023 / 02 / 18).

WALLACE, Chris (2021): *Countdown 1945: The Extraordinary Story of the Atomic Bomb and the 116 Days that Changed the World*. New York: Simon & Schuster.

WHALEN, Andrew (2020): "Despite Hiroshima Polling Trends, Doubting the Bomb Isn't New". In: *Newsweek*, 08 / 06; https://www.newsweek.com/hiroshima-bombing-anniversary-justified-polling-public-perception-debate-1523419 (last access 2023 / 02 / 18).

YAVENDITTI, Michael J. (1974): "The American People and the Use of Atomic Bombs on Japan: The 1940s". In: *The Historian* 36 / 2: 224–47.

Internet Resources

ANONYMOUS (2009): "Bombing Hiroshima Was Right, American Voters Say 3–1"; https://poll.qu.edu/Poll-Release-Legacy?releaseid=1356 (last access 2023 / 02 / 18).

DUTTON, Sarah et al. (2016): "CBS News Poll: What do Americans Think of the 1945 Use of the Atomic Bomb?". In: *CBS News*, 05 / 27; https://www.cbsnews.com/news/cbs-news-poll-what-do-americans-think-of-the-1945-use-of-the-atomic-bomb/ (last access 2023 / 02 / 18).

JAIMUNGAL, Candice (2020): "75 Years Since Hiroshima and Nagasaki—Americans Still Divided on Use of Atomic Bomb". In: *YouGovAmerica*, 08 / 10;

https://today.yougov.com/topics/politics/articles-reports/2020/08/10/75-years-hiroshima-and-nagasaki-americans-still-di (last access 2023 / 02 / 18).

KAFURA, Craig (2020): "Americans Want a Nuclear-Free World". In: *The Chicago Council on Global Affairs: Running Numbers Blog*, 08 / 06; https://globalaffairs.org/commentary-and-analysis/blogs/americans-want-nuclear-free-world (last access 2023 / 02 / 18).

KIM, Jong Un (2018): "Kim Jong Un's 2018 New Year's Address"; https://www.ncnk.org/node/1427/ (last access 2023 / 02 / 18).

MOORE, David W. (2005): "Majority Supports Use of Atomic Bomb on Japan in WWII"; https://news.gallup.com/poll/17677/majority-supports-use-atomic-bomb-japan-wwii.aspx (last access 2023 / 02 / 18).

POPE FRANCIS (2019a): "Full Text of Pope's Message in Hiroshima"; https://english.kyodonews.net/news/2019/11/117d6ed36af7-full-text-of-popes-message-in-hiroshima.html (last access 2023 / 02 / 18).

POPE FRANCIS (2019b): "Full Text of Pope Francis' Message in Nagasaki"; https://english.kyodonews.net/news/2019/11/8591e9ea1921-full-text-of-popes-message-in-nagasaki.html (last access 2023 / 02 / 18).

STOKES, Bruce (2015): "70 Years After Hiroshima, Opinions Have Shifted on Use of Atomic Bombs". In: *Pew Research Center*, 08 / 04; https://www.pewresearch.org/fact-tank/2015/08/04/70-years-after-hiroshima-opinions-have-shifted-on-use-of-atomic-bomb (last access 2023 /02 / 18).

THE WHITE HOUSE (2016): "Remarks by President Obama and Prime Minister Abe of Japan at Hiroshima Peace Memorial"; https://obamawhitehouse.archives.gov/the-press-office/2016/05/27/remarks-President-obama-and-prime-minister-abe-japan-hiroshima-peace (last access 2023 /02 / 18).

WEAVER, Al (2016): "Bobby Knight: Trump Would Have 'Guts to Drop the Bomb' Like Truman". In: *Washington Examiner*, 04 / 28; https://www.washingtonexaminer.com/bobby-knight-trump-would-have-guts-to-drop-the-bomb-like-truman (last access 2023 / 02 / 18).

ZAKARIA, Fareed (2016): *Transcripts* (CNN), 05 / 15; https://transcripts.cnn.com/show/fzgps/date/2016-05-15/segment/01 (last access 2023 / 02 / 18).

Splitting the Atom:
Nishioka Yuka's Manga Rebuke of the Nuclear-Weapon / Power Divide

Michele M. Mason

1 Rejecting "Atoms for Peace" Mythologies

In the wake of Japan's 2011 triple disaster—earthquake, tsunami, and nuclear power plant meltdowns—the world was reminded of the ever-present dangers of atomic energy. In Japan, despite the upheaval and trauma, there quickly emerged an outpouring of creative commentary, ranging from fiction and film to twitter poetry and dramatic works, each documenting and sharing the experiences of individuals, families, and communities in the affected region. Notably, manga also contributed significant observations on the immediate experience and subsequent challenges of finding family, evacuating, procuring water and food, and, eventually, the clean-up efforts. These texts naturally touch on fear, death, and despair but also include inspiring stories of care, cooperation, and generosity.[1]

The historical trajectory that led to this most recent nuclear power accident started with the successful detonation of an atomic bomb in 1945, which would later be heralded as a monumental scientific feat and establish possessing nuclear weapons as the principal factor for determining global power. The United States' deployment of two nuclear bombs on civilian populations in Japan served to not only communicate the nation's technological superiority but also its willingness to unleash unmitigated misery. The Cold War and global arms race have since been defined by the repeated "use" of nuclear weapons in the form of threats as a means to negotiate political, ideological, and military aims (Gerson 2007: 2, 12). Additionally, more than 2000 nuclear tests by nuclear weapons states have ensured that no place on the planet is left untouched by the radioactive fallout in an era now known as the Atomic Age. Also called the Anthropocene for the consequential, lasting impact made by mankind, this new geological epoch's dates are confirmed by spikes in uranium and plutonium deposits in peat bogs, coral reefs, and polar ice cores across the globe.

In the post-WWII era, as a means to counter growing, world-wide resistance to nuclear projects, politicians, scientists, and industrialists began crafting and normalizing messaging regarding nuclear power generation. Crucially, the U. S.-led "Atoms for Peace" campaign (1953–61) promoted an incontrovertible split between "dangerous" nuclear weapons and "safe" nuclear energy. Travelling exhibitions, legislative acts, juvenile literature, and promotional advertisements fostered a flight of fancies of what the little atom could accomplish, and narrative and visual depictions were without imaginative or earthly limits. Technocratic promises included not only endless energy production but also atomic technologies that would create enough food to feed the entire world, provide revolutionary medicine, propel airplanes, boats, and submarines, and launch futuristic spacecrafts into our galaxy. While some of the extravagant promises have dropped out of today's political playbook, the received narrative of "safe" nuclear power still retains its persuasiveness hold in the 82nd year of the Atomic Era. In fact, recently, the influential nuclear lobby and its adherents seeks to further strengthen this conceptual partition by packaging nuclear power in the latest environmental-friendly parlance. They argue it is not only safe and cheap, but also "clean" and "green".[2] This dangerous divorce of nuclear power's reliance on the very same ingredients that makes atomic weapons so fearsome—as well as the entire nuclear fuel cycle's reliance on fossil fuel and the attendant contamination of ecologies and humans—prompts a clear-minded challenge to such entrenched and perilous obfuscations.

In Japan's postwar years, the atomic bombings of Hiroshima and Nagasaki prompted manga artists to produce powerful works that depicted the abject human misery of nuclear weapons, but those that highlighted the dangers of nuclear power would come later and were rare.[3] KATSUMATA Susumu 勝又進 (1943–2007) was the first to publish critical comic works on nuclear power, generally, and the precarious lives of nuclear power plant workers, specifically.[4] KATSUMATA earned a degree in physics but rejected a career in the field to become a manga artist. He initially gained a reputation for his four-panel comic strips and later for the edgy collection of story-length manga, "Red Snow" (*Akai yuki* 赤い雪, 2005). 15 years after Japan's first nuclear reactor went online and soon after the U. S. accident at Three Mile Island, KATSUMATA contributed cover art for the book entitled "Why Nuclear Power is Scary" (*Genpatsu wa naze kowai ka* 原発はなぜこわいか, 1980). He continued to support projects that challenged the nuclear mythos of safe nuclear power with illustrations and expertise in the coming decades. Despite being best known for his works published in the avant-guard manga journal, *GARO* ガロ, KATSUMATA also penned two realistic pieces—"Deep Sea Fish" (*Shinkaigyo* 深海魚, 1984) and "Devil Fish (Octopus)" (*Debiru fisshu (tako)* デビルフィッシュ(蛸), 1989)—which treat the inner workings of a power plant and the after-hours of the unskilled workers with a fruitful mix of lighthearted banter and gallows humor. Relying on on-site visits, interviews, and his own understanding of nuclear science, Katsumata details the minutia of safety protocols, deploys innovative stippling to depict radiation, and depicts both the men's abject fear and stoicism in the face of ever-present contamination, offering a window into a little known and seen world. In a short commentary, originally published in the magazine *Comix Box* in 1990, Katsumata wrote, "What I really want to use manga to show are the horrors of radiation. Alas, nuclear power is not that interesting a theme for manga. [...] That's no reason to ignore it, however" (Katsumata 2018: xxx). Still, he did not create another nuclear-related manga nor live to witness the 2011 triple disaster.

It took a catastrophic nuclear accident on Japan's soil to prompt another manga artist to grab the baton from Katsumata and pen a clear rebuttal to the mythos of nuclear power.[5] Nagasaki-based Nishioka Yuka 西岡由香 (*1965) was profoundly affected by the unfolding calamity. Within three months of the nuclear meltdowns in Fukushima, she created an illustrated Radiation Q & A for her website's landing page, which spontaneously inspired an online community for learning and teaching about issues related to nuclear technologies. In July 2011, Nishioka visited Fukushima, where she listened to testimonies and began forming a more expansive critique of nuclear technologies. Within a year, she had transformed her Q & A into "Goodbye, Atomic Dragon: The Story of Atomic Weapons and Nuclear Power" (*Sayonara atomikku doragon: Kaku to genpatsu no ohanashi* さよならアトミック・ドラゴン：核と原発のお話), a multi-faceted rebuke of the specious split between nuclear weapons and nuclear power. Published in 2012, this manga weaves detailed scientific explications, sobering historical facts, and moving fictionalized personal narratives into a volume that entertains as it educates. With a pedagogical aim, Nishioka speaks to and depicts disturbing realities while endeavoring to foster a critical awareness.

"Goodbye, Atomic Dragon" emerged out of Nishioka's life-long anti-nuclear activism and established record of creating youth-oriented, civic-minded graphic novels. For decades, she has been a dedicated peace worker, nuclear abolitionist, and lecturer at the University of Nagasaki. Nishioka is also well-known for her participation in the international NGO Peaceboat and Nagasaki's Testimony Committee (Nagasaki no shōgen no kai 長崎の証言の会). Her commitment to such causes was officially recognized by an award from the Peace and Cooperative Journalist Fund in 2015.

In the years before the nuclear disaster at the Fukushima Daiichi Power Plant, Nishioka produced numerous manga aimed at young people that combined interesting plot lines and educational materials related to Nagasaki's atomic history. An important example is her 2008 manga, titled "A Summer's Afterimage: Nagasaki, August 9" (*Natsu no zanzō: Nagasaki no hachigatsu kokonoka* 夏の残像：ナガサキの八月九日), which includes five separate vignettes highlighting the pain and suffering of Nagasaki victims while maintaining the complexity of history. For instance, in one chapter, NISHIOKA recognizes the humanity of Los Alamos Laboratory scientists who worked on the bomb and even Americans who still defend their use on Japan while also highlighting the voice of a Nagasaki survivor who gives testimony at this now-famous tourist site. In another, she carefully depicts two unsettling postwar historical realities in South Korea; one, that the U. S. atomic bombings were celebrated for their role in ending Japan's brutal colonial rule in Korea, and two, that Korean victims of the atomic bombings have endured long-term neglect and discrimination from their fellow citizens.

NISHIOKA has also depicted the specific experiences of Catholics in the Urakami district whose stately cathedral was the epicenter of the Nagasaki atomic bombing before and after 2011. These includes her 2010 manga titled "The Santa Claus of August 9th: Nagasaki's A-Bomb and the Survivors" (*Hachigatsu kokonoka no Santakurōsu: Nagasaki genbaku to hibakusha* 八月九日のサンタクロース：長崎原爆と被爆者), which was first serialized in newspapers and magazines, and "The Prayer of the A-Bombed Maria: Manga Testimonies of Three Survivors" (*Hibaku Maria no inori: Manga de yomu sannin no hibaku shōgen* 被爆マリアの祈り：漫画で読む三人の被爆証言, 2015). In the former, NISHIOKA depicts the journey of a teenage girl who moves to Nagasaki when her father relocates to his hometown for his job, and, in the latter, she depicts the stories of three *hibakusha* based on their personal testimonies and interviews with NISHIOKA. These two works embody this manga artist's personal journey. The first represents her early commitment to create narratives that have young characters listen to and learn from fictional *hibakusha*, reserving depictions of the day of and following the bombing to only when survivors in the storyline are sharing aspects of their lived-experience. The second was the first time NISHIOKA committed to penning non-fiction testimonies of August 1945, being extremely conscientious about what it means to look through the "window" (*mado* 窓) of a *hibakusha* and depict their indescribable experiences with integrity (NISHIOKA 2015: 150).

Elsewhere (MASON 2009), I have written about NISHIOKA's vanguard historical and educational manga on the Nagasaki atomic-bomb experience—specifically as they compare to KŌNO Fumio's こうの史代 Hiroshima-focused manga. However, herein I highlight how NISHIOKA pushes the envelope by being the first post-Fukushima comic-creator to clearly articulate the false division between dangerous / safe forms of atomic technologies. She has always been deeply committed to moving beyond rehearsed rhetoric, witnessing the pain of others, and acknowledging various, sometimes conflicting, national perspectives. In "Goodbye, Atomic Dragon", NISHIOKA moves across grand time scales and international borders while integrating the mind-bending details of nuclear science and history into a multi-generational narrative.

2 Lively Matter and Technocratic Fantasies

Importantly, in this work, the figure of the dragon serves to contest the assumption that humans can reign complete control of natural matter. Specifically, NISHIOKA takes to task the entrenched notion that humans are capable of managing nuclear materials, which are mistakenly understood to be inert. Her critique resonates with the work of political scientist and philosopher Jane BENNETT on "vibrant materiality". In the celebrated work *Vibrant Matter: A Political Ecology of Things*, BENNETT (2010: 21) asks "how would an understanding of agency as a confederation of human and nonhuman elements alter established notions of moral responsibility and political accountability?" BENNETT's formulation of "distributive agency" constitutes a powerful critique of classical notions of matter, foregrounding the potentiality of a wide variety of "static" matter when placed into assemblages. Notably, an assemblage is "never a stolid block but an open-ended collective, a 'non-totalizable sum'" (BENNETT 2010: 24). The diversity of material and non-material elements makes possible emergent outcomes that defy tidy fantasies of control by human agents. BENNETT elaborates:

> Assemblages are not governed by any central head: no one materiality or type of material has sufficient competence to determine consistently the trajectory or impact of the group. The effects generated by an assemblage are, rather, emergent properties, emergent in that their ability to make something happen [...] is distinct from the sum of the vital force of each materiality considered alone. (BENNETT 2010: 24)

In what turns out to be a particularly apt comparative example, BENNETT exemplifies this notion of "conjoint action" with a discussion of the 2003 power blackout in North America, which crossed national boundaries and affected 50 million people. Journalist James GLANZ dubbed the North-American power grid the "biggest gizmo ever built" in 2003. BENNETT builds on this image and goes into detail about how the power grid is made up in part by "a volatile mix of coal, sweat, electromagnetic fields, computer programs, electron streams, profit motives, heat, lifestyles, nuclear fuel, plastic, fantasies of mastery, static, legislation, water, economic theory, wire, and wood" (BENNETT 2010: 25). BENNETT's theory of the distributive agency of a federation of actants in this example is instructive for the case of nuclear power. BENNETT's salient framing of "conjoint action" pertains to not only the volatility of nuclear materials in and of themselves but them within the web of the technological, governmental, legislative, and discursive apparats, which point to key components of a confederation of power well beyond the actual power plant. NISHIOKA also warns readers of the folly of presuming a strict life-matter binary, which ascribes "agency" only to humans, by formulating her notion of the "atomic dragon". The personification of atoms—the basic building blocks of all matter in the universe—through the figure of the dragon forces readers to view these isotopes as lively material.

As NISHIOKA troubles notions of agency and responsibility, she also seeks to question the irrational devotion to and exuberance for highly complicated technologies (or, in this case, extremely simple processes made dangerously complicated). Her collective body of work, and this manga in particular, calls into question the ways technocratic fantasies of future political or energy security are deemed more rational than productive uncertainty, caution, and care for living beings and their environment. In this way, "Goodbye, Atomic Dragon" resonates with media theorist Joanna ZYLINKA's critiques of both "the idea that the separation of Man from Nature signifies teleological maturation" (ZYLINKA 2018: 34) and the glorification of masculinist and technicist solutions for the challenges humankind faces in what she dubs the Manthropocene (ZYLINKA 2018: 2). In her short work, *The End of Man: A Feminist Counterapocalypse*, ZYLINSKA deftly unpacks how technofixes are filtered through a variety of systems of knowledge and governance that skew our understanding of our world and the ways in which we are an inextricable part of it. While NISHIOKA's manga may not have an explicit feminist commitment, her critique of the unquestioned panegyrics of nuclear power is unmistakable. Her depictions concretize the ways the accidents foist the repercussions and responsibilities of unrestrained technical fancies onto innocent populations that pay the corporeal and fiscal price. In addition to illustrating how atomic half-lives beggar human comprehension, NISHIOKA reconnects nuclear power and weapons by describing the harm done to U. S. soldiers and Iraqi civilians alike by the use of depleted uranium munitions, which are made by recycling waste materials from nuclear power plants. In its essence, "Goodbye, Atomic Dragon"'s overarching message resonates with ZYLINSKA's call to "[adopt] precarity as the fundamental condition of living in the globalized postindustrial world" and "contest [...] the masculinist and technicist solutions to said crisis" (ZYLINSKA 2018: 2). "Goodbye Atomic Dragon" emphasizes that ignoring the ways nuclear material functions as an actor in our nuclear assemblages will continue to wreak havoc. On the cover, NISHIOKA (2012) spells this out:

> The arrogance of people thinking they can do whatever they wish has awoken a dragon that rampages across the sky. As a result of the behavior of selfish people, animals, plants, mountains, rivers, and even the dirt and rocks are all imperiled. We stand now at a crossroads with the choice of using artificial energies in atomic power or using natural energy, living in harmony with the dragon.

By reiterating the importance of the many living bodies of our biosphere (humans, non-human animals, plants, and more) in her historical and fictional narratives, NISHIOKA articulates a succinct, clear-eyed recognition of vibrant materiality, resisting human exceptionalism and a carefree comfort with our scientific-technical knowledge and empirical prognostications.

3   TRICKY TEAPOTS AND THE ATOMIC DRAGON

"Goodbye, Atomic Dragon", aimed at readers age six and above, follows the main character, Ayumi. She is a fifth-grader who regularly visits a homey cottage-cum-science lab where a friendly professor, known only as *hakase* (doctor, Ph. D.), answers a group of youngsters' every question. Ayumi, the central character, decides to write her summer essay on "the atom" and is eager to learn. The professor uses a variety of hands on, visual, and virtual materials to teach nuclear science and history to Ayumi and her friends. The children's natural fascination and confusion propel the professor's scientific expositions of basic physics, the properties of radiation, and historical narratives that range from the role of radiation at the time of the creation of life on planet earth to the use of depleted uranium by U. S. forces in the Middle East. At the same time that NISHIOKA skillfully unpacks the minutia of nuclear matter and energy, she also incisively strips down the overwrought, glorified technicist renditions of the "miracle" of nuclear power.

Integrating a variety of visual styles, narrative-rhythms, and manga genre cues in "Goodbye, Atomic Dragon", NISHIOKA shapes her critique and sustains the reader's attention. The scientific exposition, bolstered with a variety of metaphors, is useful for any-aged reader. In one instance, board game die stamped with the equivalent of "go" and "wait" are used to illustrate the complex mechanism by which radionuclides emit radiation (see Fig. 1). Historical context is elucidated through block text, charts, and lists, highlighting the fragility of power plants near tectonic plates, the backgrounds of scientists SIEVERT and ROENTGEN, and radiation's effect on hu-

Fig. 1: Radioactivity and the Metaphorical Die that Direct Atoms (*Sayonara, atomikku doragon*, p. 20)
© NISHIOKA Yuka

man's DNA and organs. Ayumi's elementary-school teacher gives a lecture using *kamishibai* (paper theater) panels to offer historical contextualization of the "Atoms for Peace" campaign, the victims of the U. S. Bikini Hydrogen test, and the cold-war nuclear arms race. With these full- or half-page panels interspersed within more complex, conversational panel organization, NISHIOKA regulates the rhythm of what otherwise would be a dense presentation. Generous use of onomatopoeia assists in the explanations of the complicated working of atoms and the presentation of youthful reactions to the material. Deploying signature techniques of shōjo manga with requisite flower motifs, moreover, indicates Ayumi's infatuation with the professor's grandson, Marco. While roses and more dainty flowers typically adorn such crush pages, NISHIOKA doubles the metaphoric load with sunflowers, which are globally recognized as icons of peace and nuclear abolition.

"Goodbye, Atomic Dragon"'s personalized storyline is linked to Chernobyl to emphasize the frailty of humans in light of dangerous radioactive materials but, moreover, to firmly situate Fukushima within global nuclear history with an emphasis on the pathos of the intergenerational aspects of nuclear contamination. The professor's grandson, Marco, has a condition called "Chernobyl Heart", a well-documented, inherited birth defect linked to radiation. The professor feels deep regret because he brought his wife and son to Germany in the 1980s when he pursued a research position there. This fatefully put his family directly under the radioactive fallout of the 1986 Chernobyl Power Plant disaster. The professor's son was exposed and passed down his compromised DNA to his son, Marco, who developed a heart with holes in it.[6] NISHIOKA shares the anguish of bequeathing such a dark and perilous inheritance in her "Afterword" with a punchy query-cum-condemnation quoted from a Chernobyl victim. "How many children will have to die before we wise up?" (NISHIOKA 2012: 124).

In particular, "Goodbye, Atomic Dragon" deploys the figure of the dragon to critique hyperbolic promises of technofixes. This includes industry-driven analyses that overestimate mankind's ability to manage such dangerous materials and technologies as well as skewed cost assessments that prioritize profits over people and the planet. The meaning of this supernatural creature unfolds in an animated video created by Marco, titled "The Day the Dragon Returns to the Sky" (*Ryū ga ten ni kaeru hi* 竜が天に帰る日). Though described as a mystical, sacred beast that lives in our universe and governs the natural elements, in the context of nuclear power, the dragon functions as a metaphorical genie out of the bottle. The video tells a story of misguided men of politics, the military, and science who pursued technologies beyond their own control and antithetical to natural symbiosis. Now that the dragon has been unleashed, humans are helpless in the wake of the indiscriminate, persistent, and long-term consequences.

NISHIOKA emphasizes this line of critique with word play and associations in her characterization of the dragon. Most importantly, she links three specific Chinese characters, namely "thunder", "electricity", and "dragon". The opening of Marco's animation edits these in turn to make each ideograph morph into the next (see Fig. 2). In one panel, little workmen figures pull out the middle vertical line of the "rice paddy" element in the "thunder" character, converting it into an electrical cord. Now the kanji signifies the word "electricity", and the workers plug the cord into a wall to reinforce the imagery. Next, they kick the "rain" component off the top and substitute it with the necessary grapheme to create the "dragon" character. Thusly, the text implies that we humans moved from an awe of and reverence for the natural phenomenon of thunder—long associated with the "dragon god" or "dragon king" (*ryūjin* 竜神)—to man-made electricity.

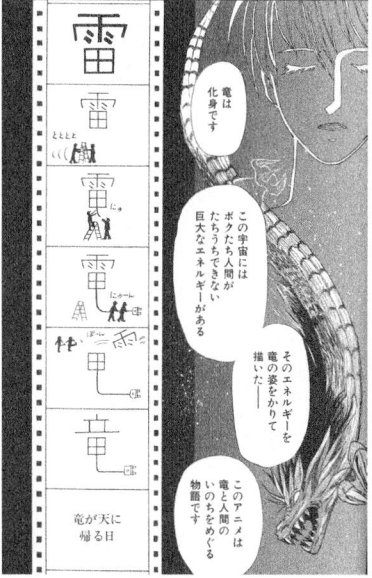

Fig. 2: Opening of Animated Video "The Day the Dragon Returned to the Sky" (*Sayonara, atomikku doragon*, p. 77) © NISHIOKA Yuka

Humanity's avaricious craving for ever more energy and comforts ultimately led to the pursuit of nuclear power, which unleashed the dragon. NISHIOKA's manga exposition of atomic science at the beginning of the work communicates the structure, reactions, and effects of the atom, which is to say it "shows and tells" what is at the heart of the atom, what it releases when it is split, and how it is connected to radioactivity. In the second and longer chapter, the focus turns more to human-centered narratives. Marco explains that the dragon is an avatar (*keshin* 化身) for the enormous energy force that exists in our cosmos, which is much too big for us to "cross swords with" (NISHIOKA 2012: 77). Throughout "Goodbye, Atomic Dragon", the energy-dragon clearly functions as an actant: that which has spontaneously created stars in the universe for millennia. In this formulation, NISHIOKA argues that the "vibrant matter" of atomic energy is beyond the control of humans.

Although a somewhat heavy metaphor, I argue the dragon is appropriate and compelling for her youth and young adult audience. In the "Afterword", NISHIOKA contextualizes the catalyzing event for the dragon motif, which happened when she visited Fukushima in July of 2011. At one point, she joins a gathering of 1,800 people, who listen to locals testify to increased deaths and illnesses in the community and their exhaustion from living in the tainted landscape. The next day, it is raining and the radiation level is "unbelievably high", and she is instructed not to touch the flowers or ground (NISHIOKA 2012: 123). NISHIOKA describes experiencing a strange and unfamiliar kind of fear there in that moment. She writes, "it was as if human-produced radiation had opened up a hole between another world and ours, and this off-limits space, into which humans had hitherto been forbidden to look, was getting larger in that particular location" (NISHIOKA 2012: 123). In this manga, NISHIOKA highlights the destruction of place and people by the "atomic dragon" through her depictions of the inescapable relationship between *hibakuchi* characters (contaminated land) and *hibakusha* characters (victims of nuclear bombs / disaster). And in doing so, she links the place and people of Fukushima with Nagasaki, Hiroshima, and Chernobyl, all connected by their tragic experiences of radiation contamination, which NISHIOKA states "is an absolute violence against life" (NISHIOKA 2012: 124). To this end, NISHIOKA depicts a scene of Ayumi visiting Marco in the hospital, where he describes what he wants to add to the unfinished video. He shows Ayumi photos of Fukushima that conjure up the people, animals, and land that have been irrevocably changed. They include images of a Geiger counter registering a high level of radiation, an elementary school with no children, masked officials in front of a "Do Not Enter" sign near a fishing supply shop, grazing cattle, and a residential home with power lines prominent in the background. Collectively, these spark image-memories that collectively hail the people, animals, land, waterways, and community that once existed here. Along with the noticeable power cables in the one photo, there is also one iconic and eerie reminder of the improvident technocratic promises: namely, the now infamous TEPCO gate in Fukushima that once proclaimed that nuclear power would bring "bright futures" (*akarui mirai* 明るい未来).

Moreover, NISHIOKA's "Goodbye, Atomic Dragon" portrays the "lively materiality" of atomic matter, contextualizing it within the larger nuclear power assemblage to question claims that tidy predictions can be totalizing. In particular, her representation of the dragon evokes the animacy of atomic particles, thus challenging the industry's exuberant confidence in emergency preparedness based on human-configured computer-generated forecasts. NISHIOKA suggests that although the technological operations of a nuclear power plant are

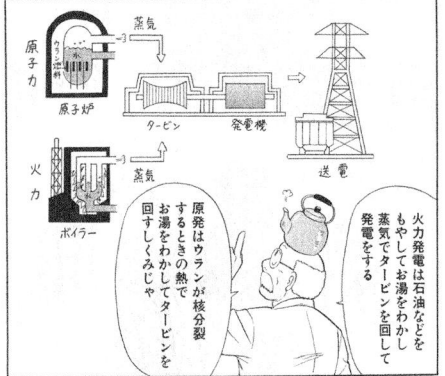

Fig. 3: Nuclear Power Plants are Water-Boiling Installation (*Sayonara, atomikku doragon*, p. 62) © NISHIOKA Yuka

analogous to heating up a "tea kettle" (NISHIOKA 2012: 62), elements / actants beyond the control of humans in their complex assemblages can have devastating consequences (see Fig. 3). In the case of the Fukushima Daiichi Power Plant disaster specifically, we could include the following inexhaustive list: nuclear fuel rods, steam, tsunamis, emergency procedures, damaged roads, radiation, skilled and unskilled workers, heat indicators, revolving door politics, alarms, nuclear reactors, AC / DC power, vacations, standing water, atomic energy laws, communication systems, S waves, ignored historical earthquake records, eight-hour batteries, and containment structures. "Goodbye, Atomic Dragon" calls out the risks of believing that humans are at the helm of any nuclear ship. Rather, it argues that the atom-dragon—envisioned not as an intrinsically malevolent creature but one bound by the laws of nature—is always an uncontrollable actant in the assemblage. In this apt analogy, she lays bare the essence of nuclear technology: it uses the most dangerous materials to achieve the simplest of tasks, namely boil water. While not named in ZYLINSKA's work, nuclear power embodies the ruse of technofix fantasies.

In keeping with her earlier youth-oriented work, NISHIOKA pulls no punches about the seriousness of the topics and our human predicament, calling her readers in to become the next generation of informed citizens and activists. In the "Afterword", she explains:

> The atomic dragon depicted in this work, is imagined as a sacred beast governing the natural elements that is turned into an evil dragon via Western technology. We must ask ourselves whether we want to keep relying on such technologies as nuclear weapons and nuclear power. We have to seriously reevaluate and reorganize our current way of life, understanding that if we do not halt what we must, the Atomic Dragon will destroy the natural world and our futures. (NISHIOKA 2012: 124)

If it seems a tall order, she pledges her own commitment to the cause. She speaks of vowing to "shed the skin of my 'pre-3 / 11' self" (NISHIOKA 2012: 124). This is a call to slough off the "Atoms for peace" rhetoric and its diverse, entrenched incarnations since the 1950s. With this turn of phrase, NISHIOKA evokes leaving behind the modern false consciousness that prioritizes economic systems and data-driven decision-making over human needs and a clear-eyed recognition of humanity's place in the natural world.

In the concluding section of "Goodbye, Atomic Dragon", NISHIOKA widens the reader's lens, capturing the sights and scenes of activism against nuclear weapons and power. These pages depict Marco and Ayumi attending a protest with a throng of people. A banner hanging over a stage marks this protest as an actual event that took place just six months after the 3 / 11 disaster, on September 19 in Tokyo. According to this rally's organizers, almost 60,000 people of all ages and backgrounds participated (GOPALAKRISHNAN 2011). NISHIOKA's panels showcase individuals of every age holding signs; "Together, we can stop nuclear power!", "Protect the children", "We don't need nuclear weapons or power", and "I'm on team Fukushima" (NISHIOKA 2012: 106, 114). Notably, the internationally-recognized smiling sun antinuclear badge in its original German, "Atomkraft? Nein danke" also appears (see Fig. 4). In a hopeful gesture, the last panel depicts the dragon receding into the sky while below on earth the antinuclear movement's most famous icon, sunflowers, dominate.

Herein I have rendered NISHIOKA's Japanese title as "Goodbye, Atomic Dragon", but an important nuance is lost in this translation. It does not capture the stronger implications of the original Japanese word *sayonara*. With its inherent sense of finality and seriousness, the *sayonara* in NISHIOKA's Japanese title

Fig. 4: People's Voices Joining Together (*Sayonara, atomikku doragon*, p. 114) © NISHIOKA Yuka

clearly argues for a total eradication of the nuclear dragon. Even though the term has been adopted into the English language as an equivalent to "goodbye", the Japanese sense of a final farewell is unmistakable. Thus, this manga not only elicits a clear-eyed view of the historical evidence that nuclear technology assemblages have actants besides humans, it advocates parting ways with nuclear power for good. The protestor's sign that reads "Goodbye to Atomic Power" (*genshiroku ni sayonara* 原子力にさよなら) is as strong as the one declaring "Let's outlaw nuclear power" (*genpatsu wa keibatsu ni* 原発は刑罰に). Thusly, the work underscores that with half-lives as long as 700 million years and 24,000 thousand years there can be no half measures.[7]

5 CONCLUSION

As NISHIOKA argues in "Goodbye, Atomic Dragon", the U.S.-led "Atoms for Peace" campaign messaging and nuclear energy boosterism since the 1950s should be recognized as misguided promises whose consequences have already indiscriminately and irrevocably affected untold numbers of victims. The overarching message of this manga both echoes and hails the mindset transformation that ZYLINSKA promotes: a call for an imbricated world view that focuses on human life over technological fancies. NISHIOKA's work organically emphasizes what ZYLINSKA describes as "a more anchored, embodied, and localized sense of response to, and responsibility for, the milieu we earthlings call home" (ZYLINSKA 2018: 7). In this vein, in the "Afterword", NISHIOKA writes, "we must rethink the state of contemporary civilization and the way we live our lives. If we don't stop this madness, the atomic dragon will destroy our natural world and the future we are supposed to pass on to the next generations" (NISHIOKA 2012: 124). Cognizant that "the line between life and death turns like that of a playing card, determining one's fate in a fragile world" (NISHIOKA 2012: 121), NISHIOKA urges her audience to understand that although humans cannot completely control the natural world, we are able to rationally avoid provoking natural phenomena with technologies that are clearly beyond our capability to control.

In particular, NISHIOKA encourages contemporary modes of activism and communicating information while emphasizing the role of the youth who will inherit the burdens of our generation's misguided technofix mentality. In the final pages of "Goodbye, Atomic Dragon", NISHIOKA includes a visual postscript of an energetic Ayumi, now 14-years-old, who is a member of the Natural Energy Research Club at her middle school. She visits a small hydropower plant to collect information for an online blog, learning how modern turbines can be constructed to protect fish in the waterways. Afterwards, Ayumi recognizes that traditional modes of energy production can be ethically modernized, concluding that the way forward is to "borrow natures' gifts a little bit at a time" (NISHIOKA 2012: 118). Eschewing an overreliance on the sensational technofixes, NISHIOKA asks us to move beyond spectacular sales pitches that are couched in narratives of modern scientific surety and ground ourselves in human-scaled projects that recognize and respect the nature of our species' entanglement with our environment.

As mentioned above, other manga creators produced a wide range of works following the triple disaster. MISUKOSO's みすこそ "Someday in a Rapeseed Field: We Will Not Forget Japan's Great East Disaster" represents an early response that focuses primarily on the aftermath of the earthquake and tsunami. The volume is a collection of vignettes depicting the experiences of victims and survivors of 3/11, which the artist learned of when she volunteered in Iwate in late June 2011. Her stories portray the terror of the earthquake itself and the loss of cherished children, parents, grandparents, and pets. The work evokes the anguish of survivors and the small and extraordinary acts of kindness and selfless ministrations extended by loved-ones, neighbors, nurses, and rescue workers. MISUKOSO's minimalist line renderings and deeply distressing scenes—for instance, when soldiers discover the body of a grandmother hugging her lifeless young grandson under a pile of rubble—productively renders abject confusion, fear, and sorrow. In the additional materials toward the end of the book, she describes the conditions, her duties, and interactions with other volunteers. Notably, she relates that a fellow, especially earnest volunteer, I-san, expressed their commitment to sharing the wretched conditions in Iwate with others in their hometown. "What I fear most is that it will be forgotten. At the beginning everyone is moved by the terrible scenarios, but thinking that it's not their issue, they forget about it. For everyone to live happily, it is necessary for all of us outside of the northeast to offer support" (Misukoso 2011: 143). Misukoso hails the call and decides to create this manga, and, in her words, she "aimed to inlay I-san's values" (*kachikan wo chiribameta tsumori* 価値観をちりばめたつもり) into the project (Misukoso 2011: 144).

Still, among the chapters in Misukoso's "Someday in a Rapeseed Field: We Will Not Forget Japan's Great East Disaster", one notably addresses the nuclear fallout following the meltdowns at the Fukushima Daiichi Power Plant. In "Fukushima: The Elderly Couple's Decision", a critique of the disaster is framed within the utter devastation of familial roots and historical livelihoods, speaking broadly to uncertain and dark futures. The storyline focuses on a bedridden man and his wife whose ancestral home lies inside the evacuation zone. The couple's abject situation first focuses on the national crisis regarding the safety of food products grown in Fukushima by highlighting the couple's enjoyment of a meal made with fresh vegetables from their garden as a news program reports on the state's efforts to screen levels of radiation in leafy greens. Then, soldiers, dressed in full hazmat suits and ominous gas masks come to warn them of the dangerous contamination and offer aid in evacuating them. However, the couple refuse, stating "We were born and raised here. We only know this town / If it has become dangerous, we have no choice but to follow fate and stay here together" (Misukoso 2011: 96). The following day the woman discovers that her husband has passed during the night, and her thoughts are dominated by grief as she contemplates who will bring the couple's ashes to and take care of their ancestral graves. In the years following the 3 / 11 disaster, numerous other media creators penned important human-centered tales of incomprehensible personal loss and social and cultural disintegration.

While Nishioka is no stranger to pathos, I argue that her body of work, broadly, and "Goodbye, Nuclear Dragon", specifically, are exceptional for their deep commitment to rigorous historical and political contextualization and critique. Now, more than ten years after one of the world's three largest nuclear power disasters, in the context of Japan's increased formal and informal restrictions on public discourse on topics branded polemical by vested commercial and political interests, NISHIOKA's voice is even more important. As a life-long citizen of Nagasaki, it is not surprising that her projects are grounded in a nuclear abolition movement that calls for a global ban on the production, testing, and use of nuclear weapons. Still, under the cloud of the 2013 State Secrecy Law (Tokutei himitsu no hogo ni kan suru hōritsu 特定秘密の保護に関する法律), there continues to be informal and formal pressures on critical voices while officials continue to downplay the seriousness of the immediate and long-term environmental and human consequences of the nuclear disaster. As a particular case in point, the "2020 Olympics" functioned as a calculated, distracting spectacle, offering festive optics to sugarcoat the realities of the ongoing nuclear disaster and deflecting funds that could have been invested in addressing it. NISHIOKA continues to be one of a small number of manga artists who speaks clearly and directly to the human loss and misery caused by nuclear weapons and power.

In this context of public discourse, one also notes the closure of small presses the likes of which have carried the lion's share of making works such as NISHIOKA's available to the public. The supremacy of conglomerate publishing companies begs the question of the future of works created by NISHIOKA and fellow activists. The progressive publishing house, Gaifūsha, which produced many of NISHIOKA's educational manga on nuclear topics, folded in 2022. Founded in 1976, Gaifūsha was a well-established press that supported critical views on a wide range of topics related to Japanese colonialism, wartime history, questions of Japanese postwar responsibility, and the U. S. occupation of Okinawa. It also printed works on Korean cultural history and Chinese literature, as well as pictorial guides of Germany's Auschwitz-Birkenau Memorial and Museum. Their publications that focus on nuclear histories at home and around the globe are particularly notable, including tomes by TAKAHASHI Hiroko 高橋博子 and TAKEMINE Seiichirō 竹峰誠一郎, founders of the Global Hibakusha Study Group and related publications.[8] Political and publishing pressures are a cause for concern as they determine who gets a voice. So, it remains to be seen how NISHIOKA will continue to share her unique voice in the changing landscape of message making.

January of 2023 was the two-year anniversary of the entry into force of The United Nations Treaty on the Prohibition of Nuclear Weapons (TPNW)—signed by 122 of the 193 UN member states. With this landmark agreement, the international community has made a very important, collective stand against one deathly form of nuclear power.[9] Even if this treaty may not immediately engender practical change given the strong-hold nuclear nations still have today, it represents a resounding majority voice. NISHIOKA's "Goodbye, Atomic Dragon" expands on this declaration, calling for abolition of both nuclear weapons and power generation. That much more still needs to be done to dispel the dangerous / safe split between nuclear weapons and power has been recently made abundantly clear by not only Russia's threats to launch nuclear weapons but also the violent commandeering of and threats of bombing nuclear power plants in Ukraine. NISHIOKA offers us an approachable and compelling argument for resisting the fallacy of "dangerous" nuclear weapons and "safe" nuclear power divide, and one must hold hope that another tragedy such as the 2011 Fukushima disaster does not occur before the global community heeds her call.

NOTES

1   To name just a few: HIRAI Toshinobu's 平井寿信 "Japan's Northeastern Disaster: The Sights We Saw" (*3/11 Higashi Nihon daishinsai: Kimi to mita fūkei* 3/11東日本大震災：君と見た風景, 2011); MISUKOSO's みすこそ "Someday in a Rapeseed Field: We Will Not Forget Eastern Japan's Disaster" (*Itsuka, Nanohana hatake de: Higashi Nihon daishinsai o wasurenai* いつか、菜の花畑で：東日本大震災をわすれない, 2011); YAMAMOTO Osamu's 山本おさむ "Good Weather Today" (*Kyō mo ii tenki* 今日もいい天気, 2013); NOBUMI のぶみ / MORIKAWA Jōji's 森川ジョージ "I'll meet you there" (*Ai ni iku yo* 会いにいくよ, 2014); and ABE Kuniyuki's 阿部国之 "3/11: Disaster Testimonies Hatakeyama Takuya: Voices from Ishimaki" (*3/11 Shinsai no kataribe Hatakeyama Takuya: Ishimaki kara no koe* 3/11震災の語り部畠山卓也：石巻からの声, 2014).

2   Consider, for instance, the 2019 *New York Times* opinion piece by GOLDSTEIN / QVIST / PINKER (2019); the more recent opinion column "Is Clean, Green Nuclear Energy the Answer to EU's Energy Security Woes?" (2022); and the article "3 Reasons Why Nuclear is Clean and Sustainable" (2021) on the website of the U. S. Office of Nuclear Energy.

3   In the former category, TAKITA Hiroshi's 滝田ひろし "Aaah, The Bells of Nagasaki Ring" (*Aa Nagasaki no kane ga naru* ああ長崎の鐘が鳴る) was published as early as 1958. 15 years later the most prominent example, NAKAZAWA Keiji's 中沢啓治 "Barefoot Gen" (*Hadashi no Gen* はだしのゲン), began serialization in "Weekly Shōnen Jump" (*Shūkan shōnen janpu* 週刊少年ジャンプ, 1973–87).

4   For a post-Fukushima tome depicting the lives of workers in the age of 21st-century precarious work force, see TATSUTA Kazuto's 竜田一人 "Ichi-F: A Worker's Graphic Memoir of the Fukushima Power Plant" (*Ichi efu: Fukushima daiichi genshiryoku hatsudensho rōdōki* いちえふ：福島第一原子力発電所労働記, 2013–15).

5   It should be noted that SHIRIAGARI Kotobuki しりあがり寿 published a nuclear-themed collection of comic strips and one-shot manga in 2011, replete with his singular gag humor and incisive critique. For more, see KNIGHTON (2013).

6   For more information on this condition, see the documentary film *Chernobyl Heart* by Maryann DELEO (2003).

7   The half-life numbers correspond to the isotopes Uranium 235 and Plutonium 239.

8   These include TAKAHASHI's work on censored documents, "Classified Hiroshima and Nagasaki: U. S. Nuclear Tests and Civil Defense Programs" (*Fūin sareta Hiroshima, Nagasaki: Bei kakujikken to minkan bōei keikaku* 封印されたヒロシマ・ナガサ

キ：米核実験と民間防衛計画, 2008), and TAKEMINE's co-written book with NAKAHARA Satoe 中原聖乃, "The Marshall Islands' Atomic Age: Society, Culture, History and Hibakusha" (*Kakujidai no Māsharu shotō: Shakai, bunka, rekishi, soshite hibakusha* 核時代のマーシャル諸島：社会・文化・歴史、そしてヒバクシャ, 2013).

9 For more information on this treaty, see this UN document: https://www.un.org/disarmament/wmd/nuclear/tpnw/ (last access 2023/02/28).

REFERENCES

ABE, Kuniyuki 阿部国之 (2014): *3/11 Shinsai no kataribe Hatakeyama Takuya: Ishimaki kara no koe 3/11* 震災の語り部畠山卓也：石巻からの声 (SG Document Comics). Tōkyō: Shōnen gahōsha.

ANONYMOUS (2022): "Is Clean, Green Nuclear Energy the Answer to EU's Energy Security Woes?". In: *ESI-Africa*, 03/17; https://www.esi-africa.com/features-analysis/op-ed-is-clean-green-nuclear-energy-the-answer-to-eus-energy-security-woes/ (last access 2023/02/28).

BENNETT, Jane (2010): *Vibrant Matter: A Political Ecology of Things*. Durham, North Carolina: Duke University Press.

GERSON, Joseph (2007): *Empire and the Bomb: How the US Uses Nuclear Weapons to Dominate the World*. London: Pluto Press.

GLANZ, James (2013): "When the Grid Bites Back: More Are Relying on an Unreliable System". In: *International Herald Tribune*, 08/18: 1 (sect. 4).

GOLDSTEIN, Joshua S./QVIST, Staffan A./PINKER, Steven (2019): "Nuclear Power Can Save the World". In: *The New York Times*, 04/06; https://www.nytimes.com/2019/04/06/opinion/sunday/climate-change-nuclear-power.html (last access 2023/02/28).

GOPALAKRISHNAN, Manasi (2011): "No More Fukushimas!". In: *DW Academy*, 09/19; https://www.dw.com/en/thousands-protest-against-nuclear-energy-in-japan/a-6621588 (last access 2023/02/28).

HIRAI, Toshinobu 平井寿信 (2011): *3.11 Higashi Nihon daishinsai: Kimi to mita fūkei* 3.11東日本大震災：君と見た風景 (Bunkasha Comics ぶんか社コミックス). Tōkyō: Bunkasha.

KATSUMATA, Susumu (2018): *Fukushima Devil Fish*. Translated by Ryan HOLMBERG. London: Breakdown Press.

KNIGHTON, Mary (2013): "The Sloppy Realities of 3.11 in Shiriagari Kotobuki's Manga". In: *Asia-Pacific Journal: Japan Focus* 11 (26); https://apjjf.org/2014/11/26/Mary-Knighton/4140/article.html (last access 2023/02/28).

MASON, Michele (2009): "A New Generation of Historical Manga: Writing Hiroshima and Nagasaki in the 21st Century". In: *The Asia-Pacific Journal*

7 (47); https://apjjf.org/-Michele-Mason/3260/article.html (last access 2023/02/28).

Misukoso みすこそ (2011): *Itsuka, Nanohana hatake de: Higashi Nihon daishinsai o wasurenai* いつか、菜の花畑で：東日本大震災をわすれない. Tōkyō: Fusōsha.

Nakazawa, Keiji 中沢啓治 (1988): *Hadashi no Gen* はだしのゲン. Tōkyō: Chōbunsha.

Nishioka, Yuka 西岡由香 (2010) *The Santa Claus of August 9th: Nagasaki's A-Bomb and the Survivors* 八月9日のサンタクロース：長崎原爆と被曝者. Tōkyō: Gaifūsha.

Nishioka, Yuka 西岡由香 (2012): *Sayonara, atomikku doragon* さよなら、アトミック・ドラゴン. Tōkyō: Gaifūsha.

Nishioka, Yuka 西岡由香 (2015): *Hibaku Maria no inori: Manga de yomu sannin no hibaku shōgen* 被爆マリアの祈り：漫画で読む三人の被爆証言. Nagasaki: Nagasaki bunkensha.

Nobumi のぶみ / Morikawa, Jōji 森川ジョージ (2014): *Ai ni iku yo* 会いにいくよ (Kōdansha Comics 講談社コミックス). Tōkyō: Kōdansha.

Takahashi, Hiroko 高橋博子 (2008): *Fūin sareta Hiroshima, Nagasaki: Bei kakujikken to minkan bōei keikaku* 封印されたヒロシマ・ナガサキ：米核実験と民間防衛計画. Tōkyō: Gaifūsha.

Takemine, Seiichirō 竹峰誠一郎 / Nakahara, Satoe 中原聖乃 (2013): *Kakujidai no Māsharu shotō: Shakai, bunka, rekishi, soshite hibakusha* 核時代のマーシャル諸島：社会・文化・歴史、そしてヒバクシャ. Tōkyō: Gaifūsha.

Takita, Hiroshi 滝田ひろし (1958): *Aa Nagasaki no kane ga naru* ああ長崎の鐘が鳴る. Tōkyō: Tōkyō manga shuppansha.

Tatsuta, Kazuto 竜田一人 (2013–15): *Ichi efu: Fukushima daiichi genshiryoku hatsudensho rōdōki* いちえふ：福島第一原子力発電所労働記 (Morning KC). Tōkyō: Kōdansha.

U. S. Office of Nuclear Energy (2021): "3 Reasons Why Nuclear is Clean and Sustainable"; https://www.energy.gov/ne/articles/3-reasons-why-nuclear-clean-and-sustainable (last access 2023/02/28).

Yamamoto, Osamu 山本おさむ (2013): *Kyō mo ii tenki* 今日もいい天気 (Action Comics). Tōkyō: Futabasha.

Zylinska, Joanna (2018): *The End of Man: A Feminist Counterapocalypse*. Minneapolis: University of Minnesota Press.

Rhetorical Perspectives on Post-nuclear Japan: Embracing Hiroshima and Nagasaki Within Fukushima

Hiroko Okuda

1 Introduction

Media discourse can be conceived of as a set of interpretive packages that give meaning to an issue. Hence the news media play a complex role not merely in producing public opinion on an issue, but in framing issues for defining and constructing social reality (Berger / Luckmann 1966; Noelle-Neumann 1993). Indeed, media narratives are used both for making sense of relevant events—suggesting what is at issue and organizing the world—allowing for a degree of controversy among those who share a common narrative (Iyengar 1991: 1–16, 127–43; Iyengar / McGrady 2007: 197–236).

While living with the tragic memories of the atomic bombings on Hiroshima and Nagasaki, postwar Japan framed nuclear power in imperatives of reconstruction, nation-building, industrialization and modernization. The progress narrative that incorporated nuclear power development into postwar Japan's overreliance on technology almost inevitably creates a false sense of security. With political and economic motives, this narrative allowed the country to differentiate between military use for atomic weapons and economic use for nuclear energy (Satō 2010: 35–85, 199–242; Yamazaki 2011: 109–14, 127–44, 152–207). Such decoupling should not be yet confused with positions for or against Japan's energy policy, but rather considered as ambivalent. On the one hand, politicians and bureaucrats painted efforts to change the country's negative attitude toward nuclear power as emblematic for a broader revival after the country's defeat in the Second World War. On the other hand, many were cynical about such administrative assurances that economic growth and stability should have priority. As such, the triple disaster—earthquake, tsunami and nuclear—on March 11, 2011, could quite conceivably have turned the tables on postwar Japan and resulted in a complete loss of faith in nuclear energy as peaceful and safe, yet it did not (Yamamoto 2012: 35–73, 187–221, 297–310; Yoshimi 2012: 7–46, 122–94, 265–94).

Hiroshima was bombed at 8:15 AM on August 6, 1945, and Nagasaki at 11:02 AM on August 9, 1945. A series of nuclear explosions and meltdowns at the Fukushima Daiichi nuclear power plant were triggered at 2:46 PM on March 11, 2011. All these Japanese local times are remembered at home and abroad for just one thing: the world looked on aghast. Post-nuclear Japan has long had an ambivalent attitude toward nuclear power. The power of dissociation allowed the country to associate *atomic* (bomb) with an *evil* power of mass destruction and *nuclear* (energy) with the *positive* power of mass production. At the beginning of this century, global discourse around nuclear energy shifted, and it experienced a positive reevaluation as a clean energy source by way of a so-called *nuclear renaissance*. Providing the technology is well regulated, so the narrative goes, nuclear energy would play an important part in cutting greenhouse gases. Yet the risks of nuclear power were still acknowledged. In a single large-scale national reconstruction and revitalization operation set up in 2013, the National Resilience Council (Higashi Nihon daishinsai fukkō kaigi 東日本大震災復興会議), a government advisory panel, recognized the danger of radioactive contamination resulting from the devastating effects of tsunamis and earthquakes. In contrast, the government's efforts at reconstruction have focused merely on infrastructure and the decontamination of land, foregoing any attention to addressing public opinion or rebuilding communities affected by previous disasters. Addressing the issue of radioactive emissions, international nuclear-related agencies such as the International Atomic Energy Agency (IAEA) and the World Health Organization (WHO) have even supported the following official platitudes provided to local residents that they should learn to live with exposure to radiation, and that their greatest health risk is in fact derived from the stress produced by those psychosomatic illnesses created by their unfounded fears of radiation (HINO 2013; NAONO 2011: 230–44).[1]

The Fukushima nuclear catastrophe, which followed the magnitude 9.0 earthquake and subsequent tsunami, shook the foundation of postwar Japan's myth of nuclear safeguards. As a clean way of generating electricity, Tokyo placed its hopes on nuclear power as a semi-domestic energy source due to its efficiency and its ability to produce sufficient energy to meet the demands of modern life. Japan wanted a clean source of energy to fight against climate change, and at the same time recognized that a power shortage might not only disrupt everyday life, but also seriously slow the growth of the entire economy. Hence Japan's nuclear programs have been delayed, due more to concerns about economics than to anxieties regarding nuclear safety (TAKEDA 2011b: 3–14, 19–30, 244–69; YOSHIOKA 2011: 245–394). The Fukushima catastrophe essentially requires Japan to make a difficult choice between restoring public confidence in nuclear energy and developing renewable energy sources. In terms of energy security, Japan has also been forced to reduce its dependence on politically unreliable sources of oil and gas as the current political situation in the Middle East—from which Japan receives almost 90 percent of its fuel imports—remains unstable. In order to mitigate these hazards, the government has invested in strategies designed to solve the technological problems and to produce carbon-free electric power so as to increase energy supplies (SAMUELS 2013: 111–18). On the whole, "[n]uclear power has suddenly found itself from being arguably part of the solution for future green energy to a now dangerous relic of the cold war era" reports Deutsche Bank (MOUAWAD / JOLLY 2011: 1).

This paper shows how different outlets within the Japanese media landscape interacted in their coverage of a series of nuclear explosions and meltdowns at the Fukushima Daiichi nuclear power plant both nationally and locally. The study focuses specifically on the photo coverage offered by the four national dailies, the *Asahi shinbun* 朝日新聞, *Mainichi shinbun* 毎日新聞, *Nihon keizai shinbun* 日本経済新聞 and *Yomiuri shinbun* 讀賣新聞, and the two local dailies, the *Fukushima minpō* 福島民報 and the *Fukushima min'yū* 福島民友, from March 11 to March 31, 2011. By doing so, the study will examine an essential context for interpreting how not just Japan, but also other nations around the world still rely on media narratives regarding the peaceful use of nuclear power. In other words, it will explore the extent to which the news media could theoretically provide counter-narratives regarding technology gone mad. Such narratives could prove compatible with the narrative of progress, which is founded on a notion of nature as controlled by technology.

2   MEDIA DISCOURSE

Public officials have often served as sponsors or advocates for nuclear power. The Nuclear and Industrial Safety Agency (Genshiryoku anzen hoan'in 原子力安全保安院), the Nuclear Regulatory Commission (Genshiryoku kisei iinkai 原子力規制委員会), and the Ministry of Economy, Trade and Industry have been important supporters for the progress narrative. Their efforts have been supplemented by industry groups such as the Japan Business Federation (Nippon keizai dantai rengōkai 日本経済団体連合会), Japanese electric power companies including the Tokyo, Hokkaidō, Tōhoku, Hokuriku, Chūbu, Kansai, Chūgoku, Shikoku and Kyūshū Electric Power Companies, Japan Nuclear Fuel Limited (Nihon gennen 日本原燃), and the Power Reactor and Nuclear Fuel Development Corporation (Dōryokuro, kakunenryō kaihatsu jigyōdan 動力炉・核燃料開発事業団). In addition, such public relations actors as Advertising Council Japan (AC Japan エーシージャパン), Dentsū 電通, the above power companies and the Japanese government have actively promoted nuclear power and have also helped articulate and spread this narrative through their public relations efforts (ŌSHIMA 2012: 87–128, 149–66; YOSHIOKA 2011: 17–28, 69–177). By reconsidering such routine channels as official proceedings, press releases, press conferences and scheduled official events, most Japanese news accounts are likely to make use of official narratives as the starting point for discussing an issue. In other words, official assumptions are generally taken on good faith (FELDMAN 2004: 15–48; TAKEDA 2011a: 35–92).

In news reporting, interpretation is often provided through quotations, and balance is provided by quoting spokespersons with competing views. In addition, norms of balance are generally shared in the commentary provided by editorial boards and syndicated columnists. On the one hand, the idea of balance as provided by mutually opposed commentators is ambiguous in practice, and yet it favors certain narratives over others. This tendency to reduce controversy to two competing positions promotes an official narrative, even if alternative narratives are available. The commentators thus share the same unstated, common assumptions as officials. On the other hand, this *balance norm* is rarely mobilized to establish counter-narratives (TUCHMAN 1978: 82–132, 156–217). According to Gaye TUCHMAN (1974: 112), the television news searches for an "establishment critic" or for a "'responsible spokesman' whom they themselves have created or promoted to a position of prominence". As Japan's *postwar* period has given way to a new *post-disaster* era, however, the March 11 disaster has not changed Japan. At first, commentators spoke of the disaster as a historical turning point in a period marked by long-standing economic stagnation and demographic decline. This shared experience, they claimed, would shape a generation, it would be a shock to reform structural failings and open closed minds. It happened to have occurred, however, precisely when the Liberal Democratic Party (LDP) was not in power. The association of the Democratic Party of Japan (DPJ), in office from 2009 to 2012, with the crisis helped to discredit both the idea of reform as well as that of political alternation. Indeed, the political impetus for reform has dissipated, as the LDP has regained power, the DPJ has disintegrated and apathy has set in. As a result, while the enduring mistrust extends to nuclear power in general, the government has not given up on it.

3 RECONSTRUCTIONS OF MARCH 11, 2011

On many issues, ordinary citizens encounter relevant phenomena directly rather than through media accounts. They try to understand events in light of their own life experiences, but few individuals have direct experience with so complex an issue as nuclear power. As long as the latter is framed as a choice between atoms for mass destruction and atoms for mass production, it is hard to see who could be against progress (LIPPMANN 1922: 3–20, 214–25; EDELMAN 1985: 114–51, 195–214). On December 8, 1953, President EISENHOWER addressed the United Nations on nuclear power in what came to be labelled by the media as his "Atoms for Peace" speech. He proposed making U. S. nuclear technology available to an international agency that would attempt to develop it for peaceful uses. The "Atoms for Peace" speech came at the height of the Cold War, giving rise to the IAEA that plays a key role in distributing nuclear technology to developing nations (SCHEINMAN 1987: 56–60). This techno-political regime promoted the either / or structure of nuclear discourse, which is often represented as a road that branches off in two directions: one leading to the proliferation of nuclear weapons, and the other to the production of emission-free electricity for advancing economies (CHERNUS 2002: 3–52, 79–131). With the Cold War discourse, the narrative of progress remained dominant, and was taken essentially as a given in postwar Japan.

At this juncture, I would like to take a critical approach to the photo coverage of the Fukushima Daiichi nuclear complex by the national and local newspapers, thus rendering visible what Paul CHILTON (1987: 16) calls "critical discourse moments". The framing inherent to postwar Japan's nuclear discourse is more easily uncovered thereby, thus revealing the ways in which Tokyo sought to shape public opinion through media discourse. The overview here presented, which examines photos of the nuclear catastrophe—as covered by the national and local news reporting from March 11 to 31, 2011—is shaped by the following three characteristics: First, the ranking of photo credits proceeds as follows; DigitalGlobe (31), the Tokyo Electric Power Company (TEPCO; 30), the Ministry of Defense (MD; 25), the Fukushima Central Television and the NHK General TV (13), the Nippon Television Network (9), the Nuclear and Industrial Safety Agency under the Ministry of Economy, Trade and Industry (METI; 8), Reuters (5), the Tokyo Fire Department and the Ministry of Land, Infrastructure, Transport and Tourism (MLITT; 3) (see Figure 1). This shows the advance of information technology, the power of TV live footage, and the control of information by state and power company officials. Second, the chronological shift of visual coverage shows the change of information sources from video footage to aerial snapshots, which were then followed by the release of visual archives to the press by government ministries and the nation's privately-run electric company (see Figure 2). Finally, while the four national dailies moved the latest news about the nuclear plant from the front to the back pages, the two local newspapers continued to highlight the nuclear issue in their front-page coverage. For a short period afterwards, the national papers rode a wave of international geo-political

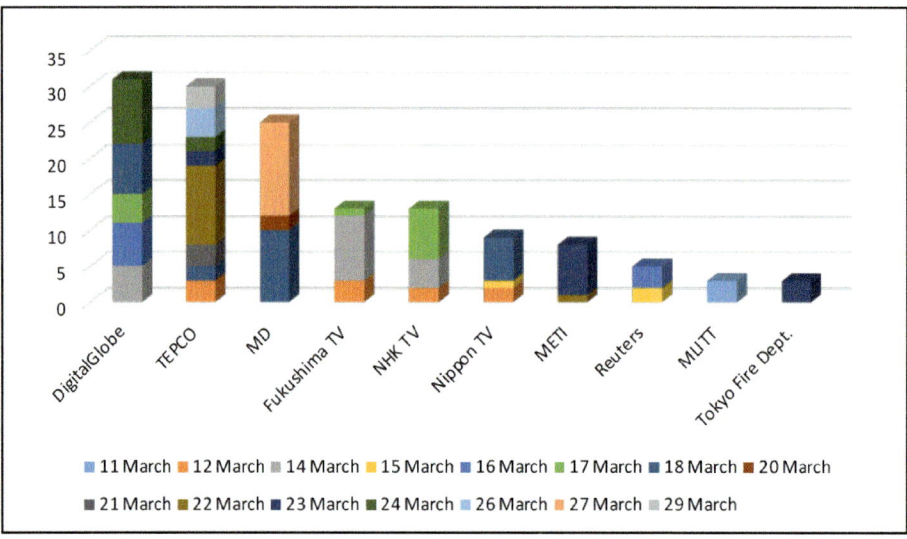

Fig. 1: A number of photo sources of national and local newspapers from March 11 to 31, 2011

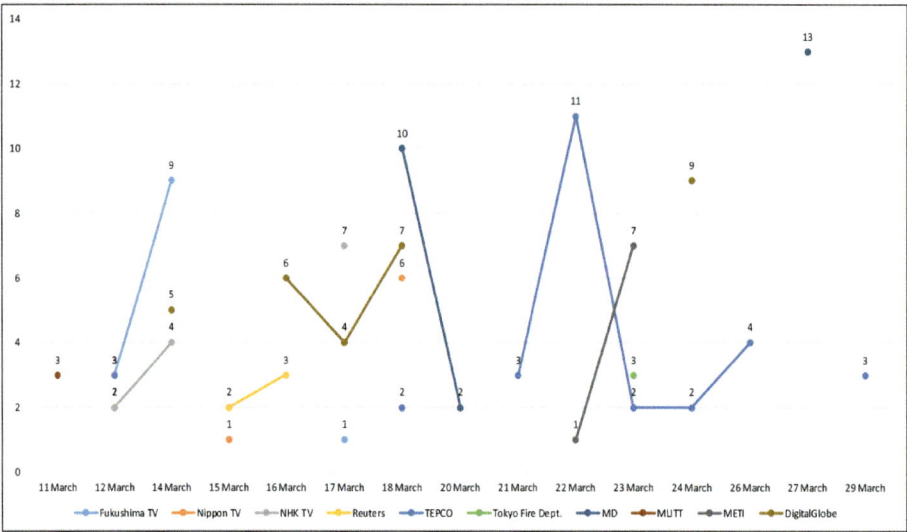

Fig. 2: A shift of photo sources from March 11 to 31, 2011

events, prioritizing the Arab Spring—the uprising in Libya after Tunisia and Egypt in particular—over the Fukushima nuclear crisis. In contrast, the local media valued subsequent reports on nuclear power—endless catastrophes—by piquing interest and assuaging fears.

Figures 1 and 2 show how the political programs, technical and scientific practices, and institutional ideology have constituted and governed postwar Japan's access to peaceful nuclear technologies. First, the number of information sources highlights the complementary relationship between technology and politics that the March 11 disaster should properly define and enact. Indeed, Figure 1 shows technological rankings such as those of U. S. over Japanese companies, or the TEPCO over the MD. Such technical and scientific hierarchies could equally constitute grounds for a narrative of inequality in terms of technologies and nuclear expertise. Beyond nuclear governance and information control, the local TV made a breakthrough in releasing footage of the first nuclear explosion of the Fukushima Daiichi nuclear site on March 12, 2011. Second, a shift of photo sources shows the degree to which information considered "nuclear" is a spectrum, not a dichotomy. The chronological change of photo credits from the broadcasting to the technical agencies indicates a pragmatic shift in deliberate attitudes toward the March 11 disaster. Media discourse sought to depoliticize the nuclear crisis by focusing specifically on technical issues, while struggles over safeguarding, by contrast, were saturated with questions of morality. Overall, the photo coverage reflected the TEPCO's privileged position as an electric supplier and its attempt to maintain the legitimacy of its place near the top of national nuclear expertise and technologies.

At present, Fukushima stands for the multifaceted complexity of the events that took place on March 11, 2011. The Great East Japan Earthquake occurred and caused triple disasters: damage from the earthquake itself, the resulting tsunami, and a string of nuclear accidents. Indeed, ultimately a fourth disaster unfolded as the first three misfortunes led to social chaos, especially at the state level. The Japanese government proved quite incompetent in its inability to respond to the disasters, and especially its failure to satisfactorily resolve the unfolding nuclear crisis. One explanation for such incompetency comes from the theory of elite panic that developed after the terrorist attacks on September 11, 2001 in the United States. The elite panic theory stipulates "relationships between elites and panic—that elites can fear panic, that elites can cause panic, and that elites can themselves panic" (CLARKE / CHESS 2008: 996).

Having critically examined visual representations of the Fukushima disaster in the media, the next section focuses specifically on the discourse in the news that surrounded the same photos covered by the national and local dailies. This approach seeks to reveal the failure of administrative officials, bureaucrats and TEPCO executives to cope with the unfolding crisis. At first, the vanishing myth of nuclear safety meant the collapse of postwar Japan's economic growth. In the crisis of social reality, then, the narrative of post-disaster Japan's recovery would appear again and again to define, explain and assimilate urgent events in the mediated discourse. It served to nationally evoke the narrative of Japan's postwar reconstruction with a view to understanding the urgency of the event, to set criteria for policy evaluation and to locate viable options for action.

4 POST-NUCLEAR JAPAN'S DISCURSIVE DUALITY

Public awareness of nuclear power in Japan began with the images of sudden, and enormous destruction, symbolized by the rising mushroom clouds of the atomic explosions over Hiroshima and Nagasaki. The blasts gave rise to fears of extinction, both in Japan and around the world. In order to assuage such lingering discomfort over the atomic bombings, the United States endeavored to transform nuclear technology from "a Force for Evil (war)" to "a Force for Good (production)" by internationalizing the great advances in technology (MACDONALD 1945: 258). In such a post-Hiroshima / Nagasaki vision, it was hard for resource-poor Japan to survive without the help of nuclear plants for the supply of electricity and for growth in the economy (BOYER 1985: 127). Here the faith that was won for the promise of "Atoms for Peace" constituted an active "part of the process by which the nation muted its awareness of Hiroshima and Nagasaki and of even more frightening future prospects" (BOYER 1985: 127). Indeed, the nuclear dualism inherent to postwar Japan's faith in progress remained largely unchallenged.

4.1 A Case Study

Both the national and the local papers used a shot from live TV footage to show how the Number One reactor at the Fukushima Daiichi nuclear power plant appeared to have been destroyed, with its outer wall apparently having collapsed ("Fukushima daiichi genpatsu de": 1; "Fukushima genpatsu de bakuhatsu: Daiichi": 1; "Fukushima genpatsu de bakuhatsu: Shūhen": 1; "Kokunai-hatsu": 1). It was shortly after 3:30 PM on March 12 that an explosion occurred and injured four workers.[2] About ten minutes later, white smoke was witnessed in the air, confirmed by the Fukushima Central Television. Here hydrogen in the outer container was combined with oxygen in the air, thus causing a hydrogen explosion, which blew up the outer concrete structure. NHK General TV highlighted the outcome of the explosion, which left only its structural frame intact. In addition, the temperature of the core and the detection of cesium led the Nuclear and Industrial Safety Agency to infer a partial meltdown of the nuclear fuel prior to the explosion. Yet, at first, neither national nor local news media made any comparison to the Chernobyl nuclear crisis in 1986. The latter had been given a level seven, a major accident, by the IAEA, making it the world's most serious nuclear accident. The public agency sought to play down this first nuclear explosion by equating it with the accident at the nuclear fuel-processing plant in Tōkaimura in 1991, which until then was the worst nuclear accident in Japanese history.[3]

Following the hydrogen explosion at the first reactor, TEPCO started to fill the reactor pressure containment vessel with seawater, claiming that the reactor containment and pressure vessels were not damaged. Experts nonetheless warned that there was a rising danger of a hydrogen explosion at the third reactor, especially given its use of plutonium-thermal power generation (also known as pluthermal generation). The reactor employed plutonium-uranium mixed oxide fuel (i. e., MOX) and had lost its ability to supply water to the core. Indeed, a core meltdown was then triggered by overheating and further abnormalities while seawater was being pumped into the reactor. At 11:01 AM on March 14, efforts to release hydrogen gas caused two explosions at the overheated Number Three reactor. With the exception of the *Asahi shinbun*, the front pages of all five dailies showed stills from the Fukushima Central Television live recording. The images were of an orange fireball and huge plumes of grey-brown smoke rising higher than those from the first explosion ("Genpatsu sangōki mo": 1; "Sangōki mo suiso bakuhatsu" (03 / 14): 1; "'Fukushima daiichi' nigōki": 1; "Genpatsu sangōki mo": 1; "Sangōki mo suiso bakuhatsu" (03 / 15): 1). In contrast, the NHK General TV footage focused on what remained after the two explosions that had been shown on the front pages of the *Asahi* and *Yomiuri* coverage ("Kōnōdo hōshanō": 1; "Nenryōbō subete": 1). This visual coverage illustrated how the explosions damaged part of an overheated reactor's protective shielding and blew the roof off the reactor. Indeed, in their hosing mission the workers were hit by flying pieces of concrete, and eleven of them were injured—four power employees, three from its partner companies, and four Self-Defense Force (SDF) personnel. As a consequence, a series of images revealed that the two reactors had already suffered partial meltdowns, and what remained not only posed a risk, but faced the imminent danger of a complete meltdown.

At around 8:30 AM on March 16, white smoke was witnessed rising out of the troubled third reactor. The local as well as the national newspapers used the aerial photograph, offered either by Reuters or by DigitalGlobe, to prove that the smoke was coming out of the temporary storage pool for spent nuclear fuel rods in the reactor's outer containment building ("Kinkyū jitai": 1; "Keishichō hōsuisha": 1; "Daiichi genpatsu": 1; "Kiki kanri": 2; "Genpatsu setogiwa": 3; "Tesaguri": 3). Three hours earlier, around 5:15 AM, the fourth reactor, where two workers were missing, was hit by a fire. This fire was also presumed to have occurred around the temporary storage pool. A separate fire that occurred on the previous day led to the collapse of side walls on the outer containment building's fifth floor, where the storage pool is located. On their first and second pages, the three national papers—the *Asahi shinbun*, *Nihon keizai shinbun* and *Yomiuri shinbun*—described the smoke as a warning sign that radioactive materials would escape from the building. It was ultimately TEPCO that took a photo identifying plumes of white vapor rising from Reactor Number 3. A fire then broke out for a second time at Reactor Number 4 ("Genpatsu nikuhaku": 1; "Hōsui inochigake": 2; "Chijō kara": 1; "Hōsui keizoku": 2; "Fukushima Daiichi genshiryoku": 1). That the steel roofs that had been blown off these two reactors was proof that radiation levels had spiked dramatically at the crippled nuclear plant, which forced helicopter pilots and firefighters to abandon their plans to cool an overheated fuel storage tank with water on that same day.

At 9:48 in the morning of the following day, two Ground Self-Defense Force (GSDF) helicopters started to dump water on the third reactor in order to quell a cloud of radioactive steam. While all four national newspapers splashed the front pages of their March 17 evening edition with the NHK footage ("Sangōki chūsui sakusen": 1; "Sangōki rikuji": 1; "Genpatsu reikyaku": 1; "Sangōki heri": 1), the *Fukushima min'yū* lead with local TV footage the following morning ("Sora": 1). The Ministry of Defense subsequently released a recording discussing the details of the SDF's water-dumping operations. By that evening, the Tokyo Fire Department had managed to deploy a water cannon truck for an hour in an attempt to contain a high level of radiation leaking. For about a week, SDF special fire and rescue engines as well as ladder trucks and chemical fire engines from the Tokyo Fire Department were deployed to cool overheated uranium fuel. These joint water-blasting operations appeared a week later not on the front-page, but in the coverage of social issues of the *Asahi shinbun* and *Yomiuri shinbun* ("Keihōon 'modoru!'": 31; "Bōgofuku": 11; "Hibiku keikokuon": 29).

After the Number 3 reactor had been soaked with water from both above and below through aerial and ground operations, radiation levels were confirmed to have declined at the Daiichi plant on March 18. According to the nuclear safety agency, however, the effect of the ground-based hosing operation was still unknown. Moreover, the decision to use two GSDF helicopters to pour tons of water on the damaged reactor was made "under strong pressure" from Washington because Tokyo turned down a U. S. technical offer to cool overheating nuclear reactors ("Govt rejected": 1). As the world watched, efforts continued to avert a full meltdown at the nuclear complex. In addition, work to restore electricity was under way in three separate reactor groups—Reactors 1 and 2, Reactors 3 and 4, and Reactors 5 and 6. As the Japanese government admitted, the highest priority was the second reactor because high radiation levels inside and outside the building made it almost certain that a partial meltdown of fuel rods had occurred. By that time, however, the IAEA had issued a statement that the situation at the first, second and third reactors was "relatively stable" (SOBLE / COOKSON / KIRCHGAESSNER 2011: 2).

From March 17 to 18, TEPCO began efforts to set up a new power line to supply electricity to the Daiichi nuclear power plant. By doing so, the power company sought to restore the emergency core cooling system (ECCS) designed to cool the reactor, thereby averting imminent risks of nuclear meltdown at Reactors 1 to 3. When the company released the photos related to the work to restore electricity about two weeks later, only the *Asahi shinbun* and the *Fukushima minpō* covered how difficult it was by focusing on a couple of unidentified engineers in a small photo ("Sagyō kankyō": 3; "Tōden, dengen": 2). By March 19, workers had almost completed laying cables that would supply external power to the second reactor, the outer building of which had not been blown off. While efforts to restore power to cool down other Reactors 1, 3 and 4 were accelerating, power was returned to the second reactor at 3:46 PM on March 20. Although occasional thin white smoke was still observed at Reactors 2 and 3, work to restore electricity resumed in order to reduce the chance of further problems. Indeed, connecting an outside power supply to the nuclear plant was essential not only for cooling pumps and instruments, but also for monitoring the reactors to be restored.

On March 22, the air conditioning system and some reactor instruments at the second reactor's central control room were restored. Electricity and lighting were at last restored to the central control room for the Number 3 reactor at 11:28 PM on March 22. Coverage provided by *Asahi shinbun*, *Mainichi shinbun*, *Nihon keizai shinbun* and *Fukushima min'yū* highlighted these developments as the first of many steps to avoid exacerbating the nuclear crisis, with the TEPCO photo given as proof ("Ponpu asu": 1; "Keikaku teiden": 3; "Sangōki reikyaku": 1; "Kiki dasshutsu": 3). The control room operates the reactors, turbines, power generators and coolant systems of the nuclear plant. As power was restored, workers were able to check meters, gauges and wiring without using flashlights in darkness. That full electricity was restored to the central control room of the Number 1 reactor just after 1:00 PM on March 24 was only reported by national outlets ("Genpatsu antei": 2; "Ichigōki: Nao": 3). In the meantime, images of black smoke rising out of the Daiichi plant were shown by the *Mainichi shinbun* and the *Fukushima min'yū*, both of whom also declared that by 4:30 AM on March 24, it had ceased altogether ("Genpatsu baishō": 6; "Kagi nigiru genpatsu": 3). Since the cause of the smoke was unknown, the contrasting photo coverage could have served to warn the country of an unpredictable situation at the Fukushima Daiichi plant. In fact, a series of visual news accounts allowed government and power company officials to offer the strongest assurances to date that the crisis was almost under control.

A deluge of contaminated water, plutonium traces in the soil, and an increasingly hazardous environment for workers at the Daiichi power station raised questions as to how long[4], and at what cost, Japan should and could maintain what nuclear technocrats called a *feed-and-bleed strategy* of keeping nuclear fuel cool with emergency measures. On March 22, TEPCO announced that seawater near an outlet at the Daiichi nuclear power plant had a concentration of radioactive iodine-131 that was 127 times higher than government standards ("Kaisui kara": 1; see also "Nenryō": 2; "Osensui": 3). This was the first time that an analysis of radioactive materials in seawater had been made public since the struggle to contain a full meltdown at the nuclear plant had begun. By seriously considering biological concentration, concern mounted in terms of the possible impact on human health and on fishery products. In other words, fears were rapidly growing and Japan now faced two further crises constituted by; the ongoing struggle to stabilize the stricken nuclear plant and secondly; the threat of increased contamination that would affect the environment and the local / national food supply. The government first responded that it was too early to evaluate the impact of the contaminated seawater on fishery products, after which it began repeatedly declaring that there would be "no immediate health hazard" (*tadachi ni kenkō ni eikyō o oyobosu mono dewa nai* ただちに健康に影響を及ぼすものではない) from radiation leakage ("Kaisui osen": 2; see also SEKIYA 2011: 155–57). While the company decided to scrap the damaged reactors at the Fukushima Daiichi plant, the government had little choice but to continue to pursue the *feed-and-bleed strategy*.

As the crippled power plant continued to leak radioactive material, the ongoing crisis underscored the unprecedented scale and complexity of the problems facing Fukushima. The Metropolitan Police Department's special squad, as well as firefighters and the SDF were all engaged in operations to cool down two overheating reactors by hosing them with water. Nonetheless, grey smoke was seen rising near the third reactor's temporary fuel storage pool around 3:55 PM on March 21. While the *Nihon keizai shinbun*, *Yomiuri shinbun* and *Fukushima min'yū* used the photo provided by the TEPCO on the front pages of their March 22 editions ("Fukushima Daiichi genpatsu": 1; "Nigōki tsūden": 1; "Daiichi genpatsu jiko": 1). Neither the local nor the national dailies referenced the white smoke rising out of the second reactor, in which overheated fuel had partially melted down shortly after 6 PM on the same day. Cracks in the suppression chamber allowed steam containing more radioactivity at Reactor 2 than at Reactors 1 and 3 to escape. At Reactor 4, TEPCO began hosing with a German-made squeeze pump that could cover greater distances using a 58-meter-long arm. Only three national dailies, the *Asahi shinbun*, *Mainichi shinbun* and *Nihon keizai shinbun*, highlighted this international, technological support by showing one of the photos released by TEPCO ("Sangōki seigyoshitsu tentō": 1; "Fukushima Daiichi genpatsu: Yongōki": 1; "58 mētoru āmu": 1).

4.2 Analysis

When it comes to public accountability for the nuclear meltdowns and explosions at the Fukushima Daiichi nuclear power plant, it is not quite so easy to benignly interpret the actions of government and private officials as having been primarily motivated by confusion. The emphasis in this case should be duly shifted from self-deception to the nuclear industry's deliberate attempts to mislead the public. A generous interpretation of events would suggest a degree of culpability on the part of company managers, with negative consequences for public consumers. A more realistic interpretation, on the other hand, would emphasize industrial profits at the expense of public safety. Moreover, government nuclear regulation is ineffective both because administrative officials double as hard-working promoters for the nuclear industry, and because industry interests work toward fully protecting themselves and providing tailor-made information to the public. In national and local media coverage, an analysis of the reasons for such internal failings—beyond general incompetence, stupidity, or carelessness—is never provided.

On the national political scene, Japan's efforts to cool the nuclear reactors showed how unprepared the government was to deal with incidents at the nuclear power plant. The police and firefighters as well as the SDF hosed the stricken reactors with water. In addition to these operations, the GSDF helicopters used large containers to dump water on the reactors. A critical insight into such photo coverage uncovers how the government tried to reassure the country that every resource possible was being deployed to resolve the problems. Hence the helicopters and water cannons would not appear unprofessional, untrained or unprepared, but rather give an encouraging sign that no effort was being spared on the part of the relevant authorities. Indeed, the Japanese government mobilized an unprecedented 100,000 SDF personnel both to conduct rescue and relief activities over a vast area along the Pacific coast of the Tōhoku and Kantō regions, and to prevent the nuclear crisis at the Fukushima Daiichi plant from developing into a nationwide disaster. In this rescue and relief effort, SDF reserves were called up for the first time. While the disaster-response measures and relief missions helped mitigate the aggressive anti-military attitudes shared by many Japanese citizens, those administrative moves were seen by some as desperate signs that the emergency would soon spiral out of control ("Kiki Dakai": 3; "Oikomareta": 3).

What is more, TEPCO started a press conference with a stock Japanese phrase to apologize to the public for the "trouble and worry" (*taihen na go-shinpai to go-meiwaku o o-kake shite iru* 大変なご心配と迷惑をおかけしている) caused by breakdowns at the nuclear power plant following the natural disaster, and then gave confusing explanations as to why the company had failed to repair the damaged Fukushima Daiichi plant. While the unfolding crisis cast renewed doubt on the safety of nuclear energy, such crisis communiqués left the country, and indeed the world, confused and incredulous. At a press conference on March 13, when the crippled nuclear power station had been plunged into crisis, its president, Shimizu Masataka 清水正孝, placed the blame squarely on mother nature by describing the scale of the tsunami as "unprecedented or unexpected" (*sōtei o ōkiku koeru tsunami datta* 想定を大きく超える津波だった) ("Kyō kara": 1). While the company was careful to offer no public comparisons to past nuclear accidents other than the Tōkaimura incident, a critical analysis of the admission that the earthquake and tsunami had been "beyond our expectations" (*sōteigai* 想定外) reveals that the company had routinely underestimated the risk of earthquakes and tsunamis (Fukushima genpatsu jiko dokuritsu kenshō iinkai 2012: 312–20). Since then, preparations for potential disasters have been improved. Nonetheless, the same flaws that handicapped the response to the disaster—lack of coordination among ministries, poor public communication, intractable bureaucratic inflexibility—persist.

Unable to contain the nuclear crisis on its own, TEPCO successfully petitioned the Metropolitan Police and the Tokyo Fire Departments as well as the Ministry of Defense for help. Those public servants worked together with the power company's engineers at the stricken Daiichi nuclear complex, often in darkness as the tsunami washed away almost all the backup systems. In order to allow repair work to continue, the government raised the legal limit on radiation exposure twice—from 50 millisieverts to 100, and then to 250. While news coverage presented those police, firefighters and SDF personnel as heroes for their willingness to fight the threat of radiation exposure for the sake of the nation, almost none of them appeared in person on national or local news. In contrast to the executive officials, those workers remained faceless and unnamed, even obliged to remain out of sight. What is more, their efforts to prevent full meltdowns were described as volunteer work, thus obscuring the workers' obligation to pump seawater onto dangerously exposed nuclear fuel, already thought to be partly melting and emitting radioactive material.[5]

This conventional pattern of storytelling sought to strengthen Japanese notions of how people are expected to quietly perform their duties without seeking attention. Yet the lack of crisis communication also meant an absence of leadership in the midst of utter chaos—the disaster both undercut trust in the authorities and diminished confidence in expert opinion. In national coverage, selflessness, stoicism and discipline in Japan are epitomized by those workers at the earthquake-stricken nuclear power plant, anonymously exposing themselves to dangerous doses of radiation as they struggle to prevent a complete meltdown. Here the general storyline was akin to an adventure tale in the tradition of *disaster averted* dramas, in which people on site take concerted action to bring an unfortunate situation under control. In this dramatic narrative, the 2011 Great East Japan Earthquake (Higashi Nihon daishinsai 東日本大震災) would rekindle confidence in the ability and fortitude of the Japanese people themselves. Japanese traits of endurance, perseverance and grace appeared repeatedly in media discourse. On reflection, however, those same perceived values led the country to the Fukushima nuclear catastrophe in the first place. To wit; postwar Japan's blind pursuit of material wealth and comfort despite the country's singular history as the target of atomic bombs.

The idea of risk aversion is often presented in the two central narratives. In the overconfidence narrative, administrative and executive officials in charge of nuclear energy may think they have it under control, where in fact they do not. Indeed, neither the government nor the power company seriously considered facts and figures of historic tsunamis and earthquakes in order to reduce costs. As such, political and business elites began to panic when Japan faced the unfolding crisis. This elite panic resulted in a failure to act in terms of crisis management. Moreover, the nuclear crisis, considered *unprecedented* or *unexpected*, shows that Japan's safeguards are not suitable for such unforeseen contingencies since the latter require a lot more risk assessment measures. In the *hidden danger* narrative, radiation is invisible and delayed, so that one may not know the true harm done until many years later. While the damage from the earthquake and tsunami was instantly visible, the nuclear impact would take days or even years to be empirically confirmed, it could affect far larger areas of Japan and neighboring countries. In the news coverage, aerial photographs of ravaged coastal areas on March 11, 2011, revealed a string of cities and villages leveled by the power of the tsunami and plumes of black smoke rising from burning industrial plants. In contrast, the nuclear crisis developing at the Fukushima Daiichi plant, consisted of multiple failures, fires and radiation leaks from at least four separate reactors and could not be visibly communicated in the same way. In addition to the euphemistic language used by government and power company officials, the conflicting reports and a constant refusal to confirm the most basic facts all served to deepen public frustration and anxiety in Japan and around the world. Indeed, many began to wonder if crucial information had been withheld.

The rhetorical power of dissociating atoms for peace from atoms for war is no longer enough to resolve a dilemma in terms of how to develop nuclear power for peaceful purposes, while at the same time restraining its development as an instrument of mass destruction. Media discourse clearly recognizes the dualism of nuclear power as controversial, yet the progress narrative is strengthened not merely in the call for energy independence to ensure national security, but in the acceptance of promoting nuclear energy as necessary and inevitable given the fight against global warming. In the fatalism narrative, however, an uncontrollable fact of life along with an anxiety regarding unknowable disasters must simply be accepted. The fatalistic narrative thus represents its stance as resigned rather than opposed. This discursive potential of broad antinuclear movements helps encourage a way of life to conserve energy, to develop sources of energy that are ecologically safe and renewable, and to decentralize production. Such a didactic conflict narrative offers little commentary regarding the costs or benefits of nuclear power.

5 Conclusions

Interpreting the March 11 disaster shows how a cascade effect threatens to make a bad situation even worse. At first glance, it seems apolitical: pointing to environmental factors can naturalize the event of urgency. Nonetheless, conflicts of interest are indeed central to Fukushima then and now. In particular, the false sense of objectivity that requires technical explanations itself evades fundamental questions about postwar Japan's pursuit of progress—economy over safety. On the whole, the story of Fukushima's nuclear pollution is likely to transition into an embryonic global reflection on the costs of unrestrained economic growth. In post-disaster Japan, however, it has fundamentally produced a breakdown of rationality.

In the name of risk communication, the Japanese government fell back on simply reiterating three words: "security" (*anshin* 安心), "safety" (*anzen* 安全) and "recovery" (*fukkō* 復興), while offering reassuring messages in optimistic tones. In the tutelage of the "nuclear safety myth" (*genpatsu anzen shinwa* 原発安全神話), the long-standing administrative efforts to produce perfectly safe systems continue to create new and unforeseen problems. Despite such dangers of "safetyism" (Lukianoff / Haidt 2018: 30), the government urges local people to take responsibility for their own health, to restore the bonds of solidarity with their shattered community, and to participate in the national reconstruction plan to revitalize their resilient nation (Takahashi 2012: 15–81). Such post-March 11 safety campaigns are seemingly associated with the concept of "ancestral hometowns" (*furusato*)—the spelling of which is distinguished as 古里 in Chinese characters and at odds with the common spelling of 故郷—as well as nostalgia for *Japaneseness*. Examining the discursive formation of *furusato* uncovers how the declining rural communities could work as repositories of amorphous *Japaneseness* in order to enhance the more traditionally Japanese lives of people in rural communities. In the national and local coverage of the ravaged communities, no antinuclear protestors appear. It was not simply that too little public deliberation was made in the mediated discourse, but that nuclear technology remained a "black box". What is more, the media failed to cover what was going on inside the six nuclear reactors, focusing only on the effects outside the reactors. Given that "the echo effect" had also occurred in the case of the 1979 Three Mile Island incident, the Japanese administration, its bureaucracy, and the electronic company all talked up the same phenomena at the Fukushima nuclear crisis, though with slightly different formulations. As a result, different stories emerged in the media, causing not only a proliferation of technical commentary, but also considerable disagreement as regards technical solutions (FARRELL / GOODNIGHT 1981: 272–73, 286).

The general idealization of rural Japan asks the country to re-discover a sense of Japanese identity, underlining a growing gap between stereotypes of rural areas and the actual reality of the country. This idealization potentially presents these devastated small towns as inherent to Japan's recovery and progress post-March 11, an act that obscures both the issue of national government subsidies, as well as the interventions of TEPCO into the everyday lives of the Japanese. The deferral of reconstruction in post-disaster Tōhoku is arguably connected to the notion of "enduring suffering" (*gaman suru* 我慢する). Indeed, the narrative of post-disaster reconstruction takes advantage of the reenforcing language of "don't give up" (*ganbare* がんばれ), which is abundant in media discourse and in government materials. The saying *ganbare* is the imperative form of "to persevere or endure suffering" (*gaman suru* 我慢する) to encourage cooperative activities. Such nationalist overtones in the crisis of social reality can potentially fuel a turn away from local communities. In the context of eco-friendly talk the phrase *ganbare* can, on the one hand, be associated with the popular notion of "resilience" (*kaifuku suru chikara* 恢復する力) which is considered as a positive way of enduring and of surviving. On the other hand, resilience also promotes adaptability: life may go on regardless of the situation. By taking into account such a complicated set of meanings that imply an enduring of suffering and adversity, the story of Tōhoku's post-disaster reconstruction can and has been used to justify endurance to radiation exposure.

Since March 11, 2011, the national and local governments have been promoting a "practical radiation protection culture" (*hōshasen kara mi o mamoru taisaku* 放射線から身を守る対策) (MINISTRY OF EDUCATION AND TECHNOLOGY 2011), deploying forward-looking messages in repeated efforts to reassure the public that there is no immediate health risk. Along with the "national plan of revitalizing a resilient nation" (*kokka kyōjinka keikaku* 国家強靱化計画), the Japanese government began asking the whole country to counteract the negative impact of baseless rumors and gossip (*fūhyō higai* 風評被害) of radiation by buying and consuming Fukushima agricultural produce and seafood as well as by visiting Fukushima itself and the surrounding areas (MIURA 2014: 58–68). In order to encourage such moral and economic support, the national government coined the slogan "Don't give up, Fukushima!" (*Ganbare Fukushima* がんばれ福島) in parallel with the encouragement given from other countries immediately after the March 11 disaster "Don't give up, Japan! Don't give up, Tōhoku!" (*Ganbare Nippon! Ganbare Tōhoku!* がんばれニッポン！がんばれ東北！). By shifting slogans from "Don't give up, Fukushima" to "Hang in there, Fukushima!" (*Ganbarō Fukushima* がんばろうふくしま) in the early 2020s, the local government sought to strengthen reassurance and asked residents to live with radioactive contamination. By running a *public acceptance* campaign in a bid for information and communication control, governments seek to trivialize, and even normalize, health risks from radiation exposure by rephrasing public anxiety as merely the result of the psychological impact of stress.

Since its inception, atomic / nuclear energy has been considered a symbol of Japan's postwar technological and economic success. In the media coverage at home and abroad, the Fukushima nuclear crisis has been used to prove to the world that a nuclear disaster of significant scale can be overcome and that Fukushima residents can survive and take back their normal lives in contaminated land. While former Prime Minister ABE Shinzō 安倍晋三 provided an optimistic vision of Fukushima's post-disaster future in which "[r]econstruction of infrastructure is nearly complete, while the rebuilding of homes and the revitalization of industries and livelihoods is proceeding step by step" (OSAKI 2017), the Japanese government has concentrated on proving that it is safe for the Olympics, safe for tourism, safe to consume local food, and safe to restart nuclear reactors. Atomic energy may have destroyed the nation in 1945, but it indeed brought about postwar Japan's progress. In other words, most Japanese seem to have accepted the logic that the benefits of nuclear power outweighed the risks. Nevertheless, in contrast to ABE's vision of "shifting to a new stage" (MCCURRY 2015), the inherent claim of his narrative that the nuclear crisis is "under control" contains one conspicuous absence. Namely, that Fukushima's misfortune in being struck by a quake-tsunami-meltdown is ultimately secondary to the welfare and centrality of a single dominant city—Tokyo (MCCURRY 2013; FACKLER 2015: A6).

The Great East Japan Earthquake devastated a thinly populated region far from Tokyo, causing nuclear meltdowns and explosions at the Fukushima Daiichi nuclear power plant. In contrast to the nuclear explosions of August 6 and 9, 1945, and despite the psychological shock to the nation in the immediate aftermath of March 11, 2011, these events have not been commemorated as a historical turning point. They are instead compared with January 17, 1995—the Great Hanshin-Awaji Earthquake (Hanshin Awaji daishinsai 阪神・淡路大震災)—which came to be seen as symbolic moments signifying the end of the Japanese asset price bubble. Rather than spurring lasting change, it gave rise to an enduring loss of faith in institutions in general. Japan stopped believing in an economic miracle that had carried it from the wreckage and poverty of 1945 to the verge of overtaking the United States and becoming the world's second largest economy. The two great earthquakes were used as calls to restore, but not to reform Japan. Unfortunately, the March 11 disaster also failed to change Japan. In order to avert further potential danger, the risk-averse country needs to begin reflecting upon a culture in which safety was once a sacred value.

While the fatalistic narrative came to predominate in media discourse after March 11, 2011, postwar Japan's faith in pro-nuclear progress is still beleaguered and defensive. In terms of a broader energy issue, any overall characterization of media discourses as either for or against the use of nuclear power necessarily obscures the promotion of much cheaper renewable resources such as solar and wind power. In addition to fatalism, predominant narratives within post-nuclear Japan make it more difficult to develop and share counter-narratives to the official discourse. In the case of Fukushima, the national government continues to pursue efforts to improve community life in contaminated areas, to encourage proactive self-responsibility (for example, self-monitoring), and to attempt to purge the social stigma attached to the residents of Fukushima, the area and its produce. Furthermore, not only does the government attack what it labels radiophobia, it also encourages the evacuees to return to their homes after and in some cases even prior to decontamination (YAMASHITA/ICHIMURA/SATŌ 2016: 19–39; see also SATŌ 2019: 222–24; SEKIYA 2011: 95–199). The unfolding nuclear disaster required the country, still facing economic uncertainty, to take up this serious challenge. Ironically, the current national pledge to make Japan carbon neutral by 2050 frames the country as forced to choose between climate disaster and nuclear disaster. Furthermore, the closing of the Fukushima Daiichi and Daini nuclear power plants resulted in approving only nine reactors to restart with stricter safety inspections, thereby providing just six percent of the country's electricity. Such a political move, regardless of the lack of public consensus, uncovers the national priority of economic growth over public safety (BLAIR 2021: 32). While embracing Hiroshima and Nagasaki within Fukushima, the crisis of the current moment challenges Japan to resist making the trade-offs demanded by the social, political and economic benefits of nuclear power in its pursuit of progress.

Notes

1 As stipulated in the Agreement referred as "WHA12-40" signed by the WHO with the IAEA on May 28, 1959, the WHO is mandated to report all data on health effects from radiation exposures to the IAEA which controls publications. In addition, on no other medical health issue is the WHO required to defer publication responsibilities to another institution.
2 Immediately after the hydrogen explosion, the Fukushima prefectural government and the nuclear safety agency confirmed that 22 people were exposed to radiation because of radioactive substances temporarily released from Reactor 1. The prefecture also announced that about 80,000 people were subjected to its evacuation order.
3 The IAEA categorized the Tōkaimura accident as level four—accidents with local consequences—because the only people exposed to serious radiation were three workers within the plant.
4 On the night of March 29, 2011, the power company stated that small amounts of Plutonium-238, 239 and 240 were detected in soil sampled at five locations inside the Fukushima Daiichi site. This is quite serious since it indicates fuel rods have melted to some degree.
5 While their bosses appeared daily in blue work coats in order to apologize to the public, TEPCO employees working to restore and reestablish order remained almost entirely anonymous. In many cases, those line workers wanted the public to know not only that they felt remorse for the nuclear crisis, but also that they tried their best to fix it. It is clear that the power company actively censored its workers, particularly if we acknowledge that it was undoubtedly too early at this juncture to ask said workers to refrain from criticism.

References

Monographs and Articles:
BERGER, Peter L./LUCKMANN, Thomas (1966): *The social construction of reality: A treatise in the sociology of knowledge*. Garden City, NY: Doubleday.
BOYER, Paul (1985): *By the bomb's early light*. New York: Pantheon.
CHERNUS, Ira (2002): *Eisenhower's atoms for peace*. College Station: Texas A&M University Press.
CLARKE, Lee/CHESS, Caron (2008): "Elite and Panic: More to Fear than Fear Itself". In: *Social Force* 87 (2): 993–1014.

CHILTON, Paul (1987): "Metaphor, euphemism, and the militarization of language". In: *Current Research on Peace and Violence* 10 (1): 7–19.

EDELMAN, Murray J. (1985): *The symbolic use of politics*. Champaign, IL: University of Illinois Press.

FARRELL, Thomas B. / GOODNIGHT, G. Thomas (1981): "Accidental rhetoric: The root metaphor of Three Mile Island". In: *Communication Monographs* 48: 271–300.

FELDMAN, Ofer (2004): *Talking politics in Japan today*. Brighton: Sussex Academic Press.

FUKUSHIMA GENPATSU JIKO DOKURITSU KENSHŌ IINKAI 福島原発事故独立検証委員会 (2012): *Fukushima genpatsu jiko dokuritsu kenshō iinkai: Chōsa, kenshō hōkokusho* 福島原発事故独立検証委員会：調査・検証報告書. Tōkyō: Discover Twenty-one.

HINO, Kōsuke (2013): *Kenmin kenkō chōsa no yami* 県民健康調査の闇 (Iwanami shinsho 岩波新書). Tōkyō: Iwanami shoten.

IYENGAR, Shanto (1991): *Is anyone responsible? How television frames political issues*. Chicago: The University of Chicago Press.

IYENGAR, Shanto / MCGRADY, Jennifer A. (2007): *Media politics: A citizen's guide*. New York: W. W. Norton.

LIPPMANN, Walter (1922): *Public opinion*. New York: Harcourt, Brace.

LUKIANOFF, Greg / HAIDT, Jonathan (2018): *The coddling of the American mind: How good intentions and bad ideas are setting up a generation for failure*. New York: Penguin Books.

MACDONALD, Dwight (1945): "The bomb". In: *Politics* 2 (9): 257–88.

MINISTRY OF EDUCATION AND TECHNOLOGY (2011): *Hōshasen tō ni kan suru fukudokuhon* 放射線等に関する副読本, www.mext.go.jp/b_menu/shuppan/sonota/attach/1313004.htm (last access 2022 / 08 / 23).

MIURA, Kōkichirō 三浦耕吉郎 (2014): "Fūhyō higai no poritikusu: Nazuke no <gōman-sa> o megutte 風評被害のポリティクス：名づけの<傲慢さ>をめぐって". In: *Kankyō shakaigaku* 環境社会学 20: 54–76.

NAONO, Akiko 直野章子 (2011): *Hibaku to hoshō: Hiroshima, Nagasaki, soshite Fukushima* 被ばくと補償：広島、長崎、そして福島 (Heibonsha shinsho 平凡社新書). Tōkyō: Heibonsha.

NOELLE-NEUMANN, Elisabeth (1993): *The spiral of silence: Public opinion, our social skin*. 2nd ed. Chicago: The University of Chicago Press.

ŌSHIMA, Ken'ichi 大島堅一 (2011). *Genpatsu no kosuto: Enerugī tenkan e no shiten* 原発のコスト：エネルギー転換への視点 (Iwanami shinsho 岩波新書). Tōkyō: Iwanami shoten.

SAMUELS, Richard J. (2013): *3.11: Disaster and change in Japan*. Ithaca: Cornell University Press.

SATŌ, Takumi 佐藤卓己 (2019): *Ryūgen no media-shi* 流言のメディア史 (Iwanami shinsho 岩波新書). Tōkyō: Iwanami shoten.

SATŌ, Toshiki 佐藤俊樹 (2010): *Shakai wa jōhōka no yume o miru* 社会は情報化の夢を見る. Tōkyō: Kawade shobō.

SCHEINMAN, Lawrence (1987): *The International Atomic Energy Agency and world nuclear order*. New York: Routledge.

SEKIYA, Naoya 関谷直也 (2011): *Fūhyō higai: Sono mekanizumu o kangaeru* 風評被害：そのメカニズムを考える (Kōbunsha shinsho 光文社新書). Tōkyō: Kōbunsha.

TAKAHASHI, Tetsuya 高橋哲哉 (2012): *Gisei no shisutemu: Fukushima, Okinawa* 犠牲のシステム：福島・沖縄 (Shūeisha shinsho 集英社新書). Tōkyō: Shūeisha.

TAKEDA, Tōru 武田徹 (2011a): *Genpatsu hōdō to media* 原発報道とメディア (Kōdansha gendai shinsho 講談社現代新書). Tōkyō: Kōdansha.

TAKEDA, Tōru 武田徹 (2011b): *Watashi-tachi wa kō shite "genpatsu taikoku" o eranda* 私たちはこうして「原発大国」を選んだ (Chūkō shinsho rakure 中公新書ラクレ). Tōkyō: Chūō kōron shinsha.

TUCHMAN, Gaye (1974): *The TV establishment*. Englewood Cliffs, NJ: Prentice-Hall.

TUCHMAN, Gaye (1978): *Making news: A study in the construction of reality*. New York: The Free Press.

YAMAMOTO, Akihiro 山本昭宏 (2012): *Kaku enerugī gensetsu no sengoshi 1945–1960: "Hibaku no kioku" to "genshiryoku no yume"* 核エネルギー言説の戦後史1945–1960：「被爆の記憶」と「原子力の夢」. Kyōto: Jinbun shoin.

YAMASHITA, Yūsuke 山下祐介 / ICHIMURA, Takashi 市村高志 / SATŌ, Akihiko 佐藤彰彦 (2016): *Ningen naki fukkō: Genpatsu hinan to kokumin no "furikai" o megutte* 人間なき復興：原発避難と国民の「不理解」をめぐって (Chikuma bunko ちくま文庫) Tōkyō: Chikuma shobō.

YAMAZAKI, Masakatsu 山崎正勝 (2011): *Nihon no kakukaihatsu: 1939–1955—genbaku kara genshiryoku e* 日本の核開発：1939〜1955—原爆から原子力へ. Tōkyō: Sekibundō shuppan.

YOSHIMI, Shun'ya 吉見俊哉 (2012): *Yume no genshiryoku* 夢の原子力: *Atoms for Dream* (Chikuma shinsho ちくま新書). Tōkyō: Chikuma shobō.

YOSHIOKA, Hitoshi 吉岡斉 (2011): *Genshiryoku no shakaishi: Sono Nihon-teki tenkai* 原子力の社会史：その日本的展開 (Asahi sensho 朝日選書 883). New ed. Tōkyō: Asahi shinbun shuppan.

Newspaper articles:

BLAIR, Gavin: "Nuclear reactor's operator hid fault line danger". In: *The Times* (London), 2021 / 08 / 20: 32.

"Bōgofuku no taiin: Kinpaku hōsui 防護服の隊員：緊迫放水". In: *Yomiuri shinbun* 讀賣新聞 (Evening ed.), 2011 / 03 / 23: 11.

"Chijō kara mo hōsui 地上からも放水". In: *Nihon keizai shinbun* 日本経済新聞, 2011 / 03 / 18: 1.

"Daiichi genpatsu 第1原発". In: *Fukushima min'yū* 福島民友, 2011 / 3 / 17: 1.

"Daiichi genpatsu jiko: Sangōki hatsuen de sagyō chūdan 第1原発事故：3号機発煙で作業中断". In: *Fukushima min'yū* 福島民友, 2011 / 03 / 22: 1.

FACKLER, Martin: "Japanese balk at returning to disaster area". In: *The New York Times*, 2015 / 08 / 09: 6 (Section A).

"Fukushima Daiichi genpatsu de bakuhatsu 福島第一原発で爆発". In: *Fukushima min'yū* 福島民友, 2011 / 03 / 13: 1.

"Fukushima Daiichi genshiryoku hatsudensho no genjō to taisaku 福島第一原子力発電所の現状と対策". In: *Yomiuri shinbun* 讀賣新聞, 2011 / 03 / 18: 1.

"Fukushima Daiichi genpatsu, tsūden sagyō, hōsui o chūdan 福島第1原発、通電作業・放水を中断". In: *Nihon keizai shinbun* 日本経済新聞, 2011 / 03 / 22: 1.

"Fukushima Daiichi genpatsu: Yongōki chūsui saikai 福島第1原発：4号機注水再開". In: *Mainichi shinbun* 毎日新聞, 2011 / 03 / 23: 1.

"'Fukushima Daiichi' nigōki「福島第1」2号機". In: *Mainichi shinbun* 毎日新聞, 2011 / 03 / 15: 1.

"Fukushima genpatsu de bakuhatsu: Daiichi, ichigōki 福島原発で爆発：第一・1号機". In: *Yomiuri shinbun* 讀賣新聞, 2011 / 03 / 13: 1.

"Fukushima genpatsu de bakuhatsu: Shūhen de 90-nin hibaku ka 福島原発で爆発：周辺で90人被曝か". In: *Asahi shinbun* 朝日新聞, 2011 / 03 / 13: 1.

"Genpatsu antei 'saitei ikkagetsu' 原発安定「最低1カ月」". In: *Asahi shinbun* 朝日新聞, 2011 / 03 / 25: 2.

"Genpatsu baishō: Kuni futan mo 原発賠償：国負担も". In: *Mainichi shinbun* 毎日新聞, 2011 / 03 / 24 : 6.

"Genpatsu nikuhaku 30-ton hōsui 原発肉薄30トン放水". In: *Asahi shinbun* 朝日新聞, 2011 / 03 / 18: 1.

"Genpatsu reikyaku: Heri kara hōsui 原発冷却：ヘリから放水". In: *Nihon keizai shinbun* 日本経済新聞 (Evening ed.), 2011 / 03 / 17: 1.

"Genpatsu sangōki mo bakuhatsu 原発3号機も爆発". In: *Nihon keizai shinbun* 日本経済新聞 (Evening ed.), 2011 / 03 / 14: 1.

"Genpatsu sangōki mo bakuhatsu 原発3号機も爆発". In: *Fukushima minpō* 福島民報, 2011 / 03 / 15: 1.

"Genpatsu setogiwa no chūsui 原発瀬戸際の注水". In: *Nihon keizai shinbun* 日本経済新聞, 2011 / 03 / 17: 3.

"58 mētoru āmu de hōsui 58メートルアームで放水". In: *Yomiuri shinbun* 讀賣新聞, 2011 / 03 / 23: 1.

"Govt rejected U. S. offer to help cool damaged reactors". In: *The Yomiuri news*, 2011 / 03 / 19: 1.

"Hibiku keikokuon 'jikan nai zo' 響く警告音「時間ないぞ」". In: *Yomiuri shinbun* 讀賣新聞, 2011 / 03 / 24: 29.

"Hōsui inochigake 放水命がけ". In: *Asahi shinbun* 朝日新聞, 2011 / 03 / 18: 2.

"Hōsui keizoku: Tanomi no tsuna 放水継続：頼みの綱". In: *Nihon keizai shinbun* 日本経済新聞, 2011 / 03 / 18: 2.

"Ichigōki: Nao takai atsuryoku 1号機：なお高い圧力". In: *Nihon keizai shinbun* 日本経済新聞, 2011 / 03 / 25: 3.

"Kagi nigiru genpatsu 'shinzōbu' 鍵握る原発「心臓部」". In: *Fukushima min'yū* 福島民友, 2011 / 03 / 23: 3.

"Kaisui kara hōshasei busshitsu 海水から放射性物質". In: *Yomiuri shinbun* 讀賣新聞 (Evening ed.), 2011 / 03 / 22: 1.

"Kaisui osen 'reisei taiō o' 海水汚染「冷静対応を」". In: *Yomiuri shinbun* 讀賣新聞 (Evening ed.), 2011 / 03 / 22: 2.

"Keihōon 'modoru!' hōsui kinpaku 警報音「戻る!」放水緊迫". In: *Asahi shinbun* 朝日新聞, 2011 / 03 / 24: 31.

"Keishichō hōsuisha tōnyū e 警視庁放水車投入へ". In: *Yomiuri shinbun* 讀賣新聞, 2011 / 03 / 17: 1.

"Keikaku teiden: Gogo no ichibu jisshi 計画停電：午後の一部実施". In: *Mainichi shinbun* 毎日新聞, 2011 / 03 / 23: 3.

"Kiki dakai: Saigo no shudan 危機打開：最後の手段". In: *Fukushima minpō* 福島民報, 2011 / 03 / 18: 3.

"Kiki dasshutsu: Mittsu no kadai 危機脱出：三つの課題". In: *Fukushima min'yū* 福島民友, 2011 / 03 / 24: 3.

"Kiki kanri 'gote' 危機管理「後手」". In: *Fukushima minpō* 福島民報, 2011 / 03 / 17: 2.

"Kinkyū jitai: Chie atsumeru toki 緊急事態：知恵集める時". In: *Asahi shinbun* 朝日新聞, 2011 / 03 / 17: 1.

"Kokunai-hatsu no roshin yōyū 国内初の炉心溶融". In: *Fukushima min'yū* 福島民友, 2011 / 03 / 13: 1.

"Kōnōdo hōshanō hōshutsu 高濃度放射能放出". In: *Asahi shinbun* 朝日新聞, 2011 / 03 / 15: 1.

"Kyō kara rinban teiden きょうから輪番停電". In: *Asahi shinbun* 朝日新聞, 2011 / 03 / 14: 1.

Mouawad, Jad / Jolly, David (2011): "Natural gas now viewed as safer bet". In: *The New York Times*, 03 / 22: 1 (Section B).

McCurry, Justin (2013): "Tokyo 2020 Olympics: Hugs, tears and shouts of 'banzai' greet news of victory". In: *The Guardian*, 09 / 08, https://www.theguardian.com/sport/2013/sep/08/tokyo-2020-olympics-jubilation-relief (last access 2022/09/17).

McCurry, Justin (2015): "Japan remembers the 18,000 victims of 2011's triple disaster". In: *The Guardian*, 03/11, https://www.theguardian.com/world/2015/mar/11/japan-remembers-the-18000-victims-of-2011s-triple-disaster-fukushima (last access 2022/09/17).

"Nenryō reikyaku-yō ponpu no dengen fukkyū 燃料冷却用ポンプの電源復旧". In: *Asahi shinbun* 朝日新聞, 2011/03/27: 2.

"Nenryōbō subete roshutsu 燃料棒全て露出". In: *Yomiuri shinbun* 讀賣新聞, 2011/03/15: 1.

"Nigōki seigyoshitsu mo tentō 2号機制御室も点灯". In: *Yomiuri shinbun* 讀賣新聞, 2011/03/27: 1.

"Nigōki tsūden sagyō chūdan 2号機通電作業中断". In: *Yomiuri shinbun* 讀賣新聞, 2011/03/22: 1.

"Oikomareta seifu no kake 追い込まれた政府の賭け". In: *Fukushima minpō* 福島民報, 2011/03/18: 3.

Osaki, Tomohiro: "Eight years on, Abe says 3/11 recovery nearing 'final stages'". In: *The Japan Times*, 2019/03/19, https://www.japantimes.co.jp/news/2019/03/11/national/eight-years-abe-says-3-11-reconstruction-nearing-final-stages-though-half-public-unconvinced/#.XXg4eY2P4uQ (last access 2022/08/23).

"Osensui: Jōkyō kōten habamu 汚染水：状況好転阻む". In: *Nihon keizai shinbun* 日本経済新聞, 2011/03/27: 3.

"Ponpu asu nimo fukkyū: Sangōki ポンプ明日にも復旧：3号機". In: *Asahi shinbun* 朝日新聞 (Evening ed.), 2011/03/23: 1.

"Sagyō kankyō, hitoguri ga kadai 作業環境・人繰りが課題". In: *Asahi shinbun* 朝日新聞, 2011/03/31: 3.

"Sangōki chūsui sakusen 3号機注水作戦". In: *Asahi shinbun* 朝日新聞 (Evening ed.), 2011/03/17: 1.

"Sangōki heri de mizu tōka 3号機ヘリで水投下". In: *Yomiuri shinbun* 讀賣新聞 (Evening ed.), 2011/03/17: 1.

"Sangōki mo suiso bakuhatsu 3号機も水素爆発". In: *Asahi shinbun* 朝日新聞 (Evening ed.), 2011/03/14: 1.

"Sangōki mo suiso bakuhatsu 3号機も水素爆発". In: *Fukushima min'yū* 福島民友, 2011/03/15: 1.

"Sangōki seigyoshitsu tentō 3号機制御室点灯". In: *Asahi shinbun* 朝日新聞, 2011/03/23: 1.

"Sangōki reikyaku ponpu asu shidō mo 3号機冷却ポンプあす始動も". In: *Nihon keizai shinbun* 日本経済新聞 (Evening ed.), 2011/03/23: 1.

"Sangōki rikuji heri hōsui 3号機陸自ヘリ放水". In: *Mainichi shinbun* 毎日新聞, 2011/03/17: 1.

SOBLE, Jonathan / COOKSON, Clive / KIRCHGAESSNER, Stephanie: "Workers battle to regain control of overheating Fukushima plant". In: *The Financial Times (Asia Edition)*, 2011/03/18: 2.

"Sora to riku: Hōsui sakusen 空と陸：放水作戦". In: *Fukushima min'yū* 福島民友, 2011/03/18: 1.

"Tesaguri kenmei no hōsui 手探り懸命の放水". In: *Mainichi shinbun* 毎日新聞, 2011/03/18: 3.

"Tōden, dengen fukkyū shashin o kōhyō 東電、電源復旧写真を公表". In: *Fukushima minpō* 福島民報, 2011/03/31: 2.

Responsibility for the Fukushima Nuclear Disaster in the Japanese Media: TEPCO and the "Nuclear Village" or Prime Minister Kan and the DPJ?

Tobias Weiß

1 INTRODUCTION: THREE CORE POLITICAL QUESTIONS ABOUT THE ATTRIBUTION OF RESPONSIBILITY AND THE SEQUENCE OF EVENTS

Who was responsible for the ways in which the 2011 nuclear catastrophe in Fukushima unfolded? The answer to this question varies enormously depending on which political actor or media outlet one endorses. There is, however, one very pervasive narrative that prevails among a plethora of key stakeholders, including: the Tokyo Electric Power Company (TEPCO), which was legally responsible for running the six reactors in the Fukushima 1 (F1) plant; its allies in the national bureaucracy, most importantly the Ministry of Economy, Trade and Industry (METI); the other energy companies; and rest of the nuclear industry. These actors have been bolstered in their assertions by multiple control authorities affiliated with METI and other ministries. Most important among these are: the Nuclear Industry Safety Agency (NISA), which is directly responsible for overseeing the safety of nuclear reactors and the handling of accidents; the Japanese Nuclear Safety Commission (JNSC), which principally functions as an advisory board; and the Atomic Energy Commission (AEC), a forum involving representatives of various actors from private and public bodies and which is responsible both for making decisions on nuclear policy, and overseeing regulations. Together with the conservative LDP (Liberal Democratic Party, Jimintō 自民党)—which had been in government almost without interruption from 1955 until two years before the accident in 2009—these actors are frequently understood as constituting the "nuclear village" (*genshiryokumura* 原子力ムラ), an industrial complex that effectively controls government energy policy (YOSHIOKA 2011) while remaining impervious to outside influence. TEPCO and the "nuclear village" were, thus, the most obvious bearers of responsibility for the Fukushima accident. They had long maintained the impossibility of any large-scale accident in Japan's nuclear industry given the highly organized and technically sophisticated nature of the latter. They had, moreover, an extensive history of both evading and subverting effective control measures, as well as hiding safety problems (YOSHIOKA 2011).

As the accident ran its course, another, partially substitutive narrative began to emerge that took the place of the first. It put responsibility firmly in the hands of KAN Naoto 菅直人—who had been Prime Minister at the time of the accident—and his affiliated party, the DPJ (Democratic Party of Japan, Minshutō 民主党). The DPJ had only recently emerged as a challenger of the LDP and was in government from 2009 to 2012. In the media discourse regarding the management of the accident, the term "the Prime Minister's Office" (Kantei 官邸) is often used to refer to DPJ politicians close to KAN as well as their aides from outside of the traditional bureaucratic hierarchy. The alternative account of responsibility asserts that KAN and his aides mismanaged the accident by too strongly interfering with the micro-management of the accident by TEPCO. I will specifically focus on three questions, each of which was hotly debated in terms of the attribution of political responsibility in the discussions that followed:

a) If there were indeed delays in venting gas from reactor 1—a measure that would have served to prevent leakage of radiation[1] on March 11 and 12—who was responsible?
b) After it became clear on March 12 that seawater needed to be injected into reactor 1, were there, in fact, delays in initiating the procedure? If so, who was responsible?
c) Once the nuclear crisis reached its peak from the evening of March 14 into the early hours of March 15, did the TEPCO leadership deliberately withdraw personnel from the plant?

With a view to providing some background information regarding the relevance of these questions and the accompanying discussion, the following Table 1 gives a rough overview of the sequence of events and the accident centered on the three questions sketched above.

Time and Date	On Site (Fukushima Plant)	Tokyo (Kantei and TEPCO)
March 11, 3:27 PM	Tsunami hits, subsequent loss of cooling (reactor 1) and electricity supply	
March 11, from around 7:30 PM	Start of meltdown (reactor 1), rise of pressure in containment	
March 11, from around 12 PM		TEPCO asks for permission to vent reactor 1
March 12, 3 AM		Venting is announced at a joint press conference of TEPCO and METI
March 12, 6:50 AM		METI (KAIEDA) issues a directive ordering TEPCO to begin venting

Responsibility for the Fukushima Nuclear Disaster in the Japanese Media

Time and Date	On Site (Fukushima Plant)	Tokyo (Kantei and TEPCO)
March 12, 7:10 AM	Prime Minister KAN arrives at F1 to inspect the plant (and enquire as to why venting has not yet begun)	
March 12, around 11 AM	After problems with the venting mechanism and high radiation levels, the on-site-team claims that venting may have been successful	
March 12, 3:36 PM	Hydrogen explosion at reactor 1	
March 12, 4:00 PM	Start of freshwater injection into reactor 1	
March 12, 5:55 PM		METI (KAIEDA) issues a directive ordering TEPCO to begin seawater injection
March 12, 7:04 PM	Start of seawater injection	
March 12, 7:25 PM		The TEPCO Manager at Kantei (TAKEGURO Ichirō) orders plant manager, YOSHIDA Masao, to cease seawater injection (the directive is not carried out, however)
March 14, 11:01 AM	Hydrogen explosion at reactor 3 damages equipment for seawater injection and electricity lines of reactor 2	
March 14, afternoon	Drop of water level in reactor 2, rise of containment pressure	
March 14, 4:00 PM		TEPCO CEO, SHIMIZU Masataka, orders that a worst-case scenario be outlined for reactor 2
Afternoon March 14 to morning hours of March 15	Venting and seawater injection fails to achieve a drop in containment pressure, perilous situation in reactor 2	TEPCO CEO, SHIMIZU Masataka, contacts politicians at Kantei regarding evacuation
March 15, 4:17 AM		SHIMIZU arrives at Kantei, agrees to create joint headquarters with DPJ politicians at TEPCO headquarters
March 15, 5:35 AM		KAN arrives at TEPCO HQ, addresses TEPCO managers and employees, joint headquarters is created
March 15, 6:14 AM	Sound of explosion at reactor 2 (hydrogen explosion and subsequent fires)	
March 15, 8:29 AM	Evacuation of the bulk of workers from F1, about 70 core personnel remain	

Table 1: Overview of the sequences of events

Over the pages that follow, I intend to describe: how different political actors and different media attributed responsibility to the two groups delineated in the introduction; where the information underpinning these two narratives of responsibility came from as well as where it was published; and how the overall attribution of responsibility changed over time. I will refer to newspaper coverage and various documentation of the crisis published by different political actors and journalists. One's choices in terms of attribution of responsibility depend on one's choice of time frame.[2] For my part, I will exclusively focus on discussions regarding the immediate handling of the accident from March 11 until March 15, which was arguably the most dramatic period of the crisis. I make this choice a) due to space restrictions and b) because the specific handling of the accident during this particular time frame has become a key political issue in the attribution of responsibility overall.

2 CHANGES IN THE ATTRIBUTION OF RESPONSIBILITY IN DIFFERENT MEDIA AND THE SHIFT OF POLITICAL COMMUNICATION REGARDING RESPONSIBILITY

The answers to the three questions above determine—at least to some extent—the attribution of accountability for the nuclear accident. In this section, I will outline how different claims regarding accountability played out in different newspapers. I will attempt to trace the changing flow of information from the political and bureaucratic milieu (Kantei, NISA, JNSC, TEPCO, the LDP) to different media and sketch its development. In general, it can be said that the narrative of DPJ politicians and outsiders of the "nuclear village", such as the administrative personnel of Kantei—as well as external advisors summoned by KAN—was spread via the progressive newspapers *Asahi shinbun* 朝日新聞 and *Mainichi shinbun* 毎日新聞. The TEPCO narrative, on the other hand, was mainly produced by insiders of the "nuclear village" (bureaucrats affiliated with METI, other agencies responsible for managing the nuclear industry, nuclear power experts integrated into the "nuclear village", and TEPCO managers) and was significantly reinforced by the conservative / right wing media outlets *Yomiuri shinbun* 讀賣新聞 and *Sankei shinbun* 産經新聞.[3] In the journalistic field as a whole, initial coverage underlined responsibility of TEPCO and the "nuclear village". Once Prime Minister KAN decided to shut down the Hamaoka nuclear power plant in early May 2011, however, criticism of his management at the time of the accident intensified. This attests to the strongly politicized character of this narrative.

In terms of venting and seawater injection, neither TEPCO nor Kantei can be said to be entirely at fault. As we shall see, however, there is no obvious justification for the emphatic shift in the attribution of responsibility toward KAN and the DPJ. The publication of three major investigations concerning the management of the crisis further contributed to the spread of the TEPCO narrative. Ultimately, the story settled as one of Kantei unduly interfering with crisis management by the "heroic workers" on site (*genba* 現場).

2.1 Venting

Initial information regarding the development of events on site was very scarce. When government agencies and TEPCO announced that venting was to commence, this was picked up in television and newspaper reports (for TV see ITŌ 2012; for newspapers: WEISS 2014). Yet there was, and is, scant information available on the progress of the venting and seawater injection processes during the most critical moments of the crisis. Conflicting reports regarding delays in the venting process first emerged during the peak of the crisis in March. The *Mainichi shinbun* published a report citing high government officials (*seifu kōkan* 政府高官) who claimed that TEPCO had initially delayed venting for fears of compromising its reputation (MS 2011/03/16). Earlier still, the conservative *Yomiuri shinbun* published a report citing anonymous DPJ politicians stating that "Kantei lacked any real conscience for the crisis, [...] fearing damage to the image of nuclear power, which it had made a pillar of its growth strategy" (YS 2011/03/13).[4] The *Yomiuri shinbun* also cited JNSC chairman MADARAME Haruki, who had accompanied KAN on his March 12 visit to the plant and who, in turn, cited KAN as expressing a desire to "study nuclear power" (*genshiryoku o benkyō shitai* 原子力を勉強したい). This fed into the narrative that KAN— an engineering school graduate—had overestimated his own competence and impeded TEPCO and responsible experts in their management of the crisis (this narrative was reproduced both in the *Yomiuri shinbun* and the *Sankei shinbun*; e. g. YS 2012/05/29, ABIRU 2011: 259, 264).

From very early on, contrasting narratives were fed to the media from diverse corners of the political spectrum. At the end of March, LDP politicians made accusations against KAN during a diet session, with conservative as well as progressive media reporting on the ensuing debate (AS 2011/03/30; YS 2011/03/29). The *Yomiuri shinbun* cited a source from the "nuclear village" who claimed that "there was criticism that Prime Minister Kan's visit to the Fukushima site had delayed venting". The paper further quoted MADARAME as saying that delays to venting had resulted in delays of the subsequent injection of seawater. The article was so formulated as to give the impression that MADARAME himself had blamed the delay on KAN's visit, which was not the case.

LDP politicians cited by the *Yomiuri shinbun* claimed that TEPCO had delayed venting to ensure the safety of the Prime Minister's helicopter (YS 2011/04/01). In another article from the same issue, the *Yomiuri shinbun* described mutual distrust between KAN and the nuclear bureaucracy, taking up the latter's discontent concerning KAN's increasing reliance on external experts. KAN addressed these concerns in the diet claiming that he had visited the plant in a bid to convince TEPCO to conduct the venting quickly. He further argued that the visit had helped establish a line of communication with the people on site (AS 2011/03/30). As of early April, more details emerged, with the *Asahi shinbun* publishing additional information from anonymous sources regarding the venting process. It claimed that normal venting methods proved ineffective due to the loss of electricity and damage to electric control devices. Moreover, attempts to open the valves by hand proved fruitless due to high radiation. The article further asserted that KAN's visit was intended to counteract what was felt to be TEPCO's sluggish, hesitant, and insufficient disclosure of information to the Kantei (AS 2011/04/10), particularly with regard to the venting process. In late April, KAN again responded to critics in the diet by arguing that TEPCO's explanations had been unsatisfactory and that his visit had aided in the subsequent management of the crisis (AS 2011/04/19).

A few days later, articles in the *Asahi shinbun* (AS 2011/04/26) and (a significantly smaller entry) in the *Yomiuri shinbun* (YS 2011/04/19) presented a statement made by TEPCO president, SHIMIZU Masataka. SHIMIZU refuted the LDP's accusations and claimed that no delays had been caused by the Prime Minister's visit. In late May, the LDP briefly shifted its focus and alluded to "Kan wearing neither a helmet nor radiation protection during his visit", an angle which was soon followed by the *Yomiuri shinbun* (YS 2011/05/17).

By June 2011, the *Yomiuri shinbun* had taken up its previous line of criticism of KAN. It cited TEPCO's claims that although the reasons why venting was not immediately initiated were unclear, progress had been partially hampered by unsuccessful attempts to evacuate the local area. Indeed, the required radius of three kilometers was not cleared until midnight on March 11. The article went on to claim that: TEPCO had said venting did not proceed smoothly because, among other reasons, the evacuation of the citizens in the surrounding areas had not succeeded, pointing out that the evacuation of a three kilometers radius had been completed by midnight of March 11 and that the reason for not venting immediately was unclear. This was supplemented by a paragraph claiming that "a photograph of a whiteboard in the control room published by TEPCO showed a blank space from the time KAN's helicopter landed at 6:29 AM until 9:04 AM". An anonymous TEPCO worker was also quoted as claiming that the plant manager—YOSHIDA Masao—had to accompany the Prime Minister at all times and that it was not clear whether the hydrogen explosion that occurred later the same day could actually have been prevented had he been at his post. Regardless, the worker stated, the whole endeavour was clearly a waste of time (YS 2011/06/08b). That the *Yomiuri shinbun* intended to uphold the narrative of the "nuclear village" by bolstering an image of KAN as having actively hampered the management of crisis should, at this juncture, be quite clear. When a government investigation into the accident was later announced (Seifu jikochō 政府事故調), the *Yomiuri shinbun* claimed that KAN's role in hampering crisis management with his visit would be a central concern (YS 2011/06/08b). The CEO of TEPCO later made it clear that this claim was unfounded. Undeterred, *Yomiuri shinbun* issued further accusations in subsequent articles (e. g. YS 2011/06/19).

KAN had stepped down as Prime Minister in September 2011, after which additional details regarding the unfolding of the accident sequence of events were published, for example, including detailed descriptions regarding high radiation levels, which impeded venting (AS 2011/09/11). At this juncture, *Yomiuri shinbun* changed its editorial line slightly, printing accusations made by the former Prime Minister according to which TEPCO had failed to respond to Kantei's requests for an explanation as to why venting had been delayed. KAN further claimed that his visit had been necessary to exact information directly from TEPCO (YS 2011/05/25a). The paper went on to criticize TEPCO some days later for inadequate crisis management. Citing information that had emerged during the initial hearings of the official investigation into the accident, the paper confirmed that TEPCO had committed various errors, to wit: sending necessary materials to the wrong location; maintaining no inventory for material such as batteries and cables stored on site; having failed—for three hours—to register that an emergency cooling system in reactor 1 had been stopped by a worker. Curiously, however, the paper continued to give credence to TEPCO claims that KAN's visit to the reactor had been a waste of time (YS 2011/05/25a). In addition to the official investigation first commissioned by the Japanese government, a further two investigations were carried out. Firstly, the Diet Investigation Commission (Kokkai jikochō 国会事故調) and secondly, one that was carried out by a Private Investigation Commission (Minkan jikochō 民間事故調). Each of the three reached their respective conclusions in 2012 and each confirmed that there was no connection between KAN's visit and delays in the venting process. Yet in spite of this evidence, occasional accusations were still made thereafter, such as those that appeared in the *Asahi shinbun* in May 2012 (AS 2011/05/27).

In the course of each of the investigations, members of the TEPCO management team—particularly SHIMIZU Masataka, and the Board Chairman, KATSUMATA Tsunehisa— shifted from first negating the impact of KAN's visit on the crisis management tout court to then condemning his visit as a significant nuisance during an extreme crisis (AS 2012 / 05 / 27).

The investigation commission rejected accusations against KAN, emphasizing instead a narrative of mutually insufficient communication between the two parties. This is the narrative forwarded in two of the three commissions, the Government Investigation Commission and the Diet Investigation Commission (AS 2012 / 05 / 27; AS 2012 / 07 / 23). With the release of information from the Government Investigation Commission, interest in the issue of venting increasingly waned. Nonetheless, the *Yomiuri shinbun* continued to push TEPCO's narrative and insinuated, for example, that the visit must necessarily have been negative given that it had been criticized by the opposition (YS 2011 / 12 / 27).

The TEPCO narrative equally continued to make appearances on the pages of the *Asahi shinbun*, where it was accompanied, however, by one which was more critical of the organization. In one such example, an employee was depicted as trying, in vain, to convince YOSHIDA Masao to start venting. This contrasts, of course, with the notion that Kantei had been interfering in the management of the plant (AS 2012 / 09 / 05). This perspective equally appears in a series of articles written by the Special Reporting Unit (Tokubetsu hōdōbu 特別報道部) at the *Asahi shinbun*, which are based on internal video communication recordings from the time of the accident. The investigative reporters had called on TEPCO to release these video recordings and TEPCO eventually had to respond to this request because of the public pressure on the company, which rose after the nuclear accident (for details see below). The *Asahi shinbun* soon had to reverse its stance, however. In September 2014, they were forced to apologize for erroneous claims in their article on Government Investigation Commission's interview with the Fukushima plant manager, YOSHIDA Masao (for details, see the section on evacuation of the plant below). In point of fact, the evacuation of the plant ought really to have been the central focal point. Instead, as a result of the media buzz surrounding the *Asahi*'s reversal and apology, print media gave extensive coverage to YOSHIDA's views on crisis management and on the accident, more generally, as well as on assertions from TEPCO that a) venting had mainly been delayed by conditions brought on by the tsunami and b) interventions on the part of Kantei had simply been a nuisance (e. g. YS 2014 / 09 / 12). Though YOSHIDA did not lend any specific support to the notion that venting had been delayed by KAN's presence, TEPCO's belief that the former Prime Minister had gotten in the way of "TEPCO workers on site" (in point of fact the conflict was between the TEPCO management—most of them based in TEPCO's Tokyo headquarters—and Kantei politicians) gained significant traction in the print media more generally.

2.2 Seawater injection

Another key issue in terms of crisis management was the question of seawater injection. Generally speaking, venting is needed to maintain low pressure in the reactor containment and pressure vessel. Water injection, on the other hand, comes into play when circular cooling ceases to function. In the case of reactor 1, water was injected from the early hours of Saturday, March 12. After the first explosion, however, it became clear that there was an inadequate supply of fresh water on site and that seawater would also have to be used. Of course, seawater contains salt, and this ultimately causes damage to various parts of the reactor. Seawater is also said to possibly lead to renewed criticality (the reactor core restarting to produce energy) because it is not isolating the fuel rods as effectively as fresh water, making nuclear reactions potentially possible. Boron is typically used to prevent the nuclear chain reaction from reaching critical levels in such a case. The injection of seawater into reactor 1 was ordered by YOSHIDA at 3:10 PM once supplies of fresh water had been exhausted. Before this could successfully be carried out, however, an explosion occurred and the use of seawater injection tout court was hurriedly discussed by politicians, experts, TEPCO managers, and bureaucrats at Kantei.

The operation ultimately went ahead at 7:04 PM, three and a half hours after the first explosion, and without the knowledge of those at Kantei who had been informed by TEPCO staff at Kantei during the preceding discussions that a further two hours of preparation would be needed before injection could begin (KIMURA 2012: 140). During his visit to the plant, former Prime Minister KAN asked a number of experts present—JNSC chairman MADARAME in particular—how strong the chances of material damage and a corollary critical status in the reactor were, to which MADARAME answered: "not zero". MADARAME's careful choice of words was surely informed by his having erroneously replied to a question of KAN on the previous evening that a hydrogen explosion was out of the question (KIMURA 2012: 145). KAN had, as such, become understandably skeptical about advice issued by the "nuclear village". Having first availed of the preparation time for additional consideration of the possible risks, he ordered the use of seawater injection. He would later approve injection with the politicians concerned being led to believe that seawater injection had started at 8:20 PM (YS 2011/04/01).

The narrative that first emerged in the media was based on this sequence of events. On March 19, *The Wall Street Journal* claimed that TEPCO had hesitated in initiating seawater injection for fears that the reactors would be damaged by the salt, a claim which was rejected by TEPCO managers and which did not appear in the Japanese media at that time. The decision to cool the reactor using seawater injection was not subject to criticism in the diet until early May when METI Minister, KAIEDA Banri, stated that there had been repeated calls for TEPCO to start seawater injection (AS 2011/05/11). The *Asahi shinbun* subsequently published an article based on internal TEPCO documents, which stated that there had been a test injection of seawater at 7:04 PM followed by full injection as of 8:20 PM (AS 2011/05/13). KAN's role in the decision to use seawater injection was then addressed by LDP diet member—and former, as well as subsequent, Prime Minister— ABE Shinzō, who conducted a press conference in Kyoto on May 20, claiming that KAN had angrily intervened to have seawater injection discontinued after its initial activation at 7:04 PM. ABE claimed that KAN's intervention had aggravated the crisis and that any assertion of KAN's role in ordering the use or seawater injection was false (*detchiage* でっち上げ). KAN's error, ABE stated, was "worth a thousand deaths" (*banshi ni atai suru* 万死に値する) and the onus, as such, was on the Prime Minister to step down. Though ABE's claim was not directly taken up by the progressive newspapers, it was certainly parroted by the *Sankei shinbun* and the *Yomiuri shinbun* (SS 2011/05/21; YS 2011/05/21). On May 21, HOSONO Gōshi—secretary for TEPCO's joint headquarters with the government and a subordinate of KAN's—responded to this criticism in the diet by revealing that the government had had no knowledge of either seawater injection or the alleged cessation of the injection process until that point. As it turned out, TEPCO had in fact withheld this information from the Kantei. According to HOSONO, MADARAME expressed fears that seawater injection could lead to another critical state in the reactor (*sairinkai no kikensei wa aru* 再臨界の危険性はある), an assertion that MADARAME refuted the following day in statements given to both the *Yomiuri shinbun* and the *Asahi shinbun* (YS 2011/05/22). On May 23, the government amended HOSONO's initial claims, however, acknowledging that MADARAME had rather stated: "The chances of renewed criticality are not zero" (YS 2011/05/23). Though the wording here is more delicate, the cautionary tone remains the same. At this juncture, the leader of the Kokumin shintō party, with whom the DPJ was in coalition government, called for the removal of MADARAME, referring to him as a "nonsense chairman" (*detarame iinchō* デタラメ委員長) (YS 2011/05/24). On May 24, MADARAME himself went on to state at the diet that by claiming "the probability is not zero, he had actually meant it is like zero" (YS 2011/05/25b). For its part, TEPCO then announced that it had, in fact, sent a fax to NISA communicating its plans for seawater injection on the afternoon of March 12 (YS 2011/05/25c). The fax was apparently ignored by the NISA office as it was declared "additional information" and, as such, not immediately relevant (*sankō jōhō* 参考情報).

Speaking in the diet, KAN refuted in the strongest terms ABE's claims regarding his intervention in the seawater injection process. Nonetheless, the LDP decided to make KAN their scapegoat and, thus, signalled their desire for a vote of no confidence. On May 26, TEPCO revealed that there had been no such cessation in the seawater injection process. Until this date, Kantei had been kept in the dark with regard to this information, as well as both the fact that seawater injection had actually begun earlier than 8:20 PM, and the fact that a TEPCO manager had attempted to bring the injection operation to a premature close. Having become aware of ongoing seawater injection during an internal hearing with plant manager, YOSHIDA Masao, TEPCO publicly announced that a manager "had sensed in the atmosphere at Kantei that seawater injection would not be welcomed by the Prime Minister" (*shushō no ryōkai ga nai to chūsui ga dekinai to iu kanteinai no fun'iki* 首相の了解がないと注水ができないという官邸内の雰囲気; YS 2011/06/08b). The LDP and the conservative media persisted with their agenda of scapegoating KAN by decrying the loss of government credibility and the lack of strong leadership. ABE claimed he had "revealed the lie and everything was Kan's fault" (YS 2011/05/27a). In the *Yomiuri shinbun*, the confusing revelations were discussed at length and generally portrayed as a failure on the part of the KAN government to correctly communicate necessary information. YOSHIDA, on the other hand, was praised as having shown great leadership (YS 2011/05/27b).

On June 1, the LDP voted on a motion of non-confidence, which was rejected in the diet. The *Asahi shinbun*—in contrast to the *Yomiuri shinbun*—foregrounded TEPCO's role in omitting the fact that seawater had been injected earlier than was generally known. They also emphasized TEPCO's failure to inform the government for over two months that seawater injection was ongoing despite attempts to bring the operation to a close. For the *Asahi shinbun*, this obviated TEPCO's "predilection for covering up" (*inpei taishitsu* 隠蔽体質; AS 2011/05/27). For its part, the *Asahi shinbun* was the first publication to put TAKEGURO Ichirō in the spotlight. A TEPCO manager, it was TAKEGURO who had contacted TEPCO headquarters asking that seawater injection be suspended. For the *Asahi shinbun*, this had been entirely his own decision (*jibun no jishuteki na handan* 自分の自主的な判断; AS 2011/05/22). The *Yomiuri shinbun*, by contrast, did not place TAKEGURO center stage until much later. Indeed, the paper cast him in a very different light. He had claimed—the paper wrote in early June—that "there was an atmosphere in the Kantei that KAN would not allow seawater injection". As such, he had called YOSHIDA directly and urged that an end be put to seawater injection (YS 2011/06/08a). Precisely why a TEPCO manager would first try to cover up his call to YOSHIDA and why, indeed, TEPCO would continue to give fake testimony to the government for two months was not a question with which the *Yomiuri shinbun* was overly concerned. Instead, it decried the Prime Minister's lack of leadership and his loss of credibility (see, for example: YS

2011/05/27b). In December 2011, the Government Investigation Commission published its mid-term report in which it was confirmed for the first time that although YOSHIDA had clearly called for the cessation of seawater injection during the video conference with TEPCO headquarters—thus giving the impression that he had followed their orders—he had, in fact, quietly advised his subordinates to "never stop seawater injection" just prior to the call (YS 2011/12/27). For the *Yomiuri shinbun*, KAN remained the sole culprit. Completely uncritical of ABE's accusations, the paper consistently sought to shift blame away from TEPCO and back to the former Prime Minister. The order to cease seawater injection, for example, was thus frequently described as "originating from Kantei" (*kanteihatsu* 官邸発) (YS 2011/12/27, but the formulation can also be found in other media). The *Asahi shinbun*, on the other hand, put greater critical focus on the role of TEPCO, with accusations against KAN discussed firmly within the context of TEPCO's mismanagement of the accident. In January 2012, the paper published a series of articles based on independent investigations carried out by the Special Reporting Unit (on the structure of the *Asahi shinbun*'s team compare WEISS 2019: 437–90). The articles clearly sympathized with the perspective of KAN and the DPJ, criticizing instead the roles of NISA, TAKEGURO, and MADARAME.

It was pointed out that KAN had called neither for cessation nor delay of seawater injection. Rather, the former Prime Minister had been informed that it would take some time to prepare the operation and had, thus, called for adequate risk assessment to be carried out in the intervening period. Tellingly enough, not a single article in the *Yomiuri shinbun* makes any allusion to this fact (AS 2012/02/03; AS 2012/02/04).

The pattern of accusations was repeated in the *Yomiuri shinbun* when the public hearings of the Diet Investigation Commission were conducted in mid-2012 (YS 2012/05/29) and the commission's final report was published subsequently (YS 2012/07/24). It continued in 2014, when the original of YOSHIDA's testimony was released (for details, see section 2.3). From beginning to end, the *Yomiuri shinbun* persisted in its accusations against the former Prime Minister. The approach taken by the *Asahi shinbun*, by contrast, was much more even-handed. When the Diet Investigation Commission's final report was published, the paper provided critical appraisals of the committee's relatively strong support for TEPCO's claims (AS 2012/06/09b; AS 2012/06/28). Indeed, articles on the hearings and final report gave ample space to KAN's counterclaims (e. g. AS 2012/06/09b). When the scandal surrounding YOSHIDA's testimony broke in 2014, the paper relayed the statements published in a relatively neutral fashion. In particular, a great deal of space was afforded to YOSHIDA, who had testified on the question of seawater injection that any statement asking to stop the injection of seawater merely sounded to him like "noise" (*zatsuon* 雑音, thus underlining the uselessness of Takeguro's and presumably also Kan's interventions). The critical perspective on TEPCO, however, had disappeared along with the team of investigative reporters.

2.3 The evacuation plans on March 14 / 15

Probably the most prominent political discussion regarding the immediate management of the accident was the question as to whether TEPCO had intended to abandon the plant. While many facts regarding the other two issues have become relatively clear, there remains some doubt as to what happened from the evening of March 14 to the morning of March 15. Following the explosion on March 13 at reactor 3, it was reactor 2 that became the biggest threat. On March 14, the emergency cooling system deteriorated and, thus, the pressure as well the temperature in the reactor began rising unchecked. High pressure ensured that any attempt to restart seawater injection either failed or had no effect. At around 4 PM, the CEO of TEPCO instructed his staff to create a worst-case scenario including procedures for evacuation of the plant (Kimura 2012: 211). At roughly 6 PM, the fuel rods were said to be fully exposed. Venting and seawater injection were both resumed and the water level appeared to recover. However, the water level subsequently began to drop once again with venting no longer having the desired effect. At 4 AM, levels in the pressure vessel surpassed the critical limit and the team on site felt that they had exhausted all available means of stabilizing the reactor (AS 2012 / 09 / 05).

The CEO of TEPCO, Shimizu Masataka, had repeatedly tried to reach various DPJ-politicians asking for permission to begin evacuating (*taihi* 退避) staff beginning from the evening of March 14. The politicians in question were shocked to hear that TEPCO staff were apparently intending to abandon the plant. Kan summoned Shimizu to the Kantei, affirming that there would be no evacuation and accepting Kan's request to create a joint team at TEPCO headquarters. At 5 AM, Kan went to the TEPCO headquarters—accompanied by a handful of politicians and staff—where he briefly addressed the TEPCO employees. He emphasized the fact that "abandoning the plant was not an option" (*tettai nado arienai* 撤退などあり得ない) and that TEPCO would go one hundred percent bankrupt if the plant was abandoned (*tettai shitara Tōden wa 100 pāsento tsubureru* 撤退したら東電は100％潰れる; YS 2011 / 03 / 17). From that time onwards, DPJ politicians, Hosono Gōshi, and a handful others were stationed at TEPCO headquarters in an office next to that of the CEO and the chairman of the board.

About an hour later, an explosion was heard near unit 2 while pressure in the suppression chamber dropped to zero. Shortly thereafter, there were reports of a fire and ensuing damage at unit 4, with radiation around the plant reaching unprecedented levels. It was first thought that there had been a crack in the containment of unit 2 but it was later asserted—an explanation that prevails to this day—that the silicone parts of the reactor's venting and cooling systems had melted. This meltdown at unit 2 sent temperatures rising, which, in turn, led to the release of radioactive materials into the atmosphere. With the evident damage to unit 4, suspicions grew that the water tanks containing older nuclear fuel rods had been empty, leading to a gas leak that eventually caused the explosion. The most recent analysis, however, has shown that the tanks had been sufficiently cooled and that the explosion had, in fact, been the result of hydrogen that had escaped from unit 2. Following the explosion, electricity was gradually restored and seawater injection begun anew. Nonetheless, radiation levels on site and in the surrounding areas rose sharply on March 15, and the date is widely considered to be the peak of the Fukushima crisis.

As compared with the specifics of the events outlined above—about which much of the information did not emerge until several months after the fact—the news of a potential full-scale retreat (*tettai* 撤退) from the plant became public relatively early by way of a speech to TEPCO management and employees held by KAN when he arrived at the TEPCO headquarters in the early morning of March 15. It was transmitted via the internal video conference system used by TEPCO for communication and thus heard by numerous TEPCO employees. In this way, KAN's announcement that a joint headquarter would be created was quite the revelation and news of KAN's speech appeared already in the evening editions of most newspapers.

In the following section, I will sketch how the *Asahi shinbun*, which exposed various critical information regarding the role of TEPCO, covered the evacuation issue, and how it eventually became engulfed in a scandal regarding its coverage of the plant manager YOSHIDA Masao. I will outline how this scandal surfaced and spread in other media and how this changed the reporting of the *Asahi shinbun*, which had until then been a driver of the TEPCO-critical narrative.

The first article to appear in the *Asahi shinbun*, for example, gave details of KAN's speech at TEPCO headquarters while pointing out that TEPCO's explanations had changed several times before (*niten santen* 二転三転). The article equally claimed that it had evidently been a mistake to put the management of a nuclear accident completely in the hands of a private company (AS 2011 / 03 / 15). As to the immediate motivation behind KAN's statement, however, the public was at a loss. The following day, the *Asahi shinbun* published another article in which KAN explained his actions and alluded to the fact that TEPCO workers had requested permission to abandon the plant. The same article reproached the former Prime Minister for his "lack of a sense of crisis" (*kiki ishiki ga usui* 危機意識が

薄い). Both TEPCO's tendency to conceal bad news and Prime Minister KAN's distrust in TEPCO were explained from TEPCO's previous failures to communicate the explosion at reactor unit 1 and the subsequent escalation of the crisis (AS 2011 / 03 / 16).

On April 10, more than three weeks later, the *Asahi shinbun* published a report focussing on what had happened on the morning of March 15. It explained that SHIMIZU had been called into the Kantei where he claimed that he would "protect Fukushima with all we have". A bureaucrat from METI (possibly a NISA official whose authority KAN had effectively side-lined) was quoted as claiming that KAN had begun to claim all authority for decision-making, and that information could not be released until after it had been first conveyed to Kantei (AS 2011 / 04 / 10).

From April until August, no new information regarding the evacuation emerged. In late August 2011, KAN Naoto officially stepped down as Prime Minister and was followed by the more conservative NODA Yoshihiko. Shortly after his departure was announced, a short article in the *Asahi shinbun* claimed KAN had prevented the retreat of TEPCO staff from the power plant (AS 2011 / 08 / 13). In September, KAN gave an interview with the paper claiming that KAIEDA had informed him of TEPCO's intention to withdraw their staff. He, thus, called the CEO of TEPCO but was unable to get a clear response as to whether or not a retreat was on the cards (*tettai shinai no ka, suru no ka hakkiri shinai* 撤退しないのか、するのかはっきりしない). He equally emphasized the fact that the creation of the joint headquarters at TEPCO had served to guarantee a free flow of information between the two bodies (AS 2011 / 09 / 06). In October, the *Asahi shinbun* published the account of a Hitachi engineer who had been at the plant during the accident, attending to repairs on the electricity lines. He claimed that the team had been reduced to four persons on March 14 and that much of the staff had chosen not to go to the plant. Indeed, he went on, he and his team had initially relocated to a neighboring prefecture on March 15 only to find themselves subsequently called back to the plant. The paper went on to claim that the total workforce at the plant had been reduced to 70 at this stage, which was, more or less, when KAN had proclaimed there would be no evacuation (AS 2011 / 10 / 04).

The overall public image of what had happened with regards to the evacuation changed with the appearance of the Government Investigation Commission's mid-term report in December. For the *Asahi shinbun*, the poor management of TEPCO, Kantei, and the other institutions involved was clearly the most relevant factor in terms of determining responsibility for the consequences of the accident. This they typically supported by pointing to the damage caused by Kantei's control of information (*jōhō tōsei* 情報統制), for example.

As for the evacuation, the claims made by TEPCO management were simply regurgitated. According to the paper, SHIMIZU claimed he had called the NISA head and informed that evacuation would have to be prepared without explicitly mentioning that a minimum number of workers would be left at the plant. This had been interpreted by Kantei as full retreat (*zen'in tettai* 全員撤退). SHIMIZU claimed he had been asked whether they planned to abandon the plant, and had answered "no, clearly not" (*sonna koto wa kangaete imasen* そんなことは考えていません; AS 2011 / 12 / 27).

In January, the *Asahi shinbun*'s special reporting unit painted a very different picture. Based on a detailed reconstruction of the events at the Kantei, it claimed that multiple DPJ politicians had been contacted directly and indirectly by TEPCO with regard to the evacuation. Cabinet Secretary EDANO Yukio had called plant manager YOSHIDA and was told "they could still go" (signalling that the TEPCO team on site did not yet consider abandoning the plant inevitable; AS 2012 / 01 / 07). According to the article, certain members of the Kantei had deemed communication between the TEPCO headquarters and the site insufficient. The current TEPCO claim, the team at the special reporting unit wrote, was that SHIMIZU asked for permission that "workers not directly related to the handling of the accident be temporarily evacuated" (*sagyō ni chokusetsu kankei no nai ichibu no shain o ichijiteki ni taihi saseru koto ga izure hitsuyō to naru tame kentō shitai* 作業に直接関係のない一部の社員を一時的に退避させることがいずれ必要となるため検討したい). The article went on to claim that five people at Kantei had confirmed this story. TEPCO's SHIMIZU Masataka, for his part, declined to comment (AS 2012 / 01 / 13). For the *Asahi shinbun*, a de facto evacuation had, to a certain degree, already occurred at this point, given that most of the TEPCO workers had been ordered to vacate the premises.

In February, the *Asahi shinbun* extensively covered findings released by the Private Investigation Commission. Though critical, to some extent, of KAN's crisis management, the report emphasized TEPCO's mishandling of the situation. It highlighted, among other things: the fact that trucks had been sent by Kantei in order to supply electricity but TEPCO was not in possession of the necessary cables; the failure of TEPCO staff to realize that the emergency cooling of reactor 1 had failed; the fact that neither the TEPCO CEO nor the chairman had been present at the TEPCO headquarters during the most important initial hours of the disaster. Indeed, the chairman was on a TEPCO-financed tour of China playing host to a group of journalists and senior media staff (NAKANO 2012: 14). It also pointed to the arrogance of the nuclear power bureaucracy. In 2007, for example, the International Atomic Energy Agency documented a lack of clarity with regards to areas of responsibility for the JAEC and NISA. TEPCO responded to the appraisal by merely stating that "Japanese regulatory agencies were internationally highly regarded" and the possibility of reform was rejected. For the Private Investigation Commission, TEPCO's claims that it had never intended on instigating an evacuation were baseless given that SHIMIZU had never declared any intention to leave workers at the plant. Indeed, the Private Investigation Commission judged that KAN's management of the situation had prevented a (possible full) evacuation (AS 2012 / 02 / 28; AS 2012 / 02 / 29).

The *Asahi shinbun* went on to publish a review in early March in which both TEPCO's and the government's respective positions were presented. The authors pointed out that TEPCO had sent a fax to NISA alluding to "the relocation of the crisis management". The sentence was crossed out and replaced with "one part [of the staff] will temporarily evacuate", which may indicate that a decision to leave a core team at the plant was taken possibly after Kantei had emphatically rejected the request (AS 2012 / 03 / 04). In another article, which gave a preview of the Private Investigation Commission's final report, the *Asahi shinbun* reviewed claims made by different parties regarding the evacuation, alluding once again to TEPCO's insistence that it had not even considered, let alone made any indication of, evacuating. The Government Investigation Commission, by contrast, asserted that the lack of any real explanation on TEPCO's part—specifically, their failure to specify that the planned evacuation would be only partial—was the reason for the mutual distrust. The Private Investigation Commission supported Kantei's claim, adding that during his call to EDANO, SHIMIZU had stated that those on site could no longer hold out (*totemo genba wa kore ijō mochimasen* とても現場はこれ以上持ちません). Of the multiple phone calls made by SHIMIZU, this was the only one of which he later claimed he "could not remember anything" (AS 2012 / 05 / 01). In mid-May, the *Asahi shinbun* published a long review of the evacuation in which former police chief, ITŌ Tetsurō, claimed to have been told by TEPCO employees at Kantei that they were considering abandoning (*hōki* 放棄) F1, and could ultimately abandon the Fukushima 2 plant too (AS 2012 / 05 / 16). Various public hearings were conducted in May as part of the Diet Investigation Commission and the *Asahi shinbun* relayed the positions of both TEPCO and the DPJ politicians, though giving clear precedence to the latter. From the perspective of political communication, however, the DPJ's story's coherence was damaged because a number of his DPJ colleagues, including KAIEDA and deputy finance Minister, IKEDA Motohisa—who had been at the off-site center, the personnel of which were evacuated to Fukushima-city on March 15—openly criticized KAN both for his interrogation of TEPCO managers and for his allegedly excessive intrusions into the management of the crisis (*kajō kainyū* 過剰介入). Attention was thereby shifted away from events that reflected positively on DPJ politicians toward KAN's supposed failures (AS 2012 / 05 / 18).

In June, the Diet Investigation Commission's final report was published, including its hearing with TEPCO CEO, SHIMIZU Masataka. SHIMIZU claimed that the accident was continually dealt with on site and there had, as such, evidently never been any plans for a full evacuation (AS 2012/06/09b). Asked if it were true that only ten persons would remain should a worst-case scenario have played out (*saiaku no baai wa jūnin kurai to ninshiki sarete ita ka* 最悪の場合は十人くらいと認識されていたか), however, SHIMIZU answered affirmatively. These assertions were addressed in a separate front page *Asahi shinbun* article on June 9 (AS 2012/06/09a). The Diet Investigation Commission argued that a full evacuation had not been planned, yet the impossibility of effectively containing six reactors with a team of only ten people cannot be stressed enough. It is also unclear as to whether TEPCO had envisioned this particular worst-case scenario for the night of March 14/15. To judge from both the TEPCO and Kantei accounts, however, it seems rather likely. In a separate article in the same edition, the *Asahi shinbun* stated that while the Diet Investigation Commission had foregrounded miscommunication as a central factor in the disaster, uncertainties regarding TEPCO's contingency plans had not been laid to rest. The paper went on to relay doubts expressed both by the investigation chair and by DPJ politicians as to SHIMIZU's supposed inability to recall his phone call with Cabinet Secretary, EDANO, as it was during this call that his recommendation for a withdrawal was most clearly expressed. One member of the *Asahi shinbun*'s special reporting unit concluded that a plan for the abandonment of the plant could now clearly be delineated (AS 2012/06/09b). The paper also published a short article reporting that KAN had criticized the Diet Investigation Commission's conclusion as one-sided and published a rebuttal on his blog (AS 2012/06/12). A subsequent editorial (*shasetsu* 社説) took the same line, arguing that the Diet Investigation Commission's conclusions lacked empirical evidence. The paper continued to actively seek out additional information, criticizing TEPCO's refusal—supposedly on grounds of privacy—to make public the recordings of its internal video conference (AS 2012/06/28; AS 2012/06/30).

In an article dated July 6, the *Asahi shinbun* also pointed out that as an official government inquiry, the Diet Investigation Commission was the only investigative body that could oblige those under investigation to release material. Yet, the body had not forced the release of video conference recordings of TEPCO's internal communication, indeed it had barely availed of this judicial authority at all (AS 2012/07/06). The Government Investigation Commission's final report was released the following month and was, essentially, inconclusive, stating only that TEPCO may have considered a full evacuation (AS 2012/07/23). In September, redacted footage of TEPCO's video conference was first shown to a number of journalists at TEPCO headquarters before being released online. The content clearly supported the narrative upheld by DPJ politicians hitherto. On two occasions, TEPCO managers are seen discussing the evacuation of everyone from the plant. The footage also makes clear that during the crisis on March 14/15, they had considered setting concrete thresholds (of radiation) that, when reached, would trigger an evacuation. It was (and remains) suspicious, however, that TEPCO deleted certain conversations while claiming that the audio recording for the most important period from around midnight of March 14 had been lost. The audio that was released provided further glimpses into the dramatic scenes of the crisis, such as heated exchanges between YOSHIDA and other TEPCO managers at headquarters as well as an argument between MADARAME and YOSHIDA as to how the crisis at reactor 2 should be handled. Indeed, due consideration was given to the notion of calling in the armed forces to blow off the roof of a reactor building in order to avoid a second explosion (AS 2012/08/07).

Following the publication of the Government Investigation Commission's final report, there were no further significant revelations on the topic of evacuation until May 2014 when the *Asahi shinbun*'s special reporting team managed to lay its hands on a copy of the testimony given by plant manager, YOSHIDA Masao. An article was subsequently published by the paper on May 20 in which it was claimed that YOSHIDA's orders—according to his testimony—had clearly been disobeyed. On the night of March 14/15, YOSHIDA had testified, the situation in reactor 2 was dangerously approaching a worst-case scenario and he had, thus, called on headquarters to consider evacuation, while preparing to move the emergency response center (*kinkyū taisakushitsu* 緊急対策室) to Fukushima plant 2. He equally claimed to have "heard that CEO Shimizu had asked Prime Minister Kan for permission to withdraw staff from the plant" (AS 2014/05/20). Indeed, the preparation of busses and drivers for evacuation was also ordered by YOSHIDA. Following the sound of an explosion near reactor 2 and a drop of pressure in its suppression chamber (there was still pressure in the upper part of the reactor), YOSHIDA judged that there must have been a leak at the reactor. Given the consistency of radiation levels in the control room, however, he concluded that the containment had not been fully destroyed and, thus, ordered his staff to first relocate to a safer part of the plant. The article used the term retreat (*tettai*) to refer to a withdrawal to F2 while remaining at a safer location within F1 was referred to as evacuation (*taihi*). The article asserted that 90 percent of the staff ignored YOSHIDA's orders and relocated to F2. Among them were the general management (GM), consisting of essential workers including section leaders whose job it was to give advice during the emergency. Faced with the mass refusal to comply with his instructions, YOSHIDA declared that he had "accepted the fact [that they had left]" (*shō ga nai kara* しょうがないから) before eventually recalling the staff, beginning with the GM. Responding to the *Asahi shin-*

bun's queries as to its position on the "disobeying of an order" (*meirei ihan* 命令違反, quotation marks in original), TEPCO maintained that no orders had been disobeyed. Towards the end of the article, the author states that preparations for withdrawal had clearly been made and that on the morning of March 15, most of the remaining personnel had moved from the control room of the reactor building (*chūō seigyoshitsu* 中央制御室) to the emergency room located in the administration building (*menshin jūyōtō* 免震重要棟). Staff were, thus, not fully able to control the plant.

At the end of the article, the author speculates as to whom responsibility for dealing with a nuclear accident falls when there is a risk that should things come to the worst, the company to whom the plant belongs will simply abandon the premises. This question, the author insisted, had not been discussed publicly, and yet it was a precondition for restarting other nuclear reactors in the country that had been shut down after the crisis in Fukushima. The article equally cited the head of the Government Investigation Commission stating that he had promised not to publish any of the testimony accrued during the course of the investigation. Nothing would be published but the final report. This was clearly at odds with what YOSHIDA had been told at the beginning of his interview, namely that the interview itself could, ultimately, be published (AS 2014/05/20). The *Asahi shinbun* published various articles discussing the withdrawal of staff from the power plant including expert opinions/debates and a request made by former Prime Minister KAN to the current government that they disclose the testimony (e.g. AS 2014/05/28; AS 2014/06/10).

The topic was taken up in the diet on May 21, 2014. The new CEO of TEPCO was questioned on the issue and stated that YOSHIDA's remarks with regard to remaining at F1 were, in fact, not a direct instruction. Rather, it included the possibility of further withdrawing to F2 should radiation levels at F1 be unacceptably high. On May 31, a statement was published by journalist KADOTA Ryūshō. KADOTA had written a book on the accident based on multiple interviews with YOSHIDA and other TEPCO workers from the Fukushima 1 site (the access to multiple TEPCO workers from the plant together with the positive portrayal of TEPCO in his book indicate the cooperation of the TEPCO public relations department; see KADOTA 2012). The journalist claimed that there had been no "disobedience of an order" and that the *Asahi shinbun* coverage was deliberately misleading and an insult to TEPCO workers. KADOTA's claim was then taken up in June by various outlets, firstly by weekly magazines and then in online media platforms. In July, the newswire supplying Japan's local newspapers, Kyōdō tsūshin, published a feature series based on interviews with multiple workers who had been present on site. Not one made any mention of YOSHIDA's supposed instructions to remain at F1. On August 18, the *Sankei shinbun* claimed to have obtained a copy of YOSHIDA's testimony, equally affirming that there had been no "disobedience of an order". Shortly thereafter, having supposedly gained access to YOSHIDA's testimony, both the *Yomiuri shinbun* and the *Mainichi shinbun* published articles criticizing the *Asahi shinbun*'s reporting. This led the *Asahi shinbun* to retract the May 20 article and hold a press conference during which the CEO issued a public apology (AS 2014/11/13). The issue came to constitute a veritable scandal that was widely covered in national media and on numerous online platforms. It also happened to coincide with a further two scandals regarding the *Asahi shinbun*.[5] Needless to say, the paper became the subject of fierce criticism and the incident is said to have triggered a substantial decline in its readership (ASAHI SHINBUN KISHA YŪSHI 2015: 210).

The coverage was examined by a commission of external experts for the *Asahi shinbun*, which stated that mid-level management had not communicated YOSHIDA's order to the staff and that YOSHIDA himself had later confirmed having no significant objection to workers relocating to F2. The formulation used in the *Asahi shinbun* article, "disobeying of an order", was thus inappropriate. As for KADOTA and the *Sankei shinbun*'s assertion that the *Asahi shinbun* had damaged both the reputation of TEPCO workers and Japan's "national pride", the heavily nationalistic tone should be noted (SHINODA 2014: 52–55). There ensued a heated debate occurring on a national level as to whether or not an order had been disobeyed. Indeed, this came to constitute the fundamental issue with regards to evacuation/withdrawal. In the wake of the scandal, large parts of YOSHIDA's testimony—as well as that of the DPJ politicians, which the government subsequently made public—were covered by every major newspaper in the country, with YOSHIDA's views clearly the greatest focus of attention. YOSHIDA's testimony included heavy condemnation of KAN for his March 15 speech at the TEPCO headquarters and saw the plant manager argue that he had never intended to abandon the plant (SEIFU JIKOCHŌ 2011b: 30). None of the other TEPCO managers interviewed by the commission consented to disclosing their testimony.

3 CONCLUSION

What conclusions can be drawn from the public discussion of responsibility for the Fukushima accident? First of all, information on the actual events on site was very slow to enter the public sphere. Only very gradually between April to May 2011 did competing claims regarding venting and seawater injection surface, and it took even longer for a more detailed picture to emerge. As for the evacuation, the specifics were even slower to come to light. When they did, the Japanese public found itself confronted with two competing narratives, one based mainly on the claims of DPJ politicians, and the other on those of TEPCO and its allies in the "nuclear village".

During this initial period immediately following the disaster, from roughly April to December 2011, the *Asahi shinbun* gave more prominence to the narrative favored by the DPJ[6], while the *Yomiuri shinbun* attempted to shift blame from TEPCO to the Prime Minister. Indeed, the paper was the first to make serious claims aimed at discrediting the Prime Minister, alleging, namely, that his visit to the TEPCO plant had delayed venting. Though contradictory testimony would subsequently render this claim untenable, it continued, nevertheless, to be reproduced by conservative media and was later revived, in a slightly different guise, by TEPCO board chairman, KATSUMATA Tsunehisa, among others.

As for ABE Shinzō's claims that KAN had impeded seawater injection, it became clear that though these claims may arguably have contained some grain of truth, they nonetheless constituted fundamental misrepresentations of the actual events. It was also revealed that, for over two months, TEPCO management and workers on site obscured—from both the government and the public more broadly—the fact that seawater injection had begun a full hour earlier than had been officially declared. It also came to light that one Kantei-based TEPCO manager had attempted—while giving no notice of his actions to Kantei— to halt the injection of seawater, a fact that attests to TEPCO's obfuscation of any information that may have proven inconvenient for them.

The *Yomiuri shinbun* and the *Sankei shinbun* sought to offer up KAN as a scapegoat for almost every ill-fated event at the plant (as can be seen in the claims in ABIRU 2011, for example). This served to obscure the decisive behavior of the TEPCO manager on site, TAKEGURO Ichirō, and was closer to propaganda than anything that might constitute responsible journalism. A similarly cynical approach can be seen in the wide variety of accusations levelled against KAN, including: "Kan intimidated TEPCO employees and nuclear power experts and this is why the crisis escalated" (ABIRU 2011: 230, 285); "Kan visited the plant without safety gear and helmet" (YS 2011 / 05 / 17); "Kan overestimated his own competence and made the wrong decisions" (ABIRU 2011: 250); "Kan has hidden negative information because he wanted to continue to promote Japanese construction of nuclear power plants abroad" (YS 2011 / 03 / 13); "Kan has stopped the Hamaoka power plant with no scientific or legal basis for doing so" (YS 2011 / 06 / 01); "Kan has peddled the idea that the regions affected by the disaster will be unhabitable for 20 years" (ABIRU 2011: 243–45); "Kan has appointed a large number of external advisors to the nuclear power, complicating and hampering the crisis management and interfering with the handling of the crisis on site" (ABIRU 2011: 238–40). Accusations of this order appear in inordinate quantities in two books in particular, YOMIURI SEIJIBU (2011) and ABIRU (2011), which were published by *Yomiuri* and *Sankei* editors and dedicated to the sole purpose of discrediting KAN. That these claims were political in nature is evident in their timing, given that the bulk were made following KAN's decision to shut down the Hamaoka nuclear power plant run by the Chūbu Electric Power Company (early May 2011). It was this decision as well as his increasingly concentrated efforts to reduce Japan's future reliance on nuclear power that invoked the rage of the "nuclear village". His actions equally brought him into conflict with certain members of his own party including KAIEDA Banri, who supported NISA and METI's attempts to push through a quick restart of nuclear power plants that had been undergoing maintenance (AS 2011/09/06).

The political influence permeating the news was evident in the many accusations made against KAN and the DPJ and which were the result of the "nuclear village" insider information, namely from JNSC chairman MADARAME Haruki, for example, or other anonymous METI, NISA—and, of course—TEPCO sources. Many other accusations were advanced, or supported, by nuclear engineers significantly invested in the propagation of the "nuclear village". Take, for example, MADARAME and MIYA Kenzō, who the *Yomiuri shinbun* cited as accusing KAN of hampering the venting process for reactor 1. They, and numerous others, received sizeable donations and research funding from the nuclear industry.

DPJ-favorable news appearing in progressive media, by contrast, tended rather to have at their root certain DPJ politicians, their aides, and the external advisors called in by KAN. It is, thus, no wonder that the appointment of external advisors—which could well have diminished the "nuclear village's" voice in the formulation of nuclear policy—was the target of criticism from both the *Yomiuri shinbun* and the *Sankei shinbun* (e.g. YS 2012/05/29; ABIRU 2011: 238–43). That the DPJ narrative lost out to that of TEPCO and its allies is surely due to the DPJ's lack of internal cohesion. Moreover, key individuals such as KAIEDA and his deputy, IKEDA Motohisa, ultimately turned on KAN, criticizing him for his inflammatory treatment of TEPCO managers and experts, such as had served to galvanize the TEPCO narrative. The DPJ itself enjoyed the full support of the energy company labor unions (including TEPCO's) and had, among its ranks, some of the most ardent advocates of nuclear power.

Ultimately, it was the TEPCO narrative that took precedence over that of the DPJ in public discourse, partly because substantial elements thereof were supported by two of the three major reports. It became clear that the Diet Investigation Commission, in particular, was, to a considerable extent, politically motivated given that investigators were very much invested in foregrounding KAN's personal misconduct. Significantly less attention was paid to why TEPCO had been so unwilling to disclose information, both to the government and to the public more generally.⁷ The three major investigations carried out went some way to clarifying certain issues related to the disaster, such as interactions between Kantei, broader government administrative bodies, and TEPCO. It was revealed, for example, that TEPCO did not intentionally delay venting or seawater injection. However, the idea of simply accepting the delays as unavoidable given the chaos of the tsunami also proved to be untenable. The implausibility of these two notions ought to have pointed to a more serious lack of contingency planning prior to the tsunami. Yet, the fact that the possibility of a severe accident had been drastically underestimated went fundamentally unacknowledged. It is a realization that was, ultimately, lost when public discourse finally settled around the image of "brave workers" on site (symbolized by YOSHIDA) versus allegedly incompetent Kantei politicians. Precisely why TEPCO appears to have concealed what was happening at the plant during the initial phase of the accident—and why both NISA and MADARAME himself completely ignored the gravity of the situation—are questions that remain largely unanswered.

How could Kantei have been expected not to interfere with the handling of the accident when the operator of the plant had been actively trying to conceal the reality of events on site? For KIMURA Hideaki (2012: 42), the complete failure of NISA—as well as other organs of the nuclear bureaucracy whose remit it was to manage the crisis and advise the Prime Minister—amounted to outright sabotage in certain instances.⁸ As for TEPCO's attempts to obfuscate what was happening at the plant, they are coherent with the company's history of effectively shaping public discussion of nuclear power over an extended period of time. TEPCO was notorious for trying to conceal bad news and suppressing unfavorable coverage (for one account thereof, see WEISS 2019: 81–322, 369–540). The electricity companies have also been a pillar of support for the LDP's long term rule in Japanese politics. Using their regional monopolies, they have channeled large sums of political donations to LDP politicians (ASAHI SHINBUN TOKUBETSU HŌDŌBU 2014). It is, thus, no wonder that TEPCO was hesitant to communicate its failures to Kantei, especially given that the latter was in the hands of a government outside of the traditional power circle and a Prime Minister who had a history of activism before entering formal politics. Public debate regarding responsibility for the crisis has ultimately crystallized around YOSHIDA's testimony, a fact that is further proven by the narrative portrayed in the recently released blockbuster movie "Fukushima Fifty". This has served to consolidate a politically distorted image of the events of the crisis while concealing many of the true issues surrounding the accident.

Notes

1 The reactor contains nuclear fuel which is cooled by a constant circulation of water. The fuel is located in a pressure vessel. The pressure vessel, in reactors like the one in Fukushima, is placed within a solid containment to hold radioactive materials and radiation from escaping.
2 An inquiry taking into account the complete history of nuclear policy in Japan will naturally put stronger emphasis on the role played by the LDP and the "nuclear village", while a focus on the immediate handling of the accident will put more emphasis on how DPJ politicians and TEPCO handled the crisis.
3 In this paper, I will mainly use the newspapers *Asahi shinbun* and *Yomiuri shinbun* and only occasionally add articles from *Mainichi shinbun* and *Sankei shinbun*. A full systematic analysis of *Sankei* and *Mainichi* as well as other newspapers is beyond the scope of this paper.
4 The translations of newspaper headlines and content throughout the text are my own.
5 One scandal concerned the newspaper's coverage of the issue of wartime sexual slavery sponsored by the Japanese Imperial Army during World War II. Another concerned the critical opinion of a commentator regarding this issue that was not published in the *Asahi shinbun*.
6 This may equally apply to other progressive newspapers such as *Mainichi shinbun* and the Chūnichi Tokyo Group (*Tōkyō shinbun*, *Chūnichi shinbun*). A detailed and systematic analysis of the wider journalistic field including these and other newspapers would be desirable.
7 The political constellation in place during the installation of the Diet Investigation Commission as well as—to some extent—the Government Investigation Commission may well explain their intensely political thrust. The push to oust KAN was in full swing both inside and outside the DPJ while the party remained in power. The LDP and its allies controlled the upper house, the cooperation of which the party needed in order to progress with its policy plans. KAN's proposition to form an alliance with the government—which was facing a national crisis—was rejected by the LDP leadership, which had stipulated that KAN's resignation was a precondition for the party's cooperation. Certain members of the DPJ had considered joining the LDP's non-confidence vote in early June 2011 but were successfully dissuaded by the former Prime Minister. The ouster later gained momentum, however, and he was successfully forced to resign in late August. The subsequent NODA administration showed clear support of nuclear power. The investigation commissions were made up of nominally "neutral" members such as scholars, lawyers, and prosecutors etc. Furthermore, it is unclear as to whether any direct political influence was exerted. Yet, if the questions posed during the published hearings are anything to be judged by, the investigation appears to have been considerably influenced by public accusations levelled by the LDP.
8 Following arguments with KAN, NISA's head, TERAZAKA Nobuaki, disappeared from Kantei for several days. His agency failed to carry out basic tasks such as sup-

plying Kantei with necessary information or proposing possible countermeasures. Reneging on their remit to supervise the plant, its officers quickly evacuated to the off-site center and subsequently to distant Fukushima City on March 15. Over the course of the crisis, it became clear that NISA had neither had the means, nor the independence to supervise TEPCO and was excessively relying on TEPCO for information and expertise. MADARAME, for example, stated that there would be no hydrogen explosion and failed to give useful advice to deal with the crisis (KIMURA 2012: 42–44).

REFERENCES

ABIRU, Rui 阿比留瑠比 (2011): *Sōri, anata koso fukkō no shōgai desu: Kan Naoto seiken no 'taizai'* 総理、あなたこそ復興の障害です：菅直人政権の「大罪」. Tōkyō: PHP kenkyūjo.

ASAHI SHINBUN KISHA YŪSHI 朝日新聞記者有志 (2015): *Asahi shinbun Nihongata soshiki no hōkai* 朝日新聞日本型組織の崩壊 (Bunshun shinsho 文春新書). Tōkyō: Bungei shunjū.

ASAHI SHINBUN TOKUBETSU HŌDŌBU 朝日新聞特別報道部 (2014): *Genpatsu riken o ou* 原発利権を追う. Tōkyō: Asahi shinbun shuppan.

ITŌ, Mamoru 伊藤守 (2012): *Dokyumento Terebi wa genpatsu jiko o dō tsutaeta ka* ドキュメントテレビは原発事故をどう伝えたか (Heibonsha shinsho 平凡社新書). Tōkyō: Heibonsha.

KADOTA, Ryūshō 門田隆将 (2012): *Shi no fuchi o mita otoko: Yoshida Masao to Fukushima Daiichi genpatsu jiko no gohyakunichi* 死の淵を見た男：吉田昌郎と福島第一原発事故の五〇〇日. Tōkyō: PHP kenkyūjo.

KIMURA, Hideaki 木村秀昭 (2012): *Kenshō Fukushima genpatsu jiko: Kantei no hyakujikan* 検証福島原発事故：官邸の一〇〇時間. Tōkyō: Iwanami shoten.

NAKANO, Yōichi 中野洋一 (2012): "Genpatsu sangyō no kane to hito 原発産業のカネとヒト". In: *Shakai bunka kenkyūjo kiyō* 社会文化研究所紀要 70: 1–48.

SHINODA, Hiroyuki 篠田博之 (2014): "Kokuzoku, baikokudo tobikau Asahi basshingu sōdō no ijō: Shūkanshi ga iyō na jōkyō ni 国賊、売国奴飛び交う朝日バッシング騒動の異常：週刊誌が異様な状況に". In: *Tsukuru* 創 44 (9): 52–55.

TŌKYŌ DENRYOKU FUKUSHIMA GENSHIRYOKU HATSUDENSHO NI OKERU JIKO CHŌSA, KENSHŌ IINKAI (SEIFU JIKOCHŌ) 東京電力福島原子力発電所における事故調査・検証委員会(政府事故調) (2011a): *Yoshida Masao chōsa kekkasho* 吉田昌郎調査結果書 1, 07/22; https://www8.cao.go.jp/genshiryoku_bousai/fu_koukai/pdf_2/020.pdf (last access 2023/03/15).

TŌKYŌ DENRYOKU FUKUSHIMA GENSHIRYOKU HATSUDENSHO NI OKERU JIKO CHŌSA, KENSHŌ IINKAI (SEIFU JIKOCHŌ) 東京電力福島原子力発電所における事故調査・検証委員会(政府事故調) (2011b): *Yoshida Masao chōsa kekkasho* 吉田昌郎調査

結果書 11, 11/06; https://www8.cao.go.jp/genshiryoku_bousai/fu_koukai/pdf_2/350.pdf (last access 2023/03/15).

Tōkyō denryoku Fukushima genshiryoku hatsudensho ni okeru jiko chōsa, kenshō iinkai (Seifu jikochō) 東京電力福島原子力発電所における事故調査・検証委員会(政府事故調) (2012): *Kaieda Banri chōsa kekkasho* 海江田万里調査結果書 1, 02/08; https://www8.cao.go.jp/genshiryoku_bousai/fu_koukai/pdf_2/541.pdf (last access 2023/03/15).

Weiss, Tobias (2014): "Die Japanischen Medien und die Atomkatastrophe von Fukushima". In: *Japan Jahrbuch 2015*: 245–69.

Weiss, Tobias (2019): *Auf der Jagd nach der Sonne: Das journalistische Feld und die Atomkraft in Japan*. Baden-Baden: Nomos.

Yomiuri shinbun seijibu 読売新聞政治部 (2011): *Bōkoku no saishō: Kantei kinō teishi no hyakuhachijūnichi* 亡国の宰相：官邸機能停止の180日. Tōkyō: Shinchōsha.

Yoshioka, Hitoshi 吉岡斉 (2011): *Shinpan: Genshiryoku no shakaishi* 新版：原子力の社会史 (Asahi sensho 朝日選書). Tōkyō: Asahi shinbunsha.

Newspaper Articles:

Asahi shinbun 朝日新聞 (AS)

AS (2011/03/15): "Ayaui kiki kanri. Kan shushō mizukara Tōden-iri. Minkan makase itten, taisaku honbu. Higashi Nihon daishinsai, genpatsu jiko 危うい危機管理　菅首相自ら東電入り　民間任せ一転、対策本部　東日本大震災・原発事故": 2 (Morning ed.).

AS (2011/03/16): "Kiki kanri, gote. Kan seiken, Tōden, fushin no kabe. Fukushima Daiichi genpatsu jiko 危機管理、後手　菅政権・東電、不信の壁　福島第一原発事故": 3 (Morning ed.).

AS (2011/03/30): "Kan shushō no Fukushima genpatsu shisatsu de kōbō. Yatō 'haiki chakushi ni nanajikan' 菅首相の福島原発視察で攻防　野党「排気着手に7時間」": 4 (Morning ed.).

AS (2011/04/10): "'Ichi–gogōki, zendengen sōshitsu!' Kenshō, Fukushima Daiichi genpatsu jiko 「1~5号機、全電源喪失！」　検証・福島第一原発事故": 1 (Morning ed.).

AS (2011/04/19): "Kan shushō, benmei mata benmei. Higashi Nihon daishinsai, Fukushima genpatsu taiō, san'in'i de shūchū shingi 菅首相、弁明また弁明　東日本大震災・福島原発対応、参院委で集中審議": 4 (Morning ed.).

AS (2011/04/26): "Kan shushō no shisatsu de chien, Tōden shachō ga hitei. Fukushima genpatsu ichigōki no haiki 菅首相の視察で遅延、東電社長が否定　福島原発1号機の排気": 4 (Morning ed.).

AS (2011 / 05 / 11): "Osensui, kunō kyūmanton kenshō. Fukushima Daiichi genpatsu no chūsui to taiō 汚染水、苦悩9万トン検証　福島第一原発の注水と対応": 3 (Morning ed.).

AS (2011 / 05 / 13): "'Chūsui ga saiyūsen' 'bento seyo' kinpaku no sagyō, namanamashiku Fukushima genpatsu jiko, Tōden shiryō kijutsu「注水が最優先」「ベントせよ」緊迫の作業、生々しく　福島原発事故、東電資料記述": 3 (Morning ed.).

AS (2011 / 05 / 22): "Kaisui chūnyū→chūdan→saichūnyū no jōhō, seifu ni todokazu. Fukushima Daiichi genpatsu jiko 海水注入→中断→再注入の情報、政府に届かず　福島第一原発事故": 4 (Morning ed.).

AS (2011 / 05 / 27): "Genba inpei, minukenu Tōden. Chūsui chūdan, shochō to ittan gōi. Fukushima Daiichi genpatsu jiko 現場隠蔽、見抜けぬ東電　注水中断、所長といったん合意　福島第一原発事故": 2 (Morning ed.).

AS (2011 / 08 / 13): "'Taijingo wa baiomasu'. Kan shushō, ene seisaku ni wa kodawari「退陣後はバイオマス」菅首相、エネ政策にはこだわり": 4 (Morning ed.).

AS (2011 / 09 / 06): "Fukushima Daiichi genpatsu jiko e no taiō, Kan zenshushō ni kiku. 'Arayuru sōtei, umaku ikanakatta' 福島第一原発事故への対応、菅前首相に聞く「あらゆる想定、うまくいかなかった」": 3 (Morning ed.).

AS (2011 / 09 / 11): "'Genpatsu ga bakuhatsu shimasu'. Jieikan ga tsuge ni kita. Fukushima genpatsu jiko, jichitai no genba kara kenshō「原発が爆発します」自衛官が告げに来た　福島原発事故、自治体の現場から検証": 1–2 (Morning ed.).

AS (2011 / 10 / 04): "Bakuhatsu, gekishin, kao nadeta bakufū. 'Nigero' fūjō e hashitta Fukushima Daiichi genpatsu jiko, genba sekininsha ga shōgen 爆発、激震、顔なでた爆風　「逃げろ」風上へ走った　福島第一原発事故、現場責任者が証言": 39 (Morning ed.).

AS (2011 / 12 / 27): "Kanteinai no bundan shinkoku. Gokai to chika, nagarenu jōhō. Seifu jikochō no chūkan hōkoku 官邸内の分断深刻　5階と地下、流れぬ情報　政府事故調の中間報告": 2 (Morning ed.).

AS (2012 / 01 / 07): "(Purometeusu no wana) Kantei no itsukakan: 5 Mada yaremasu ne (プロメテウスの罠)官邸の5日間: 5 まだやれますね": 3 (Morning ed.).

AS (2012 / 01 / 13): "(Purometeusu no wana) Kantei no itsukakan: 11 Chō sūpāman nara　(プロメテウスの罠)官邸の5日間: 11 超スーパーマンなら": 3 (Morning ed.).

AS (2012 / 02 / 03): "(Purometeusu no wana) Kantei no itsukakan: 32 'Ro e kaisui chūnyū seyo' (プロメテウスの罠)官邸の5日間: 32「炉へ海水注入せよ」": 3 (Morning ed.).

AS (2012 / 02 / 04): "(Purometeusu no wana) Kantei no itsukakan: 33 Tsugi no te, teian ga nai' (プロメテウスの罠)官邸の5日間: 33 次の手、提案がない": 3 (Morning ed.).

AS (2012/02/28): "'Tōden no soshikiteki na taiman' 'Kisei wa Garapagosuka' Fukushima genpatsu, minkan jikochō ga hōkokusho「東電の組織的な怠慢」「規制はガラパゴス化」福島原発、民間事故調が報告書": 5 (Morning ed.).

AS (2012/02/29): "Minkan jikochō, Kan-shi no genpatsu taiō o hihan. Shushō mizukara 'batterī no ōkisa wa?' Tantōsha ni shitsumon 民間事故調、菅氏の原発対応を批判　首相自ら「バッテリーの大きさは?」担当者に質問": 3 (Morning ed.).

AS (2012/03/04): "(Higashi Nihon daishinsai ichinen: 4) Genpatsu jiko. Fukushima Daiichi genpatsu 'unmei no hi' wa sangatsu jūgonichi datta（東日本大震災1年: 4）原発事故　福島第一原発「運命の日」は3月15日だった": 36 (Morning ed.).

AS (2012/05/01): "Kakushin, doko made semareru. Kenshō, Fukushima Daiichi genpatsu jiko no chōsai 核心、どこまで迫れる　検証・福島第一原発事故の調査委": 11 (Morning ed.).

AS (2012/05/16): "Fukushima Daiichi 'tettai' shinsō wa. Seifu to Tōden, kotonaru kenkai. Kokkai jikochō, asu kara seijika chōshu 福島第一「撤退」真相は　政府と東電、異なる見解　国会事故調、あすから政治家聴取": 7 (Morning ed.).

AS (2012/05/18): "Kinkyū sengen no okure 'Kan-shi no rikai ni jikan' Kaiedashi shōgen. Kokkai genpatsu jikochō = teisei ari 緊急宣言の遅れ「菅氏の理解に時間」海江田氏証言　国会原発事故調=訂正あり": 1 (Morning ed.).

AS (2012/05/27): "Genpatsu jiko, kantei no jittai wa. Kokkai jikochō, shōchi ōzume 原発事故、官邸の実態は　国会事故調、招致大詰め": 4 (Morning ed.).

AS (2012/06/09a): "Saiakuji wa zanryū jūnin, Shimizu, Tōden zenshachō 'ninshiki'. Zenmen tettai wa hitei. Kokkai genpatsu jikochō 最悪時は残留10人、清水・東電前社長「認識」全面撤退は否定　国会原発事故調": 1 (Morning ed.).

AS (2012/06/09b): "Tōden to kantei, tairitsu no mama. Shimizu zenshachō, 'zenmen tettai' hitei no kenkai. Kokkai genpatsu jikochō 東電と官邸、対立のまま　清水前社長、「全面撤退」否定の見解　国会原発事故調": 3 (Morning ed.).

AS (2012/06/12): "Kan-shi, burogu de hanron. Kokkai genpatsu jikochō no kenkai 'ippōteki na kaishaku' 菅氏、ブログで反論　国会原発事故調の見解「一方的な解釈」": 4 (Morning ed.).

AS (2012/06/28): "Puraibashī riyū ni rokuga kōkai kobamu Tōden kabunushi, nezuyoi fushinkan. Genpatsu terebi kaigi hozen shinsei e プライバシー理由に録画公開拒む東電株主、根強い不信感　原発テレビ会議保全申請へ": 3 (Morning ed.).

AS (2012/06/30): "Edano keisanshō, hikōkai no Tōden hinan. Terebi kaigi no rokuga, kabunushira hozen kyōgi 枝野経産相、非公開の東電非難　テレビ会議の録画、株主ら保全協議": 2 (Morning ed.).

AS (2012/07/06): "'Genpatsu jiko wa jinzai' Kokkai jikochō ga saishū hōkoku. Tōden, kuni no sekinin o kyōchō「原発事故は人災」国会事故調が最終報告　東電・国の責任を強調": 1 (Morning ed.).

AS (2012 / 07 / 23): "Anzen saiyūsen no shisei ketsujo. Tōden, kyūmei no ishi fujūbun. Seifu genpatsu jikochō, saishū hōkoku de hihan 安全最優先の姿勢欠如　東電、究明の意志不十分　政府原発事故調、最終報告で批判": 1 (Evening ed.).

AS (2012 / 08 / 07): "'Bento yare hayaku' Sekasu Tōden komon. Genpatsu jiko kaigi eizō, kakō shi kaiji「ベントやれ早く」せかす東電顧問　原発事故会議映像、加工し開示": 1 (Morning ed.).

AS (2012 / 09 / 05): "Ichinichi kūhi, sangōki bōso. Fukushima Daiichi genpatsu jiko, Tōden terebi kaigi kiroku no yaritori 一日空費、3号機暴走　福島第一原発事故・東電テレビ会議記録のやりとり": 17 (Morning ed.).

AS (2014 / 05 / 20): "Hōmurareta meirei ihan 'Yoshida chōsho' kara tōji o saigen. Fukushima Daiichi genpatsu jiko=kono kiji o torikeshimasu 葬られた命令違反「吉田調書」から当時を再現　福島第一原発事故＝この記事を取り消します": 2 (Morning ed.).

AS (2014 / 05 / 28): "'Yoshida chōsho' kōkai, Kan-shi no motome kyohi. Naikaku kanbōgawa「吉田調書」公開、菅氏の求め拒否　内閣官房側": 4 (Morning ed.).

AS (2014 / 06 / 10): "(Shisō no chisō) Hō no shihai to genpatsu. Zanryū no gimu, dare ni mo nakatta Oguma Eiji（思想の地層）法の支配と原発　残留の義務、誰にもなかった小熊英二": 3 (Evening ed.).

AS (2014 / 11 / 13): "Asahi shinbunsha 'Yoshida chōsho' hōdō. Hōdō to Jinken iinkai (PRC) no kenkai zenbun (1) 朝日新聞社「吉田調書」報道　報道と人権委員会（PRC）の見解全文 (1)": 0 (Morning ed.).

Mainichi shinbun 毎日新聞 (MS)

MS (2011 / 03 / 16): "Higashi Nihon daishinsai: Fukushima Daiichi genpatsu jiko. Seifu no taiō, gote ni. Tōden to renraku honbu, setchi-okure 東日本大震災：福島第1原発事故　政府の対応、後手に　東電と連絡本部、設置遅れ": 3 (Morning ed.).

Sankei shinbun 産經新聞 (SS)

SS (2011 / 05 / 21): "Fukushima Daiichi genpatsu. Jishin yokujitsu ichigōki, 'shushō gekido' de kaisui chūnyū chūdan 福島第1原発　地震翌日1号機、「首相激怒」で海水注入中断": 2 (Morning ed.).

Yomiuri shinbun 讀賣新聞 (YS)

YS (2011 / 03 / 13): "Fukushima genpatsu bakuhatsu. Seifu no hinan shiji gote. Han'i kakudai. 'Kantei wa kikikan usui' 福島原発爆発　政府の避難指示後手　範囲拡大「官邸は危機感薄い」": 8 (Morning ed.).

YS (2011 / 03 / 17): "Genpatsu jiko. Kantei to Tōden surechigai. Sagyōin taihi-meguri oshi-mondō 原発事故　官邸と東電すれ違い　作業員退避巡り押し問答": 6 (Morning ed.).

YS (2011 / 03 / 29): "Genpatsu shisatsu wa 'jintō shiki'. Shushō shodō-okure hihan ni hanron 原発視察は「陣頭指揮」 首相初動遅れ批判に反論": 1 (Evening ed.).

YS (2011 / 04 / 01): "Genpatsu jiko sanshūkan kenshō. Tōden shodō tsumazuku. Kaisui chūnyū okure. Kiki rensa maneku 原発事故3週間検証 東電、初動つまずく 海水注入遅れ 危機連鎖招く": 3 (Morning ed.).

YS (2011 / 04 / 19): "Kokkai ronsen no shōhō. Jūhachinichi san'in yosan'i shūchū shingi 国会論戦の詳報 18日参院予算委集中審議": 11 (Morning ed.).

YS (2011 / 05 / 17): "Yatō, genpatsu shodō o tsuikyū. Shūin yosan'i 'jiko wa jinzai da' 野党、原発初動を追及 衆院予算委「事故は人災だ」": 4 (Morning ed.).

YS (2011 / 05 / 21): "Shushō ikō de kaisui chūnyū chūdan. Fukushima Daiichi shinsai yokujitsu 首相意向で海水注入中断 福島第一震災翌日": 1 (Morning ed.).

YS (2011 / 05 / 22): "'Ichigōki sairinkai no osore'. Kanteigawa 'Madarame iinchō ga shiteki'. Tōnin 'itte nai'「1号機再臨界の恐れ」官邸側「班目委員長が指摘」 当人「言ってない」": 1 (Morning ed.).

YS (2011 / 05 / 23): "Seifu 'Madarame shingen' o teisei. 'Sairinkai kikensei aru' → 'zero dewa nai' 政府「班目進言」を訂正 「再臨界危険性ある」→「ゼロではない」": 1 (Morning ed.).

YS (2011 / 05 / 24): "Detarame iinchō da. Kamei-shi, Madarame-shi no kōtetsu yōkyū でたらめ委員長だ 亀井氏、班目氏の更迭要求: 4 (Morning ed.).

YS (2011 / 05 / 25a): "Kan-shi 'genpatsu jiko wa jinzai' 'shimurēshon kinō sezu'. Honshi to kaiken 菅氏「原発事故は人災」「シミュレーション機能せず」 本紙と会見": 2 (Morning ed.).

YS (2011 / 05 / 25b): "'Kanōsei, jijitsujō zero no imi'. Madarame-shi 'sairinkai zero dewa nai' o shakumei「可能性、事実上ゼロの意味」 班目氏「再臨界ゼロではない」を釈明": 4 (Morning ed.).

YS (2011 / 05 / 25c): "Tōden chūsui sanjikanhan mae. Hoan'in ni fakkusu 東電注水3時間半前 保安院にファクス": 1 (Evening ed.).

YS (2011 / 05 / 27a): "Kaisui chūnyū mondai. Shushō sekininron sara ni. 'Hiraita kuchi fusagaranai'. 'Nani ga hontō ka' 海水注入問題 首相責任論さらに 「開いた口ふさがらない」「何が本当なのか」": 4 (Morning ed.).

YS (2011 / 05 / 27b): "'Kaisui chūnyū' niten santen. Seifu wa 'Tōden makase'. Kokusai shakai sara ni kōka mo 「海水注入」二転三転 政府は「東電任せ」 国際社会さらに硬化も": 3 (Morning ed.).

YS (2011 / 06 / 08a): "Higashi Nihon daishinsai sankagetsu. Kenshō, shoki taiō. Kantei, shodō kara meisō 東日本大震災3か月 検証・初期対応 官邸、初動から迷走": 34 (Morning ed.).

YS (2011 / 06 / 08b): "[Kenshō, genpatsu kiki] (1) bento. Kūhaku no nijikanhan (rensai) [検証・原発危機] (1) ベント 空白の2時間半 (連載)": 1 (Morning ed.).

YS (2011/06/19): "Fukushima Daiichi shinsai chokugo. Bento-meguri seigyo-shitsu konran. Tōden shiryō. Shushō taizaichū no kisai nashi 福島第一震災直後　ベント巡り制御室混乱　東電資料　首相滞在中の記載なし": 1 (Morning ed.).

YS (2011/12/27): "Henshū techō 編集手帳": 1 (Morning ed.).

YS (2012/05/29): "[Sukyanā] Kan-zenshushō jōzetsu ni hanron. Jikochō chōshu [スキャナー]菅前首相冗舌に反論　事故調聴取": 3 (Morning ed.).

YS (2012/07/24): "Yottsu no jikochō hōkoku. Hikaku 'tsunami' fubi de itchi= tokushū sono ichi 4つの事故調報告　比較「津波」不備で一致＝特集その1": 12 (Morning ed.).

YS (2014/09/12): "Genpatsu jiko chōsho no yōshi sono ni 原発事故調書の要旨その2": 13 (Morning ed.).

Living with the Nuclear: Life Experiences of Farmers in Fukushima Prefecture

Anna Wiemann

Today, the Fukushima disaster became ordinary. It no longer captures the public's attention. [...] Fukushima gradually became a thing of the past, a memory. (BENSAUDE-VINCENT / BOUDIA / SATO 2022: 1–2)

1 INTRODUCTION

The Fukushima Daiichi nuclear power plant borders the small towns of Ōkuma and Futaba on the coastline of Fukushima prefecture in Northeastern Japan. Just over a decade has now passed since the explosions that occurred there in March 2011, and yet the nuclear disaster appears to have been normalized. Contradictory though it may seem, the previously "unimaginable" nuclear disaster is now considered by the European nuclear sector to be a largely "closed file", from which all necessary lessons have already been learned. In Japan, after a temporary halt of all nuclear reactors in the country following the disaster in 2011, nuclear reactors are consecutively back up and running already since 2015[1] (ARNHOLD 2019; OECD 2021: 38–42). Meanwhile, in Fukushima prefecture, evacuation orders have been lifted, other than for a very small area to which return will continue to be difficult for a long time to come. In 2022, this area accounts for no more than 2.4 % of the prefectural land (RECONSTRUCTION AGENCY 2022).

Yet, Fukushima Daiichi continues to emit radioactive material and farmers in Fukushima prefecture still face the task of reducing radiation levels in their fields and produce despite far-reaching efforts to decontaminate the soil (MCCAULEY 2021; NUCLEAR REGULATION AUTHORITY 2022: 3). The nuclear disaster, which unfolded in the aftermath of the Great East Japan Earthquake and Tsunami in March 2011, is thus ongoing and will continue for decades to come.

This paper focuses on the ongoing framing and narration of the nuclear disaster by farmers in Fukushima prefecture—one of the most vulnerable groups to environmental contamination. From the perspective of disaster memory studies, I present a longitudinal narrative case study of two farmers in Fukushima prefecture who were interviewed in 2012 / 2013 by a group of students from Sophia University (Tokyo) under the supervision of David SLATER and again by me in 2022, thus covering the first decade after the initial disaster. Focusing on the farmers' definition of the disaster and the way it affects their lives, I seek to carve out the social frames they draw upon in order to position themselves in the broader disaster-affected social context. This serves to hypothesize the effect of an invisible nuclear disaster on social structures in affected communities in Fukushima prefecture.

2 DISASTER MEMORY STUDIES

The region and communities adjacent to Fukushima Daiichi have "undergone significant social, political, demographic and environmental changes" (ABEYSINGHE et al. 2022: 1). In the broad field of the social sciences, much has been—and continues to be—written about mental and physical health (e. g. LOH / AMIR 2019; MAEDA / OE / SUZUKI 2018), recovery processes (e. g. GERSTER 2019; YAMAKAWA / YAMAMOTO 2017), evacuees (e. g. SHAKUTO 2022), compensation litigation (e. g. FELDMAN 2015; TOGNI 2022, WIEMANN / SAHIN 2021), citizen science (e. g. REIHER 2016; STERNSDORFF-CISTERNA 2019), social movements (e. g. HASEGAWA 2014; WIEMANN 2018), risk communication (KIMURA 2016), social media (e. g. HJORTH / KIM 2011), gender (e. g. HOLDGRÜN / HOLTHUS 2016; KIMURA 2016), nuclear safety (e. g. SAITO 2021), trust (e .g. FELDHOFF 2018), as well as disaster preparedness and risk reduction (e. g. CALLEN-KOVTUNOVA / HOMMA 2022)—to highlight just a few.

These diverse fields have recently been enriched by a developing branch of research concerned with the memory of the multiple disaster, integrating approaches from memory studies and disaster sociology. Remembering and memory is essential in keeping people oriented in time and space and it is essential to individuals as well as collective identities (EICHENBAUM 2017: 1015; EYERMAN 2004: 65). Memory is constructed and differs according to linguistic, social, historical, or national contexts (ERLL 2017: 5). The interdisciplinary field of memory studies struggles with a variety of concepts and definitions, describing "memory-related issues ranging from the neuropsychological to the cultural" (OLICK 2009: 249). Astrid ERLL (2017: 5) thus proposes broadly defining collective memory in terms of "all biological, psychological, medial, and social processes which are important in the mutual influence of past, present and future in cultural contexts."[2]

From a methodological point of view, two approaches to collective memory may be distinguished. One focuses on memory stored in the material world, such as paintings, museums, monuments, literature, symbols, practices etc. This strand of research refers mainly to theorists such as Aby WARBURG (Bildgedächtnis) and Pierre NORA (lieux de mémoire) and may be associated with the concept of "cultural memory" as coined by Aleida ASSMANN (2006: 51–58). The other strand of research, on which I draw for the following analysis, is concerned with memory found in narratives, in communicative interaction. This body of work was particularly influenced by the writings of Maurice HALBWACHS (1992: 38) who emphasizes the fact that individual memory is inseparable from social or collective memory: "[It] is in society that people normally acquire their memories. It is also in society that they recall, recognize, and localize their memories." From there, HALBWACHS establishes the idea of social frames in which individual memory is embedded. Social frames are transmitted to us by our social environment from the day we are born. These cognitive patterns enable individuals to orientate themselves in time and social space, with social frames thus shaping cognitive perceptions and memory. When an individual recalls events from the past, they do so by referring to social frames salient in the present moment of communication. Memory in narrative thus has a function in the present, directed towards an envisaged future (LINDE 2015: 1).

In disaster sociology, disasters are defined in terms of "discrete events, concentrated in time and space, that disrupt the social order and interfere with the ability of a community or society to continue to operate" and are thus essentially social (TIERNEY 2019: 5). However, there is some discussion as to what extent the hazard agent, the extent of physical destruction, or the threat should be viewed as constitutive of disaster (PERRY 2018: 14–16). For the purposes of this paper, I adopt the view that a disaster manifests in patterns of social disruption, which are influenced by characteristics of the social system as well as of the hazard agent itself. It is empirically proven, for example, that disasters arising from armed conflict differ in their social consequences from disasters triggered by natural hazards such as environmental contamination (PERRY 2018: 16). An environmental disaster caused by radioactive particles, for example, has different social implications than the physical destruction caused by earthquake or tsunami. In a study that compares the psychosocial effects of the earthquake and tsunami with those of the nuclear accident, MAEDA / OE / SUZUKI (2018: 52) have shown that in the case of the former, the impact of trauma is rather acute, psychological acceptance relatively easy and community cohesiveness high. In the latter case, by contrast, the impact of trauma is chronic, psychological acceptance difficult, and community cohesiveness low.

Disaster research, on the other hand, seeks to enable better preparation for the future and thus naturally centers on the "lessons learned", i. e. disaster mitigation, risk reduction, community resilience, and recovery processes. Focusing on memories of disaster, however, enhances our understanding of memory, disaster, politics, and trauma as well as how it is they influence one another. This has been demonstrated quite adequately in the work of Bin Xu (2016), which focuses on a variety of cases. There has also been some academic scholarship focusing on Japan with regard to the Great East Japan Earthquake and Tsunami in March 2011 and the ways in which the disaster has found its way into cultural memory, such as through museums or commemorative practices (CREIGHTON 2014; LITTLEJOHN 2021). For instance, in his analysis of metaphorical links between war and disaster, commemoration, and newspaper reportages, Philip SEATON (2016) finds an interweaving of disaster and war memories in the context of Japan. In their analysis of 17 government-established museums, which deal with the 2011 disaster in the Tōhoku region, Julia GERSTER and Elizabeth MALY (2022: 205) confirm that the museums' framings of the disaster in terms of overcoming hardship and building a better future are in line with the framings of memorial museums on World War II. However, the Fukushima nuclear disaster is most often excluded from this presentation, thus contributing to an "othering" of the nuclear catastrophe in cultural memory. In many disaster areas, museums also organize guided tours to disaster remains accompanied by disaster storytellers (*kataribe* 語り部). In most cases—though not exclusively—such tours are given by those who have experienced the disaster and now want to contribute to disaster education (FULCO 2017; MALY / YAMAZAKI 2021; NAGAMATSU / FUKASAWA / KOBAYASHI 2021; SATO / IMAMURA 2021; THANG 2005). Disaster storytellers refer to their past experiences in communicative acts, however, their stories follow well-established patterns that cohere with the master framing of most disaster museums: "natural hazards are unavoidable, but passing on the experiences of the survivors, we can be better prepared in the future" (GERSTER / MALY 2022: 200).

This paper complements this research by exploring the social memory of the Fukushima nuclear disaster from the perspective of farmers in Fukushima Prefecture over the course of the past decade. I consider this group the most vulnerable in terms of the nuclear disaster. Vulnerability may be defined as "the combined effects of exposure to a hazard agent, susceptibility to harm from that exposure, and the ability to cope with or adjust to the effects" (BOLIN / KURTZ 2018: 183). In other words, vulnerability "arises out of differential social relations among groups in a given society" (FORDHAM et al. 2013: 4). The only way for farmers whose fields are exposed to radioactive contamination to escape personal exposure and from dealing with the contamination of their products is to move away. However, in most cases, the farmers and their ancestors lived on the land and had worked hard for many years to maintain fertile grounds; their identity is closely tied to land and community. To break with these practices and relationships would be to risk significant mental health problems. The process of receiving financial compensation for the harm suffered is also frequently a lengthy one, and the amounts given out are comparatively small (FELDMAN 2015; MC-NEILL 2012; WIEMANN / SAHIN 2021).

Following the detection of radioactive contamination of agricultural products in Fukushima Prefecture in March 2011, the Nuclear Emergency Response Headquarters (Genshiryoku saigai taisaku honbu 原子力災害対策本部; NERHQ) under the Prime Minister's Office banned certain products from Fukushima while subjecting them to tests in order to ensure they adhered to new limits for radioactive nuclides in food. Thereby, many farmers lost their crops in 2011. In addition, the government restricted the planting and sowing of rice in 2011 and farmers were asked to spread zeolites[3] and other minerals said to reduce radioactivity on their fields. Moreover, contaminated areas have been subjected to large-scale decontamination, involving measures such as removing top soil or deep plowing of farmlands, washing fruit trees or using potassium-containing fertilizer (BACHEV 2015; MINISTRY OF AGRICULTURE, FORESTRY AND FISHERIES 2022; MINISTRY OF THE ENVIRONMENT 2018). Such measures—designed to ensure food safety—were taken by the government in accordance with a risk communication strategy to reduce "harmful rumors"[4] on the side of consumers while enabling Fukushima farmers to continue their business (RECONSTRUCTION AGENCY 2020; WALRAVENS / O'SHEA / AHRENKIEL 2022). However, this strategy put farmers under great pressure to follow the rules, and to believe established radiation limits and estimations of the possible health risks of long-term exposure to low-dose radioactivity. Consequently, Fukushima farmers who remained on their farmlands after the nuclear disaster came to be particularly dependent on government policies and continued consumer trust.

Providing insight into the disaster memory of farmers in Fukushima over a decade therefore contributes to understanding an invisible and ongoing disaster—from the perspective of a particularly vulnerable group in rural areas of Fukushima Prefecture—along with its changing implications for the lives of individual farmers; the effects of the disaster on social structures; and its social meaning over the course of time.

3 DATA AND METHODOLOGY

For the purposes outlined above, I sampled interviews from the "Voices from Tohoku" (*Tōhoku kara no koe* 東北からの声) oral narrative archive from 2012 and 2013 with Fukushima farmers who were willing to sit a second narrative interview with me online in autumn 2022.[5] The "Voices from Tohoku" archive is administered by David SLATER from Sophia University (Tokyo) who went to the Tōhoku region with groups of student volunteers immediately after the disaster. While volunteering, SLATER and his students realized that many disaster victims felt unheard by the broader public. Beginning in 2012, groups of students conducted narrative interviews with different groups in the Tōhoku area (SLATER / VESELIC 2014). These comprise groups from the tsunami-affected areas in Ishinomaki, Minami-Sanriku, Otsuchi, Sendai, Yuriage, and Ogatsu, as well as antinuclear activists, mothers, Fukushima University students and fishermen. The Fukushima farmers' group consists of 46 interviews. However, many interviewees in this sample did not provide email addresses. Of those I could reach, two agreed to a second interview online. This study thus draws from two cases in-depth: a female farmer from Nihonmatsu city (M) and a male farmer from Fukushima city (Y). Nihonmatsu and Fukushima city are both in the Nakadōri region of Fukushima Prefecture between the coastal Hamadōri region and the mountainous Aizu region. The two farms are situated about 70 and 90 km from the damaged nuclear power plant respectively.

The goal of narrative research is to "generate comprehensive, layered, nuanced understandings of human experience and meaning in context". The key questions to be considered when designing a narrative analysis are: "(1) how a narrative is structured into a story; (2) what function the story serves; (3) what the story contains in terms of contents or substance; and (4) how the story is told or performed" (RAU / COETZEE 2022: 5). The aim of this research on the disaster memory of farmers is two-fold: First, I want to evaluate what the nuclear disaster means to Fukushima farmers, how they understand the disaster, and how this perception has changed over time. Secondly, I want to explore the social frames employed by the farmers and the continuities and differences of these social frames over time. This concerns their references to social structures and how they position themselves in society. Drawing on this, I derive general patterns of social disruption and vulnerability following the nuclear disaster in Fukushima Prefecture from the farmers' perspective.

To answer these two research questions—both of which concern the content of the discussion, its different layers as well as the way the content is talked about—I apply narrative sequential analysis drawing on the model proposed by FISCHER-ROSENTHAL / ROSENTHAL (1997) and ROSENTHAL / FISCHER-ROSENTHAL (2004). These authors suggest first reconstructing biographical data and then sequentially analyzing textual segments from the self-presentation in the narrative interview. They refer to this step as text and thematic field analysis. This is followed by a reconstruction of the case history in terms of "life as lived" which is contrasted with the life story "as narrated" (ROSENTHAL / FISCHER-ROSENTHAL 2004: 261). By paying special attention to social frames (as narrated), this analysis is thus focused on the experienced past (as lived), on the meaning that the narrated past carries for the present moment and, as such, its social meaning.

The two cases in the subsequent sections are presented in accordance with the following steps: I first reconstruct the two farmers' biographies over the past decade and carve out the role played by the nuclear disaster in their lives so far. This is followed by a comparative section looking at the social frames referred to by the farmers over time. The social frames the two farmers employ during their interviews and the way they refer to them exposes a great deal in terms of which aspects of life they feel were most affected by the disaster. Given that as social knowledge, frames and themes are salient in the present moment of communication, those interviewed thus imply characteristics of social structures (PFISTER 2020: 411) such as patterns of vulnerability and resilience in the present when talking about memories of disaster.

4   THE CASE OF M

In 2013, M was 25 years old. She was born into a farmer's family and has two brothers and one sister. In 2010, she began working as an organic farmer on her family's farm. She produced rice, tomatoes, radishes, and seasonal vegetables. Before becoming a farmer, she studied at a sports university in Tokyo. After graduation, however, she reconsidered which job genuinely suited her and decided on agriculture. She had first wanted to work in an agricultural company before deciding to learn the business firsthand by returning to her family's farm. During her first year, she struggled to adjust to her new environment.

On March 11, 2011, M was at her farm in Nihonmatsu city about 70 km from the Fukushima Daiichi nuclear power plant. During the days immediately following the disaster, it was unclear just how much radiation was reaching the city. Like everybody else, M tried to be careful and wore a mask, a hat, and long sleeves when she left the house. When the neighboring town Namie was evacuated, about 3,000 people came to Nihonmatsu whom M helped support by providing food and other provisions. About two weeks after the accident, the first long-distance bus arrived in Nihonmatsu, which M then took as far as Niigata and then Tokyo, where she visited for some ten days with friends who had been worried about her. During these ten days, she felt that she was unable to obtain any real information about the situation in Fukushima, despite the consistent television coverage of the disaster. When she saw the images of the destruction in the coastal areas, she felt that she still had a house and a family. As such, she went back to her family's home despite her fear of the radiation. Since then, she has interacted with radiation by measuring her food and by frequenting the whole-body counter provided by the city administration.

After returning to her hometown, she realized that many people were leaving Fukushima Prefecture and her family was unable to sell their produce. The situation struck her as one of genuine crisis and she was keen for people to come to Fukushima and see it for themselves. She hoped to assure people that local produce was being tested responsibly and thus safe to eat. In 2012, she learned of a program launched by the cabinet, which supported start-ups in the three disaster-hit prefectures (Fukushima, Miyagi, and Iwate). She applied in March 2013 and founded a company organizing farming experience events including overnight stays at farm guesthouses as well as events that facilitated contact with the local population. At the time of the interview in May 2013, she had been running this business for about two months.

In 2022, M was 35 years old. She continued to run her company actively until 2019. When the evacuation orders were lifted in 2017, she moved to the town of Namie where she lived for about two years and supported returning farmers to establish a direct selling point for their produce (*michi no eki* 道の駅). Exhausted and, indeed, disappointed that nuclear energy had not ultimately been phased out, she put her business in Nihonmatsu on hold in 2019 and moved to Miyagi Prefecture where she began working in an unrelated field. She was convinced that she would be unable to gain new perspective if she did not leave Fukushima for a time.

In 2022, M realized that her mental health had suffered significantly in the years since the accident. Astutely aware that she could not bring about significant change unless she was in good mental health, she decided to leave Miyagi Prefecture and spend some time in the U. S. Over a period of three months, she connected with members of the Japanese diaspora and, in particular, with those who had raised their voices for Fukushima. At the time of the interview, she had returned to her family's home and had just decided to dissolve her business completely and start something new. She had a continued desire to make the world a better place and to show that human beings are connected to nature. Moreover, she was considering writing a book on her experiences with the nuclear disaster from a global point of view, i.e. including issues such as nuclear weapons.

4.1 Restoring Connections and Making the Invisible Visible

In 2013, the disaster signified survival for M: survival in a contaminated environment. She was no longer able to sell her produce as she had previously. When the government asked her to treat her fields with zeolite, she did so. She was not happy with the measures but her knowledge of radiation was insufficient, both in terms of how it affected farmland and with regards to safeguarding her own health. As she saw it, consumers had lost faith in produce from Fukushima given the lack of knowledge regarding radiation and its effects. She also claimed that it was generally very unclear which information was trustworthy and that she felt a loss of connection both to those in the prefecture as well as those beyond its borders. Fukushima was, moreover, misrepresented in the media and misunderstood by the majority in Japan. She thus tried to carefully filter all of the information she received and believe only such revelations as came from those she knew she could trust.

Anxious about her health, M took radiation measuring very seriously, as did her parents. She developed strategies to reduce the contamination of vegetables she produced and thus sought to minimize her own exposure. She also frequented the whole-body counter provided by the city administration. Her first check showed levels that were slightly higher than normal but at that juncture, she was not sure how to interpret them. The medical staff at the center explained that the numbers measured were about a quarter of those found in people in Chernobyl. On her third visit to the center, she was completely clear of radioactive cesium.

The scientific data that made the invisible visible, so to speak, allowed her to sell her produce and to invite others to the region in good conscience. She coped with the social disconnection she felt by rebuilding connections and trust. As a business model, she cooperated with the travel company HIS to organize tourist stays and farming experiences in the region, thereby highlighting the many positive aspects of Fukushima and bridging the gap with Tokyo, i.e. the very gap between life in the city and life in nature. In addition, she established connections between farmers and groups of concerned mothers in the area such that trust in local produce could be restored.

4.2 Searching for a New 'Life Axis'

In 2013, M had been speaking as a representative of Fukushima, in 2022 she emphasized that she now spoke as an individual. For the past twelve years, the Fukushima disaster has been a central axis in her life. She felt it was her fate to be in Fukushima at the time of the disaster and it was equally her fate to better the situation and prove that Fukushima had a great deal to offer. Occasionally, she reflected as to what her life would have been in the absence of the disaster or if she had been elsewhere when it occurred. She began to wonder if dealing with the aftermath of Fukushima was not her fate after all:

> Well, when I think that this is now the twelfth year, it feels like it was yesterday; honestly, it really became a central axis in my life. Almost the main axis. It was 2010 that I started working in agriculture and the disaster occurred one year later. So, the Fukushima disaster naturally became the axis of my work. But now, in the end, I can see that it is not my life.[6]

M pointed out that for most people the situation had been normalized over the past decade. New farmers were now arriving and radiation was rarely a topic of discussion. Still, though she could not really talk about it, she still worried about her health. She said that she did not regret the decisions she had made at the time but:

> Among my concerns, my concern for my body was huge. The question was what kind of influence the radiation would have on my body. Of course... I measured. I measured anything I ate, but I was concerned not knowing what would happen to my body. In a way, to cultivate decent produce, I needed to face this concern on a daily basis, but I was worried the whole time.

Referring to the questions she was commonly asked when she was invited to talk about her business, she stated:

> When I was asked: Is everything really ok? I would answer that I was measuring the vegetables and my body for radiation, so I was reducing the risk of radiation exposure (*hibaku*). This is what I said out loud but inside, there was this anxiety that I couldn't express.

After 12 years of restoring trust and rebuilding relationships, M had begun to feel isolated in terms of her personal health concerns. She no longer frequented the whole-body counter, though she did have her thyroid checked regularly. She dealt with the isolation by concentrating on the fact that her life was greater than the disaster alone, by trying to connect to people outside the prefecture and to network on an international level to gain greater perspective. She also wondered how best to express what she had experienced so that others could learn from it. She sought some means of connecting her past with her present to envisage a different future.

5 The case of Y

Y was 35 by the time of the first interview in 2012. He was born and raised in Fukushima city. After his graduation from university in Kanazawa Prefecture where he studied bioresources, he hoped to do something to support his family's fruit farming business and started working at the central fruit market in Yokohama. While there, he wondered how produce from Fukushima could best be sold to Tokyo and other big cities. He inherited a fruit orchard from his father in 2009/10 and has since produced cherries, peaches, and apples. Y had fond childhood memories of enjoying—along with many others—the delicious fruit produced by his father. His determination to continue that tradition was his primary motivation for taking over the family business.

When the earthquake struck on March 11, 2011, he was in his fields. Returning to the house, he saw a lot of damage, the boiler in the bathroom, for example, had collapsed causing a large hole in the ground. During the nights following the disaster, he slept with his family in one of the greenhouses. Sales figures for food from Fukushima then fell drastically. Together with neighboring fruit farmers who were equally uncertain what was to be done next, he helped found an "earth club" to study radiation and how it could best be measured. He and his colleagues immediately began washing their trees and working to determine the levels of contamination in their orchards. At the same time, they were eager to connect more efficiently with other farmers in the prefecture.

In August 2011, his first daughter was born, and childcare kept him and his wife busy. They were worried about radiation, of course, but Y thought that it would be more dangerous if the family separated. He felt that the "balance of the heart" (*kokoro no baransu* 心のバランス) achieved by keeping the family together was most important. However, his wife joined a mother's group and Y was worried that she might want to move away from the prefecture if the opportunity arose.

Y was 45 years old at the time of the second interview in 2022. He continued living in his hometown growing fruits and his "earth club" was still active. His second daughter was born in 2012. In 2015, he published a children's picture book, which tells the story of how peaches are produced. The book was distributed throughout the country and enjoyed by a large audience, Y claimed. The book aimed to make children and adults in Fukushima proud of the fruits from their prefecture, thus motivating them once again to send Fukushima fruits as gifts. By the same token, he hoped the book would persuade young people stay in the prefecture. Over the past ten years, a program launched by the prefecture had allowed him to give occasional talks about Fukushima at middle and high schools all over Japan. Shortly before the interview, he was invited to a middle school in Osaka. The middle school children were about two years old when the disaster happened and were thus quite ignorant of the events at Fukushima. On such occasions, Y attached great importance to speaking from his own perspective, as views on the events varied greatly. He told the students what he and his farmer friends were doing and how they produced safe fruit.

Owing to substantial conflicts among the members, the mothers' group to which his wife was connected no longer existed. She was now active with a farmers' wives group, which organized activities for children. Though it was no longer mandatory, the fruit farmers continued to measure their produce. The city and the prefecture still recommend doing so and it gave them a feeling of confidence when they brought what they had produced to markets. In Y's opinion, this would continue to be necessary well into the future.

5.1 Missing Knowledge, Unclear Regulations and Reaching out to Others

In 2012, the disaster had caused Y's sales to come to a standstill. He and his fellow fruit farmers were at a loss in terms of how to cope with radioactive contamination. Moreover, he complained that regulations concerning the measurement of produce were unclear. In the interview, he alluded to different measuring instruments: One device measured contamination very accurately but was time-consuming and expensive, while another took only 20 to 30 minutes but did not measure levels below 30 Becquerel. If levels were indeed below 30 Becquerel, the result was "ND" (no data). This led to difficulties in communication with potential consumers. Which was "safer": A "no data" result from a cheaper instrument or a 2.0 Becquerel result from an expensive but more accurate instrument? These disparities, among others, led to a loss of trust between the consumer and the producer. However, Y also emphasized the misrepresentation of Fukushima in the media, as a consequence of which, ashamed locals were no longer sending produce as gifts beyond the prefecture. He also observed increased migration away from the province and personally accused the government of ignoring those affected and endlessly replacing the responsible administrative personnel. However, part of the problem was that Tōhoku was not able to raise its voice effectively. This may be attributed—at least to some extent—to the fact that making such demands would rather contravene Tōhoku culture.

Y's reaction to this was to reach out to others and to act together by founding the "earth club". When invited to do so, he would also talk about the situation in Fukushima and continued to promote local fruit. Y continues to seek transparency in his public communication, so as to allow consumers themselves to make an informed decision. Still, he thought that the Japan Agriculture Cooperative (JA) as well as the consumer cooperative were responsible for communicating well with consumers while he did his best to reduce the numbers. Moreover, he wanted the media to show that there are people in Fukushima living a normal life.

5.2 The Younger Generation, Interrupted Traditions, and the Decision to Stay

In 2022, Y expressed his worries about the impact of the disaster on the younger generation. Caring for the younger generation pushed him to continue restoring pride in produce from Fukushima. This applied, in particular, to reestablishing the old tradition of giving fruit as a gift, a tradition that Y claims was interrupted by the disaster and was returning only very slowly. Although one did not hear the term "harmful rumors" anymore, he observed an unconscious ranking of available produce. Items originating in Fukushima were sold more cheaply, despite the lack of any open discourse linking them to radiation. He felt that Fukushima was gradually being forgotten, a phenomenon he viewed with some ambivalence.

In the course of our interview, Y theorized that had he not been in Fukushima at the time of the disaster, he might not have returned. As the case was, however, he felt it was his destiny to care for Fukushima:

> If I hadn't come back [to Fukushima] at that time [just a year before the disaster] ... if, for example, the disaster had happened when I was still in Yokohama, I don't think I would have returned to Fukushima. But things just happened that way, and that's all the more reason to care for Fukushima and to do something positive for it. If I had lived elsewhere, this task may have fallen to somebody else... I am not sure I would have accepted it as my calling (*jibun no koto*).

Moreover, Y was faced with certain difficulties in terms of educating children in Fukushima about what had what happened without thereby stigmatizing them. He continues to worry for his family's health yet admits that ultimately "they are only worries". Still, he was not convinced that they had really done enough to tackle the radiation problem and reflected on the decision to stay in Fukushima:

> In truth, it's a question of whether we ever really faced up to it [the radiation problem]. I really don't know. There was so much information coming in, good and bad. As for the levels of radiation in micro sieverts, we didn't understand them at all, and I was busy on the orchard every day. I was busy, but it could also be that I just fled from the radiation problem. [...] There are many people who really thought about it and left. I think that what they decided was the right thing for them, but I also can't say that it was wrong to stay. So, we stayed. That's the result. I think, there were many 'rights'. None of us was wrong, but it is difficult.

In 2022, Y still coped with the disaster by connecting to other farmers in the area and focusing on community and family. He has reconciled himself with his decision to remain in Fukushima together with his family. He now largely believes that there are no rational grounds for continued health concerns, yet some uncertainty remains. Regarding his feelings about the government, he claims to remain neutral when it comes to nuclear energy and politics in general, as he would not be able to express his opinions without growing angry.

6 Social Frames over Time

In 2012 and 2013, both interviewees pointed out that they had to make decisions based on knowledge they did not have. There was no clear picture regarding either the impact of radiation on their fields or the long-term effects of radiation on their bodies. They each felt insecure given the sense of isolation surrounding their decisions. This insecurity affected many of their relationships, from those within the community to those with customers. Equally, both interviewees strongly identified with their hometown, the very reason for which they each decided to remain. The precarity of the situation, however, has posed a serious threat to community solidarity and stymied efforts to repair and strengthen relationships with peers. These uncertainties also heavily influenced the producer-consumer relationship, a relationship that is, of course, essential for the farmers' survival. The inability to sell their produce had direct consequences for their livelihoods, their family histories, and their communities, not least because it promotes emigration, particularly among younger generations.

Both interviewees also felt misrepresented by the mass media insofar as they were each of the opinion that interviewers generally came with their own agenda and prefabricated narratives, giving little space to the perspectives of those on the ground. As such, they felt instrumentalized for the production of a somewhat myopic perspective that invisibilized the complexity of their experience. The two also referred to children, but in different ways: While M said that it was safe for children to live in Fukushima and sought to find worried mothers whom she could assure that this was the case, Y represses his concerns for fears that his family may separate should the children become worried and decide to move elsewhere. For him, family separation would be worse than long-term exposure to low doses of radiation.

M also referred to the urban-rural divide in Japan and regretted that consumers in urban areas did not show enough appreciation for those who produce their food. It was for this reason that she hoped to make herself useful in terms of re-establishing connections between humans and nature. Y attributed responsibility for declining sales and Fukushima's public image to the Japan Agriculture Cooperative (JA). It was the responsibility of farmers to reduce radiation levels he felt. Generally speaking, he did not believe the government paid much attention to those genuinely affected. M was equally critical of the government.

Date	M	Y
2012/13	decision-making, missing knowledge, no clear information	decision-making, missing knowledge, no clear information
	insecurity concerning health effects	insecurity concerning health effects
	identification with hometown	identification with hometown
	community solidarity	community solidarity
	producer-consumer relationship (harmful rumors)	producer-consumer relationship (harmful rumors, gift culture)
	demographic decline	demographic decline
	national media vs. real stories > misrepresentation	national media vs. real stories > misrepresentation
	save the children from radiation > measuring	save the children vs. save the family
	urban-rural divide	direct sales vs. sales through JA Japan Agriculture cooperative
	connection human-nature	power / responsibility of the JA
		government vs. people in the affected regions

Date	M	Y
2022	destiny	destiny
	individual vs. society	children, next generation, population decline and aging
	silence / taboos (selfcare; *seiji-garami*)	insecurity concerning health effects
	insecurity concerning health effects	identification with hometown, traditions (fruit gift culture)
	consolidated social gap: those who stayed and those who went away	relationship producer-consumer
	save the children (but also the adults)	structural disadvantage, bad "ranking" of Fukushima products
		community solidarity

Tab. 1: Social frames over time

In 2022, both interview partners had accepted the disaster as part of their lives, as their destiny. M's health concerns and the silence/taboo around speaking about it has brought her into conflict with prevailing opinion. While self-care was not generally considered a worthwhile pursuit, M also claimed that it had become difficult for many to voice any real criticism of the government/local administration given their "political involvement" (*seiji-garami* 政治がらみ), i.e. the ways in which they were now profiting from the current system. She also observed a consolidated gap between the people who stayed in Fukushima and those who left.

M expressed concerns that though while there were many activities—such as recuperation camps outside the prefecture—that aimed to address the needs of children, there were no such programs for adults, whose needs she considered equally important. Y, on the other hand, was rather concerned for the next generation, particularly in terms of population decline and the effects that an aging population would have on his children's chances of securing a solid school education. He also worried that the future children of Fukushima would be faced with a general unconscious bias toward the area. His hope was that such children could instead be proud of their hometown, that they could identify with it, and the culture it embodies, including the fruit-giving custom outlined above for example. He felt very strongly that one ought not feel so embarrassed as to refrain from engaging in this practice. As such, he is still invested in improving relationships with potential customers. According to him, if customers, particularly those from Fukushima Prefecture, trust the producer they are more likely to buy Fukushima fruits as seasonal gifts to acquaintances nationwide. He highlighted the fact that he still benefited from the solidarity and support that had been built with other fruit farmers during the disaster, both in terms of radiation-related matters as well as childcare.

7 Conclusion: The Meaning of "Nuclear Disaster"

The foregoing in-depth longitudinal analysis of the two farmers' memory narratives allows us to hypothesize characteristics of the ongoing nuclear disaster. First, for Fukushima farmers, the nuclear disaster is essentially social. The greatest feature of the nuclear disaster is the rupture of relationships, which—in terms of the farmers—most directly applies to those with consumers, the very source of their livelihood. Beyond that, significant strain has been put on their family and community relationships as well as those with the local administration, the big agricultural players, the media and the central government. For M, the disaster also puts into question the relationship between humans and nature, leading her to criticize consumerism and politics on a global scale. Y, by contrast, opts not to voice his opinion when it comes to politics, claiming that the subject would only make him angry. The reasons for the ruptures in these various relationships are health concerns and a lack of knowledge regarding the effects that radiation has on air quality and on natural flora and fauna. Mistrust and isolation have resulted from this general insecurity though disparities in the interpretation of what little knowledge exists, and differences in the level of trust into administrative measures have also contributed. For this reason, both farmers in the sample invested a lot of energy into restoring relationships, principally with consumers though also with colleagues and peers.

Second, over the course of the past decade, the meaning of the disaster has shifted from the social to the private, which in turn feeds back into the social and political. Many social ruptures have been consolidated: some consumers could be regained, while others were lost and an "unconscious negative ranking" of produce from Fukushima lingers on. Certain people have severed ties with their former communities by moving away, while those who remained have established new group-belongings. The media no longer reports on the situation in Fukushima and for the government, the "Fukushima problem" has almost been solved. On the other hand, both farmers indicated ongoing anxiety concerning possible health effects. M decided to dissolve her business in Fukushima, while for his part Y commented in our most recent interview that he now acknowledges health concerns he previously ignored. However, these fears are generally kept private, while expressing anger towards the government has almost become a taboo. Many individuals, as such, keep a veneer of neutrality in the public sphere, which has the effect of dissociating them from society and politics in general.

It goes without saying that a nuclear disaster leaves a deep impression on affected individuals and communities while taking a heavy toll on all kinds of relationships. The continued radioactive threat requires affected people to suppress worries or to re-evaluate past decisions and thus even though normality may be restored on the surface, there is a fundamental distinction between the aftermath of nuclear incidents as compared with other types of disaster. Farmers in Fukushima Prefecture are particularly vulnerable as their livelihoods depend on nature and they are reliant on government regulations and image campaigns. Decisions on staying or leaving, worries concerning unexpected effects of low-dose radiation on health, and stigmatization were and are equally shared by other occupational groups and through all social strata in the area. Memory narratives, when self-affirming, can support community recovery after large-scale disasters (MADSEN / O'MULLAN 2013: 66). Still, within Fukushima Prefecture expressing worries about ongoing radioactive contamination appears to be morally constrained (NISHIZAKA 2017: 636). A social memory narrative including such worries and their changes over time may be difficult to find.

The in-depth longitudinal study of two Fukushima farmers' life experiences has shown that disaster memories and therefore the meaning of a nuclear disaster changes over time. Whereas in the direct aftermath, surviving by restoring relationships was the main concern, a decade later, individual concerns disturb individual relations to the community to some degree. Nevertheless, the study leaves room for additional research particularly regarding the question as to which overarching disaster memory narratives and frames influence farmers' perceptions and how other social and age groups in the area experience and narratively frame the disaster over time.

NOTES

1 According to the World Nuclear Association (2023), Japan has 33 operable nuclear power reactors, of which ten are currently in operation in January 2023. After the disaster in 2011, the first two reactors were restarted in August and October 2015 (Sendai 1 and 2). The other eight reactors were restarted consecutively (Takahama 3 and Ikata 3 in 2016, Takahama 4 in 2017, Genkai 3, Genkai 4, Ohi 3, Ohi 4 in 2018, and Mihama 3 in 2021). Before 2011, Japan had 54 operating nuclear reactors.
2 Translation by the author.
3 Because of their highly absorbing characteristics, minerals of the zeolite family are used to absorb radioactive cesium and strontium, to remove ammonia from waste and drinking waters, as filters in kidney-dialysis units and more besides (MUMPTON 1999).
4 After the Fukushima nuclear accident, harmful rumors or *fūhyō higai* 風評被害 became a term and concept to diminish many voices, particularly those of mothers, who were worried about the potential health risks of radioactive contamination in food. These mothers were construed as irrational, hysterical "dangerous fearmongers" in the sense that their scientifically unjustified fear threatened the livelihoods of people in the affected areas (KIMURA 2016: 28).

5 Due to travel restrictions to Japan caused by the Covid-19 pandemic, personal travel to Japan was impossible.
6 All citations have been translated by the author.

REFERENCES

ABEYSINGHE, Sudeepa et al. (2022): "The Reconstruction of Community and Wellbeing in Fukushima—Situating the Case in the Field". In: ABEYSIGNGHE, Sudeepa et al. (eds.): *Health, Wellbeing and Community Recovery in Fukushima*. London: Routledge, 1–16.

ARNHOLD, Valerie (2019): "L'apocalypse ordinaire. La normalisation de l'accident de Fukushima par les organisations de sécurité nucléaire". In: *Sociologie du travail* 61(1); https://journals.openedition.org/sdt/14611 (last access 2023/02/14).

ASSMANN, Aleida (2006): *Der lange Schatten der Vergangenheit: Erinnerungskultur und Geschichtspolitik*. München: C. H. Beck.

BACHEV, Hrabrin (2015): "March 2011 Earthquake, Tsunami and Fukushima Accident Impacts on Japanese Agri-Food Sector". In: *Munich Personal RePEc Archive (61499)*; https://mpra.ub.uni-muenchen.de/61499/ (last access 2023/02/03).

BENSAUDE-VINCENT, Bernadette / BOUDIA, Soraya / SATO, Kyoko (2022): "Introduction: Shaping the Nuclear Order". In: BENSAUDE-VINCENT, Bernadette / BOUDIA, Soraya / SATO, Kyoko (eds.): *Living in a nuclear world: From Fukushima to Hiroshima*. London, New York: Routledge, 1–19.

BOLIN, Bob / KURTZ, Liza C. (2018): "Race, Class, Ethnicity, and Disaster Vulnerability". In: RODRIGUEZ, Havidan / DONNER, William / TRAINOR, Joseph E. (eds.): *Handbook of Disaster Research*. Cham: Springer, 181–203.

CALLEN-KOVTUNOVA, Jessica / HOMMA, Toshimitsu (2022): "Ten years since the Fukushima Daiichi NPP disaster: What's important when protecting the population from a multifaceted technological disaster". In: *International Journal of Disaster Risk Reduction* 70.

CREIGHTON, Millie (2014): "Wasuren—We won't Forget! The Work of Remembering and Commemorating Japan's and Tohoku's 2011 (3.11) Triple Disasters in Local Cities and Communities". In: *Journal of Global Initiatives* 9 (1): 97–120.

EICHENBAUM, Howard (2017): "On the Integration of Space, Time, and Memory". In: *Neuron* 95 (5): 1007–18.

ERLL, Astrid (2017): *Kollektives Gedächtnis und Erinnerungskulturen: Eine Einführung*. Stuttgart: J. B. Metzler.

EYERMAN, Ron (2004): "Cultural Trauma: Slavery and the Formation of African American Identity". In: ALEXANDER, Jeffrey et al. (eds.): *Cultural Trauma and Collective Identity*. Berkeley: University of California Press, 60–111.

FELDHOFF, Thomas (2018): "Visual Representations of Radiation Risk and the Question of Public (Mis-)Trust in Post-Fukushima Japan". In: *Societies* 8 (2): 32; https://www.mdpi.com/2075-4698/8/2/32 (last access 2023 / 02 / 14).

FELDMAN, Eric A. (2015): "Compensating the Victims of Japan's 3-11 Fukushima Disaster". In: *Asian-Pacific Law and Policy Journal* 16 (2): 127–57.

FISCHER-ROSENTHAL, Wolfram / ROSENTHAL, Gabriele (1997): "Narrationsanalyse biographischer Selbstrepräsentation". In: HITZLER, Ronald / HONER, Anne (eds.): *Sozialwissenschaftliche Hermeneutik: Eine Einführung*. Wiesbaden: Springer, 133–64.

FORDHAM, Maureen et al. (2013): "Understanding Social Vulnerability". In: THOMAS, Deborah S. K. et al. (eds.): *Social Vulnerability to Disasters*. Boca Raton: CRC Press, 1–29.

FULCO, Flavia (2017): "Kataribe: A Keyword to Recovery: Practice of Storytelling in Post-Disaster Japan". In: *Japan Insights*; https://www.japan-insights.jp/pdf/essays/JIN_Kataribe_01.pdf (last access 2023 / 01 / 20).

GERSTER, Julia (2019): "Hierarchies of affectedness: Kizuna, perceptions of loss, and social dynamics in post-3.11 Japan". In: *International Journal of Disaster Risk Reduction* 41; https://www.sciencedirect.com/science/article/abs/pii/S2212420919309586?via%3Dihub (last access 2023 / 02 / 28).

GERSTER, Julia / MALY, Elizabeth (2022): "Japan's Disaster Memorial Museums and framing 3.11: Othering the Fukushima Daiichi nuclear disaster in cultural memory". In: *Contemporary Japan* 34 (2): 187–209.

HALBWACHS, Maurice (1992): *On Collective Memory*. Chicago: University of Chicago Press.

HASEGAWA, Koichi (2014): "The Fukushima nuclear accident and Japan's civil society: Context, reactions, and policy impacts". In: *International Sociology* 29 (4): 283–301.

HJORTH, Larissa / KIM, Kyoung-hwa Y. (2011): "The Mourning After". In: *Television & New Media* 12 (6): 552–59.

HOLDGRÜN, Phoebe / HOLTHUS, Barbara (2016): "Babysteps towards Advocacy. Mothers against Radiation". In: MULLINS, Mark R. / NAKANO, Koichi (eds.): *Disasters and Social Crisis in Japan. Political, Religious, and Sociocultural Responses*. New York: Palgrave Macmillan, 239–66.

KIMURA, Aya H. (2016): *Radiation Brain Moms and Citizen Scientists: The Gender Politics of Food Contamination After Fukushima*. Durham: Duke University Press.

LINDE, Charlotte (2015): "Memory in Narrative". In: TRACY, Karen / ILLIE, Cornelia / SANDEL, Todd (eds.): *The International Encyclopedia of Language and Social Interaction*; https://onlinelibrary.wiley.com/doi/10.1002/9781118611463.wbielsi121 (last access 2023 / 01 / 20).

LITTLEJOHN, Andrew (2021): "Museums of themselves: disaster, heritage, and disaster heritage in Tohoku". In: *Japan Forum* 33 (4): 476–96.

LOH, Shi L. / AMIR, Sulfikar (2019): "Healing Fukushima: Radiation Hazards and Disaster Medicine in Post-3.11 Japan". In: *Social Studies of Science* 49 (3): 333–54.

MADSEN, Wendy / O'MULLAN, Cathy (2013): "Responding to Disaster: Applying the Lens of Social Memory". In: *Australian Journal of Communication* 40 (1): 57–70.

MAEDA, Masaharu / OE, Misari / SUZUKI, Yuriko (2018): "Psychosocial effects of the Fukushima disaster and current tasks: Differences between natural and nuclear disasters". In: *Journal of the National Institute of Public Health* 67 (1): 50–58.

MALY, Elizabeth / YAMAZAKI, Mariko (2021): "Disaster Museums in Japan: Telling the Stories of Disasters Before and After 3.11". In: *Journal of Disaster Research* 16 (2): 146–56.

MCCAULEY, D. J (2021): "Farming in Fukushima One Decade after Nuclear Disaster". In: *CSA News* 66 (3): 14–22.

MCNEILL, David (2012): "The Fukushima Nuclear Crisis and the Fight for Compensation". In: *The Asia-Pacific Journal: Japan Focus* 10 (6); https://apjjf.org/-David-McNeill/3707/article.pdf (last access 2023 / 01 / 31).

MINISTRY OF AGRICULTURE, FORESTRY AND FISHERIES (2022): "Measures for Reduction of Radionuclide Contamination of Agricultural Produce"; https://www.maff.go.jp/e/policies/food_safety/emer/attach/pdf/202209_slide.pdf (last access 2023 / 02 / 03).

MINISTRY OF THE ENVIRONMENT (2018): "Environmental Remediation in Affected Areas in Japan"; http://josen.env.go.jp/en/pdf/environmental_remediation_1812.pdf (last access 2023 / 02 / 03).

MULLINS, Mark R. / NAKANO, Koichi (eds.) (2016): *Disasters and Social Crisis in Japan. Political, Religious, and Sociocultural Responses*. New York: Palgrave Macmillan.

NAGAMATSU, Shingo / FUKASAWA, Yoshinobu / KOBAYASHI, Ikuo (2021): "Why Does Disaster Storytelling Matter for a Resilient Society?". In: *Journal of Disaster Research* 16 (2): 127–34.

NISHIZAKA, Aug (2017): "The moral construction of worry about radiation exposure: Emotion, knowledge, and tests". In: *Discourse & Society* 28 (6): 635–56.

NUCLEAR REGULATION AUTHORITY (2022): "Results of Airborne Monitoring in Fukushima Prefecture and neighboring prefectures and the Fourteenth Airborne Monitoring in the 80km zone from the Fukushima Daiichi NPP"; https://radioactivity.nsr.go.jp/en/contents/16000/15474/24/2021_16th%20Airborne%20monitoring%20press_english.pdf (last access 2023 / 01 / 04).

OECD (2021): *Fukushima Daiichi Nuclear Power Plant Accident, Ten Years On*; https://read.oecd-ilibrary.org/energy/fukushima-daiichi-nuclear-power-plant-accident-ten-years-on_124c2774-en#page1 (last access 2023 / 01 / 04).

OLICK, Jeffrey (2009): "Between Chaos and Diversity: Is Social Memory Studies a Field?". In: *International Journal of Politics, Culture, and Society* 22 (2): 249–52.

PERRY, Ronald W. (2018): "Defining Disaster: An Evolving Concept". In: RODRIGUEZ, Havidan / DONNER, William / TRAINOR, Joseph E. (eds.): *Handbook of Disaster Research*. Cham: Springer, 3–22.

PFISTER, Sandra Maria (2020): "Deutungsmuster des Katastrophischen". In: HEINLEIN, Michael / DIMBATH, Oliver (eds.): *Katastrophen zwischen sozialem Erinnern und Vergessen: Zur Theorie und Empirie sozialer Katastrophengedächtnisse*. Wiesbaden: Springer Fachmedien Wiesbaden, 405–29.

RAU, Asta / COETZEE, Jan K. (2022): "Designing for Narratives and Stories". In: FLICK, Uwe (ed.): *The SAGE Handbook of Qualitative Research Design*. Los Angeles: Sage, 1–25.

RECONSTRUCTION AGENCY (2020): "Eliminating Negative Reputation Impact: Reconstruction from Nuclear Disaster and the History of Safety and Revitalization in Fukushima"; https://www.reconstruction.go.jp/topics/main-cat1/sub-cat1-4/fuhyou/pamphlet/latest/huhyou-higai-husshoku_E.pdf (last access 2023 / 02 / 03).

RECONSTRUCTION AGENCY (2022): "Fukushima Updates"; https://fukushima-updates.reconstruction.go.jp/faq/fk_040.html# (last access 2023 / 01 / 03).

REIHER, Cornelia (2016): "Lay people and experts in citizen science: Monitoring radioactively contaminated food in post-Fukushima Japan". In: *ASIEN* (140); http://asien.asienforschung.de/wp-content/uploads/sites/6/2017/01/140_RA_Reiher.pdf (last access 2023 / 02 / 03).

ROSENTHAL, Gabriele / FISCHER-ROSENTHAL, Wolfram (2004): "The Analysis of Narrative-Biographical Interviews". In: FLICK, Uwe et al. (eds.): *A Companion to Qualitative Research*. Translated by Bryan JENNER. London: SAGE Publications, 259–65.

SAITO, Hiro (2021): "The Sacred and Profane of Japan's Nuclear Safety Myth: On the Cultural Logic of Framing and Overflowing". In: *Cultural Sociology* 15 (4): 486–508.

Sato, Shosuke / Imamura, Fumihiko (2021): "Evaluation of Listeners Reaction on the Storytelling of Disaster Response Experience: The case of service continuity at Miyagi Prefectural office after experiencing the Great East Japan Earthquake". In: *Journal of Disaster Research* 16 (2): 263–73.

Seaton, Philip (2016): "Japanese War Memories and Commemoration after the Great East Japan Earthquake". In: Tota, Anna Lisa / Hagen, Trever (eds.): *Routledge International Handbook of Memory Studies*. New York: Routledge, 345–56.

Shakuto, Shiori (2022): "'Radiation Refugees': The Role of Gender and Digital Communication in Japanese Women's Transnational Evacuation after Fukushima". In: *Journal of Immigrant & Refugee Studies* 20 (2): 177–89.

Slater, David H. / Veselic, Maja (2014): "Voices from Tohoku. 'Public' Research, New Media Practices and the 'Archive of Hope'". In: *5: Designing Media Ecology* 1 (1): 28–41.

Sternsdorff-Cisterna, Nicolas (2019): *Food Safety After Fukushima: Scientific Citizenship and the Politics of Risk*. Honolulu: University of Hawaiʻi Press.

Thang, Leng Leng (2005): "Preserving the Memories of Terror: Kobe Earthquake Survivors as 'Memory Volunteers'". In: Tsun, Yun Hui / van Bremen, Jan / Ben-Ari, Eyal (eds.): *Perspectives on Social Memory in Japan*. Folkstone: Global Oriental, 191–203.

Tierney, Kathleen J. (2019): *Disasters: A Sociological Approach*. Cambridge, UK: Polity.

Togni, Giulia de (2022): *Fall-Out from Fukushima: Nuclear Evacuees Seeking Compensation and Legal Protection After the Triple Meltdown*. Milton: Taylor and Francis.

Walravens, Tine / O'Shea, Paul / Ahrenkiel, Nicolai (2022): "'Let's eat Fukushima': Communicating risk and restoring 'safe food' after the Fukushima disaster (2011–2020)". In: *Japan Forum* 34 (1): 79–102.

Wiemann, Anna (2018): *Networks and Mobilization Processes: The Case of the Japanese Anti-Nuclear Movement after Fukushima*. München: Iudicium.

Wiemann, Anna / Sahin, Köksal (2021): "Betroffenenbewegungen und das Recht: 'Heimatverlust' als Schaden in der Nariwai-Sammelklage gegen TEPCO und den Staat". In: *Japan Jahrbuch*, 97–118.

Xu, Bin (2016): "Disaster, Trauma, and Memory". In: Tota, Anna Lisa / Hagen, Trever (eds.): *Routledge International Handbook of Memory Studies*. New York: Routledge, 357–70.

Yamakawa, Mitsuo / Yamamoto, Daisaku (2017): *Rebuilding Fukushima*. London, New York: Routledge.

Transcending the Nuclear Fallacy: Japan's Atomic Legacy as Thematized in Hayashi Kyōko's Late Work
—In Lieu of an Epilogue—

Stephan Köhn

> It was around ten thirty or ten forty when it happened. Lying stretched on the floor, Shiro suddenly pricked up his ears and stood up. He let out a howl, as if to say "Master!" and then started to growl. There wasn't a soul on the street. 'Maybe he can hear distant footsteps' thought the father and looked out toward the street. There wasn't a sign of life. "Don't be a nuisance!" said the son and gave Shiro a slap on the nose. Shiro howled in a high-pitched voice. He steadied his legs as though to attack [an enemy]. He seemed scared. And yet, there was no sign of anyone that Shiro could had been howling at. The son and father both looked up to the sky while placing back the sweet potato tendrils that had already reached the street to the edges in the field. They looked, of course, because one can [immediately] see in the sky and the sun if a disaster on earth is occurring. But the sky was as blue as ever [...].
> "Isn't that a siren?" said the son, straining his ears. They both glanced in the direction of the fence. By all appearances, the sound of the siren came from beyond that fence. The tops of the pine trees, taller than the fence already, were rustling heavily. "There is no doubt, father, that is a siren", said the son as he bundled the tendrils so the potatoes would be easy to dig out and harvest. "Is it really an alarm?" wondered the father. The two looked toward the nuclear power plant near the coast. The sky [...] was silent. (HAYASHI 2005a: 267)

HAYASHI Kyōko's 林京子 (1930–2017) short novel "Harvest" (*Shūkaku* 収穫) was first published in the literary magazine *Gunzō* 群像 in 2002 and clearly demonstrates the effect that an "accident" at a nuclear power plant—built in the direct vicinity of a small town near the coast—can have on the everyday life of the inhabitants. The story centers upon the daily routine of a sweet potato farmer by the name of Yamada. Preparing for the harvest, as his ancestors had done for generations, Yamada hears the sudden sound of a siren. Inhabitants within a 300-meter radius of the power plant are urged to leave their homes and find shelter in a designated refuge. But the old potato farmer stays put. He and his son manage to harvest the potatoes but are subsequently kept in the dark about the extent of soil contamination, while the vegetables, ultimately, are left to rot.

One could be forgiven for thinking that this seemingly inconspicuous short novel was based on the "criticality accident" (*rinkai jiko* 臨界事故) at the Tōkaimura Nuclear Fuel Processing Plant (Ibaraki Prefecture) in September 1999. As a result of the "accident", several hundred people were exposed to radiation with three workers exposed to doses as high as those caused by the Chernobyl meltdown in 1986. The accident caused two fatalities and was considered the most severe nuclear accident in postwar Japan, at least until March 11, 2011.

HAYASHI visited the area for herself in December 1999 and spoke with inhabitants about their personal experiences and the sequence of events (HAYASHI / SHIMAMURA 2011: 49–51). "Harvest" comprises first-hand and second-hand impressions that HAYASHI gathered during her visit to Tōkaimura, whereby her literary focus is on the figure of "the old farmer". Modelled after one of HAYASHI's interviewees, the farmer is both the protagonist and the narrator of the story. In this sense, "Harvest" is a remarkable literary account on the events of September 1999. At the same time, the novel is also a timeless tale of man-made disasters (*jinsai* 人災) in Japan. The old farmer epitomizes the helplessness of everyday people suddenly confronted with the supposedly "unforeseeable" (*sōteigai* 想定外). This is demonstrated by: his 'inability' to grasp the full meaning of the "criticality accident" described by a television newscaster; his 'unwillingness' to leave behind his home (and his crop) to take refuge in a shelter; and his 'naivety' regarding the various risks of radioactive pollution empathically depicted in HAYASHI Kyōko's novel. Reading "Harvest" more than two decades after the book was first published, readers may well be puzzled by what KAWAMURA Minato (2013: 94) has referred to as its almost prophetic power. Indeed, the power plant officials in the novel try to convince the old farmer that low doses of radiation are relatively harmless for the human body and one cannot help but think of the barely believable discourse employed in official statements after March 11, 2011. In one further scene, an official performs a public ritual of ostentatiously consuming sweet potatoes from the disaster-stricken region in a bid to prove the harmlessness of low doses of radiation. As the older farmer looks on, commenting dryly "Oh, that poor guy!" one is inevitably reminded of the absurdity of analogous scenes following the Fukushima disaster, such as that of former Prime Minister, KAN Naoto 菅直人, filmed by the press as he ate cucumbers and strawberries from the Fukushima region on April 15, 2011. Fundamentally, HAYASHI's story could have been set anywhere and at any time, from Tōkaimura to Fukushima or even Chernobyl. When the author visited Tōkaimura two months after the accident, public interest had already waned significantly. HAYASHI later summarized her impressions as follows: "It really is a pity. We regular people soon acquiesce to [the situation] instead of questioning [the use of nuclear energy] (HAYASHI / SHIMAMURA 2011: 51).

"Harvest" marks a turning point in HAYASHI Kyōko's literary career. It is the first fictional work in which she explicitly thematizes the danger of the supposedly "good" side of nuclear energy, a danger epitomized by the nuclear accident at the Tōkaimura Nuclear Fuel Processing Plant. As KUROKO Kazuo (2007: 183) has pointed out, not only is "Harvest" one of the very few examples of literary works on nuclear energy and nuclear power plants to be written in the postwar period. It is, first and foremost, a fictionalized report based on personal research carried out by the author on site. It is worth remarking that HAYASHI's work was not subject to any form of criticism as her depiction remains quite faithful to the facts. Indeed, even the name, "Tōkaimura", remains unmentioned in the entire text. Nonetheless, HAYASHI's short novel creates a dense atmosphere whereby the invisible fear of radiation is almost omnipresent (WATANABE 2005: 152–58). In this sense, the work should be considered a milestone as it articulates the problem of exposure to radiation against the backdrop of Japan's "safety myth" as regards clean nuclear energy (*anzen shinwa* 安全神話).

Born in Nagasaki and raised in Shanghai, HAYASHI Kyōko repatriated to Nagasaki in February 1945. She personally experienced the atomic bombing of Nagasaki on August 9, 1945, where—along with many other female students her age—she was on duty in the Mitsubishi weapons factory. In contrast to ŌTA Yōko 大田洋子 (1903–63), who began writing about her first-hand experiences just a few weeks after the atomic bombing of Hiroshima on August 6, 1945, HAYASHI Kyōko waited almost 20 years before she finally began work on her own material on the matter. Her first piece on the aftermath of August 9—"A Flash in Summer" (*Senkō no natsu* 閃光の夏)—was published in 1964 in the magazine "Capital City of Literature" (*Bungei shuto* 文藝首都). The piece did not yet address the day of the bombing itself, however. Instead, the short novel thematized a survivor's struggle to obtain national recognition as an atomic bomb victim (*hibakusha* 被爆者) as well as access to medical treatment in postwar Nagasaki. Ultimately, it took another ten years for HAYASHI Kyōko to publish her 1975 novel, "Festival Ground" (*Matsuri no ba* 祭りの場), which was her first literary eyewitness account of August 9 and the aftermath in *Gunzō*.

It would be no exaggeration to say that literature from Nagasaki on the atomic bombings was rather scarce compared to literary productions from Hiroshima, or as John Whittier TREAT puts it: "The canon (if we may speak of one) of atomic-bomb literature is overwhelmingly dominated by Hiroshima writers and Hiroshima works. Nagasaki atomic-bomb literature, for whatever reasons […] is quantitatively smaller and critically less favored" (1995: 302). According to SHIJŌ Chie (2015: 43–51), the predominance in public discourse of physician and author, NAGAI Takashi 永井隆 (1908–51), impeded the emergence of other literary voices from Nagasaki, even after his death. A devout Catholic, NAGAI managed to reformulate—and thus to distort—the narrative surrounding the atomic bomb such that it would be understood as a blessing for humankind. He equated the devastation of Nagasaki with the Old Testament "burnt offering" (*hansai* 燔祭), by which interpretation the survivors of the bomb were no longer victims but the blessed subjects of divine providence. Needless to say, NAGAI's interpretation of the bomb as divine blessing was welcomed with open arms by both the Allied Powers and the Japanese Emperor, the latter of whom this interpretation implicitly exculpated of any wrongdoing. So positive was NAGAI's reputation that he received a visit from the Emperor during the latter's May 1949 visit to Nagasaki (YAMADA 2001: 44–46). NAGAI was the most prominent voice from Nagasaki and enjoyed a sacrosanct status in Japanese society that rendered dissenting voices with conflicting memories, more or less, speechless.

When HAYASHI Kyōko published "Festival Ground" in 1975, NAGAI Takashi's spell had, of course, already been broken, but the Japanese public had essentially lost interest in so-called atomic bomb literature (*genbaku bungaku* 原爆文学) whether from Hiroshima or Nagasaki (NAKANO / NAGAOKA 1985: 20). The situation changed dramatically, however, when HAYASHI's work was suddenly awarded both the 18th Gunzō Newcomer Prize (Gunzō shinjin bungaku shō 群像新人文学賞) as well as the most renowned literary prize in Japan, the 73rd Akutagawa Prize (Akutagawa Ryūnosuke shō 芥川龍之介賞). HAYASHI's moment of fame arrived as she became the most prominent voice from Nagasaki and, by the same token, the target of significant literary criticism.

HAYASHI's "Festival Ground" became a challenge for both literary critics and readers. The ambivalence with which her literature was generally received can be seen in the comments of the selection committee members for the Akutagawa Prize. While the "subject matter" was praised by most of the committee's nine members, the "literariness" received less favorable scrutiny. ŌOKA Shōhei 大岡昇平, for example, criticized the clumsiness and vagueness of HAYASHI's phrases and expressions, while NAKAMURA Mitsuo 中村光夫 pointed to the immaturity of her narrative techniques, and TAKII Kōsaku 瀧井孝作 referred to the incomprehensibility of several passages in her work. YASUOKA Shōtarō 安岡章太郎 aptly summarized the committee member's ambivalent impressions as follows: "I was deeply moved by the historical facts of this novel, but not by its literary realization" ("Akutagawashō senpyō" 1975: 348). To a certain extent, HAYASHI Kyōko could be said to have shared the same fate as ŌTA Yōko some 30 years prior, when she was equally criticized for lacking literariness. Yet, they differ in one fundamental respect, namely that "Festival Ground" became the first work of atomic bomb literature that was embraced by the literary establishment and officially recognized as a worthy piece of sophisticated literature (*junbungaku* 純文学).

In the years that followed, HAYASHI Kyōko published several novels and essays in which she thematized both her experiences as an eyewitness of August 9, 1945, as well as the aftermath of the bombing. She equally focused on: her crisis-ridden life as an atomic bomb survivor in postwar Japan; her childhood in colonized China during the war; and her ambivalent encounter with the United States. HAYASHI's readers considered her an uncomfortable and unrelenting voice. Her identity as an 'authentic' survivor of August 9 significantly affected her writing on the atomic bombing and the aftermath. It also strongly affected how she thought about her readership. She felt the urge to become a storyteller of the bombing (*genbaku no kataribe* 原爆の語り部), to give, for posterity, testimony on behalf of all the other victims. Yet at the same time—as John Whittier TREAT has fittingly remarked—"Hayashi, like many hibakusha, [did] not easily believe in our ability to understand what we are asked to listen to" (TREAT 1995: 318). She undoubtedly felt the limits of comprehensibility for an experience that could essentially not be adequately represented to non-victim (*hi-hibakusha* 非被爆者) readers. For HAYASHI Kyōko, there was no form of articulation in existence that could sufficiently express the unspeakable events of August 9. Wakako, the protagonist of HAYASHI's short novel "Two Grave Markers" (*Futari no bohyō* 二人の墓標) and most likely an alter ego of the author herself, puts it very well: "Only someone who was at that very place at that very time can really understand" (HAYASHI 1983: 74). Just as Wakako feels alienated from the unharmed inhabitants of her home village to which she returns after her traumatic escape from the devastated city of "N", HAYASHI felt estranged from large sections of postwar Japan.

In some regards, HAYASHI Kyōko was radical and uncompromising in her assertion that only real victims could legitimately bear witness to that day. The author was once asked during a roundtable discussion if she had ever read IBUSE Masuji's 井伏鱒二 (1898–1993) famous novel "Black Rain" (*Kuroi ame* 黒い雨, 1965–66), which focusses on the atomic bombing of Hiroshima. HAYASHI answered that she had not: "I had the feeling that if I had read this work, my mind would have been disturbed and distracted" (INOUE / KOMORI 2004: 26). This appears to be generally indicative of HAYASHI Kyōko's attitude toward accounts of August 1945 from *hi-hibakusha* writers, including IBUSE Masuji whose "Black Rain"—from which extracts are included in many textbooks for Japanese language education (*kokugo* 国語)—is probably the most widely read novel on the atomic bombings in Japan.

Needless to say, HAYASHI's monopolization of August 9 displeased some of her colleagues and literary critics. Her colleague NAKAGAMI Kenji 中上健次 (1946–92) became one of her fiercest critics. In a 1982 "joint review" (*sōsaku gappyō* 創作合評), published in the magazine *Gunzō*, NAKAGAMI attacked HAYASHI Kyōko as an "atomic bomb fascist" (*genbaku fashisuto* 原爆ファシスト). However, this was less for her alleged cynicism by way of which "she discourages readers from sympathizing with her"— as John Whittier TREAT (1995: 327) put it—but for asserting herself as 'the' authentic voice of Nagasaki. According to NAKAGAMI, HAYASHI Kyōko capitalized on her experiences as a *hibakusha* by publishing one uncritical eyewitness account after the other in the naive belief that such accounts would automatically be considered literary (KARATANI / NAKAGAMI / KAWAMURA 1982: 288). For NAKAGAMI, HAYASHI disqualified herself as an author because her "literature" lacked any form of self-reflexivity regarding the historical context of the atomic bombing. Indeed, she did not seem capable of prescinding, NAKAGAMI claimed, from her own personal history as a victim and refused to critically reflect on the course of events that eventually led to the dropping of the atomic bomb on Nagasaki (KARATANI / NAKAGAMI / KAWAMURA 1982: 290). NAKAGAMI's criticism is remarkable insofar as it questions not only the literariness of HAYASHI's personal accounts but problematizes, first and foremost, the lack of historical awareness in HAYASHI's writing. As NAKAGAMI sees it, the duty of atomic bomb literature is to establish a counter-narrative to the predominant national history narrative whereby Japan's victimhood alone is the point of focus. Those authors that do not question this national narrative, on the other hand, are little more than collaborators. Indeed, NAKAGAMI goes so far as to call them fascists. It remains difficult to say if HAYASHI Kyōko was unaware of the resentment that resulted from her rigid monopolization of August 9, or if she was simply indifferent to it. Be that as it may, she continued to think of August 9 as a personal experience dislocated from any specific time or place. HAYASHI's "Nagasaki" begins on August 9, 1945 as a dehistoricized and depoliticized abstract space.

A fundamental change in HAYASHI Kyōko's writing about August 9, 1945, occurred in 1999, however, when she decided to visit the Trinity test site in New Mexico where the first atomic bomb had been detonated on July 16, 1945. HAYASHI had been strangely fascinated by the location for some time already, given that the same kind of bomb was used in New Mexico for the nuclear test as was dropped on Nagasaki, namely a plutonium bomb. HAYASHI describes her visit to the test site as "a way for me to finally escape from August 9. I wanted, at all costs, to put an end to my own August 9 after having made a short trip to my 'birthplace', the place where all exposure to radiation had actually started" (HAYASHI / SHIMAMURA 2011: 36). In 2000, HAYASHI returned to Japan and published her impressions of the site in the magazine *Gunzō*, with a piece titled "From Trinity to Trinity" (*Toriniti kara Toriniti e* トリニティからトリニティへ). Her trip signaled the end of her former perspective on August 9. It was an eye-opener in terms of relativizing personal experiences and reconceptualizing the atomic bombing of Nagasaki in a global historical context.

Let us turn our attention now to HAYASHI Kyōko's autobiographical semi-documentary about her encounter with humankind's dystopic nuclear past and future. It begins with protagonist, "I", embarking on a trip to Los Alamos together with her *hi-hibakusha* friend Tsukiko, who was born in Nagasaki but has been living in Texas for more than 40 years. The test site opens its doors to visitors but twice a year, specifically on the first Saturday of April and October respectively, and the two thus took their opportunity on October 2, 1999. On her way to Los Alamos, "I" and her friend make a brief stop at the National Atomic Museum, which today is the National Museum of Nuclear Science & History in Albuquerque. The exhibits and audio guides force "I" to confront both her past and her present: "Here, I was entirely a *hibakusha* once again. Prior to visiting the museum, I had not felt like a Japanese woman or like a *hibakusha*. I was rather concerned about my relationship with Tsukiko, who had been living here in the United States for such a long time. But when the elderly man got up from his seat, it suddenly struck me that I was both Japanese and a *hibakusha*" (HAYASHI 2005b: 89). The museum exhibits and its souvenirs relay narratives on August 6 and August 9 that frame the nuclear "success story" from the victor's perspective and this leaves "I" feeling significantly alienated. More irritating still, however, is the observation that all the museum's visitors and staff members are, without exception, "white". "I" finds further grounds for irritation during the couple's ensuing short stop at the Bradbury Science Museum in Los Alamos. Beyond the bewildering exhibits and souvenirs, "here again, all of the museumgoers are white" (HAYASHI 2005b: 96). The topic of "nuclear racism" as regards the production, detonation, and commemoration of the atomic bombing in the United States is not one on which HAYASHI Kyōko explicitly elaborates in this work. Yet, her insistence on the "whiteness" of visitors and staff members is reminiscent of the fact that "nuclear racism" was (and is) a crucial factor in terms of U. S. legitimation of its use of the atomic bombs, as John DOWER (2010: 166–74) has pointed out.

On October 2, the two finally reach their destination: the Trinity test site in White Sand National Park. Standing at the area demarcated "Ground Zero", "I" suddenly begins to imagine the wave of devastation brought to this innocent landscape by the first plutonium bomb just as would occur in Nagasaki four weeks later: "Before coming to the 'Trinity site', I always thought that we human beings were the first nuclear victims on that planet. But I was wrong. There were older victims and they were here. They could not weep, they could not scream. They were right here" (HAYASHI 2005b: 106). To put it another way, "Ground Zero" forces "I" to both face August 9 anew—as the hour of her birth as a *hibakusha*—while at the same time helping to overcome her solitary *hibakusha* status. When "I" first tells her friend, Rui, that she plans to visit the Trinity test site, she responds by teasing her and calling her an "atomic bomb maniac". Yet "I" ultimately proves the value of the trip by putting an end to her personal August 9 (HAYASHI 2005b: 83).

"From Trinity to Trinity" became a key work in HAYASHI Kyōko's literary oeuvre. She began to historically recontextualize August 9 by embracing July 16 as the ultimate origin of all future *hibakusha* worldwide. HAYASHI arrived in the U. S. in September 1999 and saw the Tōkaimura accident on the news the night before her visit to "Ground Zero". It was this that compelled her go on a solitary trip to Tōkaimura and, ultimately, to write this short novel on the new generation of *hibakusha*. The trip created a nexus in which the vectors "past" and "future"—as well as "good energy" and "evil energy"—became irreversibly intertwined in her writing. HAYASHI's confrontation with "Ground Zero" revealed to her that though she may have been a victim, she was also a victimizer. Indeed, victim (*higaisha* 被害者) and victimizer (*kagaisha* 加害者) are two sides of the same nuclear coin. According to WATANABE Sumiko and Manuela SURIANO (2009: 291), HAYASHI began to see herself as part of the same humankind that had developed and tested the first atomic bomb at Los Alamos. As such, she began to feel that she had, ultimately, been complicit in exposing innocent mother earth to radiation and, thus, having produced the very first *hibakusha* on the planet. However, this interpretation is only half of the story. The "criticality accident" in Tōkaimura made it painfully clear to HAYASHI that as part of Japan's affluent, nuclear-based society, she was equally complicit in exposing innocent inhabitants of a small town near a nuclear power plant to radiation. HAYASHI had initially envisioned "From Trinity to Trinity" as her final work on (the aftermath of) August 9, 1945. Ironically enough, however, her trip to the U. S. marked a new starting point in the writer's recontextualization of her experiences as a *hibakusha*, as XIONG Fang (2018: 285–86) has pointed out. In the years that followed, HAYASHI capitalized on her experiences of Japan's past "nuclear nightmare" to critically reflect on Japan's ongoing "nuclear dream". She became one of the few voices in Japan that addressed both atomic bombs (*genbaku* 原爆) and nuclear power plants (*genpatsu* 原発) in her work.

Following, in particular, the triple disaster on March 11, 2011, HAYASHI Kyōko grew significantly more critical of the Japanese government. In 2012, she wrote a message to her readers for a reprint of her novel "Hope" (*Kibō* 希望) titled "With Sincere Gratitude" (*Shimijimi kansha o komete* しみじみ感謝をこめて) in which she harshly criticized the government's plan to restart the nuclear power plant in Ōi (Fukui Prefecture): "While brushing the Fukushima accident under the carpet as 'unforeseeable', the restart is forced through with the motto 'for people's everyday life'. I am fed up with this [...] sophistry. We ordinary people are not so stupid that we can be duped by the hollow rhetoric uttered by politicians, officials, power plant managers, and so-called scientists" (HAYASHI: 2012: 209). HAYASHI Kyōko's analysis of March 11, 2011, was always based on the evocation of memories around August 9, 1945, which had been—as Peter

KUZNICK (2011: 22) puts it—deliberately buried in oblivion for the promulgation and implementation of Japan's vision of clean and safe nuclear energy. She decried the "safety declaration" (*anzen sengen* 安全宣言)—that was proclaimed in the context of the power plant's planned restart—as a cheap trick to obfuscate the very real dangers inherent in the decision. For HAYASHI, history has proven that nuclear energy is neither controllable nor safe. As a fatal result of this national delusion, a new generation of *hibakusha* was born. And they faced the same social ostracism and discrimination that had been forced onto survivors of August 1945.

HAYASHI's last literary work, "To Rui, Once Again" (*Futatabi Rui e* 再びルイへ) was published in *Gunzō*, in 2013 and takes the form of a fictitious letter. It is addressed to her fictionalized *hi-hibakusha* friend "Rui"—the addressee of two letters in HAYASHI's "From Trinity to Trinity"—to develop a very personal and critical reflection on Japan's atomic legacy. HAYASHI's "letter" has no inherent plot structure *per se*. Instead, it resembles a collection of associations with and memories of Nagasaki, Trinity, and Fukushima that crisscross throughout the entire text. For HAYASHI Kyōko, these three nuclear disaster sites have irreversibly lost their chronological order and are now simultaneous and omnipresent in the author's mind.

According to HAYASHI Kyōko, the inconceivable damages caused by the March 11 earthquake and tsunami became a welcome excuse for all of those in charge to feel unaccountable for the "unforeseeable" meltdown in Fukushima Daiichi. However, it was only the natural disaster that was "unforeseeable" and not the meltdown. As HAYASHI states: "The explosion in the reactor [...] was not a natural disaster [*tensai* 天災]. It was an accident that was, of course, foreseeable" (HAYASHI 2016: 215). As an eyewitness of August 9, 1945, HAYASHI was keenly aware of the alarming similarities between August 1945 and March 2011. These similarities quickly became unignorable in the wake of the meltdown. In particular, the Japanese public again found itself dealing with a government that denied any accountability for the nuclear disaster. HAYASHI was one of the few literary voices after March 11 to deliberately employ the unpopular term, *hibakusha*, to talk about the inhabitants of the contaminated areas. In contrast to the majority of Japanese literati who adopted a rather depoliticized approach to (the aftermath of) March 11 in their works, HAYASHI continued to tell her readers bitter truths about Japan's nuclear legacy (KÖHN 2022: 135–37):

> The nuclear power plant—initially highly praised for its peaceful use—was now struck by a meltdown of the nuclear fuel rods in the reactors that caused its collapse. This was the worst accident to ever occur. Inhabitants living at a distance of less than 20 kilometers to the power plant had to be evacuated. A new danger, even more terrible than the tsunami, spread very quickly: the danger of radioactive particles. And this led to the problem of 'internal exposure to radiation' [*naibu hibaku* 内部被曝]. Perhaps, these words are unfamiliar to most people. But

for us *hibakusha*, 'internal exposure to radiation' is a problem with which we have been confronted for more than six decades: man and atom, life and radioactivity. (HAYASHI 2016: 227)

By linking August 9 and March 11 in her writing, HAYASHI Kyōko unveils the full extent of the nuclear disaster in Fukushima while, at the same time, recalling the near-forgotten effects of the nuclear disaster in Nagasaki. In a sense, HAYASHI's "To Rui, Once Again" was aimed against the oppressive power structures that had already rendered (the aftermath of) August 1945 both invisible and unrepresentable. These same structures were now taking exactly the same approach to March 2011.

Then, as now, *hibakusha* are either silenced or chose to remain silent. In the wake of March 11, radioactivity became such a sensitive topic in Japan that even *hibakusha* who spoke in public about the possible effects of exposure to radiation ran the risk of being blamed for spreading "harmful rumors" (*fūhyō higai* 風評被害) (HAYASHI 2016: 231). It seems that *hibakusha* forfeited their right to share their knowledge and experiences of the "evil" nuclear energy.

In her late work, HAYASHI Kyōko finally managed to transcend the nuclear fallacy. The discursive splitting of atomic energy into "good" and "evil" had also split Japanese society for decades. As the "accidents" in Chernobyl, Tōkaimura, and Fukushima have shown, nuclear energy can neither be controllable nor safe. As Japan's dream of nuclear-based prosperity continues unabated, it is—as HAYASHI Kyōko (2016: 216) has put it—only a question of time before a new generation of *hibakusha* is born. In this last work, the author poses the question of personal responsibility for the nuclear disaster and for the restart of power plants after the disaster. For HAYASHI, the anonymity of the actors involved—such as *the* government and *the* management team of the Fukushima Daiichi Power Plant as entities that attempt to render their individual members somewhat invisible—is the main reason that an entire nation again feels "victimized" by the "unforeseeable". The nuclear disaster is, thus, downplayed as an "accident" that occurred merely by chance. Speaking to the structural similarity between August 1945 and March 2011, YAMAMOTO Akihiro (2021: 185) alludes to the accountable actors in the following terms: "The nuclear disaster of Fukushima was the result of an 'irresponsibility' and 'recklessness' on the side of the government and the nuclear community. [Their unwillingness to stop Japan's nuclear energy policy] very much resembles the inability of the government to stop the [Pacific] War at that time."

HAYASHI Kyōko's "To Rui, Once Again" calls into question the price for Japan's bright and promising future under the spell of the nuclear. Her personal experiences of (the aftermath of) August 9, 1945 help her unmask the "unforeseeable" of March 11, 2011, for what it was: a man-made disaster. HAYASHI sensitized her readers for the "nuclear issue" by highlighting mechanisms and processes that were—though rendered invisible by powerful discursive structures—inherent to Japanese society. As a *hibakusha*, HAYASHI sees a nuclear timeline that began with the Trinity test on July 16, 1945, and stretches from Los Alamos to Hiroshima, Nagasaki, Chernobyl, Tōkaimura, and Fukushima.

In a certain sense, "To Rui, Once Again" is HAYASHI's payoff for her history within Japan's nuclear legacy. Her sweeping criticism has demonstrated, beyond any doubt, the relevance of atomic bomb literature both during its infancy and today. Its charge, as KAWAGUCHI Takayuki (2011: 45–47) formulates it, is to construct a discursive field in which the hypocrisy and the constructed nature of the system, 'postwar Japan', can be unveiled.

The nuclear disaster on March 11, 2011, came to be HAYASHI Kyōko's final turning point. The sight of people flocking to an anti-nuke-demonstration held in Yoyogi Park, Tokyo, filled her with hope that Japanese society would finally overcome the *genbaku* / *genpatsu* dichotomy that had established itself as a stable point of public consensus in the early 1950s:

> Rui, I never felt such natural and honest concern for 'life' among people as I did on my short way from the station to the park. More than 60 years have passed since the end of war and people had finally made up their mind. Those of us who survived the war could finally pass the baton to the next generation. I was deeply moved. [...] All the skepticism I have felt since the great disaster, all the confusion—gone. [...] All I have to do now is honestly live the time left to me. (HAYASHI 2016: 245)

REFERENCES

"Akutagawashō senpyō 芥川賞選評" (1975). In: *Bungei shunjū* 文藝春秋 53 (9): 344–49.

DOWER, John W. (2010): *Cultures of War*. New York / London: W. W. Norton / The New Press.

HAYASHI, Kyōko 林京子 (1983): "Futari no bohyō 二人の墓標". In: *Hayashi Kyōko* 林京子 (Nihon no genbaku bungaku 日本の原爆文学 3). Tōkyō: Horupu shuppan, 69–93.

HAYASHI, Kyōko 林京子 (2005a): "Shūkaku 収穫". In: *Hayashi Kyōko zenshū* 林京子全集, vol. 6. Tōkyō: Nihon tosho sentā, 262–76.

HAYASHI, Kyōko 林京子 (2005b): "Toriniti kara Toriniti e トリニティからトリニティへ". In: *Hayashi Kyōko zenshū* 林京子全集, vol. 6. Tōkyō: Nihon tosho sentā, 80–111.

HAYASHI, Kyōko 林京子 / SHIMAMURA, Teru 島村輝 (2011): *Hibaku o ikite: Sakuhin to shōgai o kataru* 被爆を生きて：作品と生涯を語る (Iwanami bukkuretto 岩波ブックレット 813). Tōkyō: Iwanami shoten.

HAYASHI, Kyōko 林京子 (2012): "Shimijimi kansha o komete しみじみ感謝をこめて". In: HAYASHI, Kyōko 林京子: *Kibō* 希望 (Kōdansha bungei bunko 講談社文芸文庫). Tōkyō: Kōdansha, 206–10.

HAYASHI, Kyōko 林京子 (2016): "Futatabi Rui e 再びルイへ". In: HAYASHI, Kyōko 林京子: *Tanima / Futatabi Rui e* 谷間／再びルイへ (Kōdansha bungei bunko 講談社文芸文庫). Tōkyō: Kōdansha, 205–45.

INOUE, Hisashi 井上ひさし / KOMORI, Yōichi 小森陽一 (eds.) (2004): *Zadankai: Shōwa bungaku shi* 座談会：昭和文学史, vol. 5. Tōkyō: Shūeisha.

KARATANI, Kōjin 柄谷行人 / NAKAGAMI, Kenji 中上健次 / KAWAMURA, Jirō 川村次郎 (1982): "Sōsaku gappyō 創作合評". In: *Gunzō* 群像 37 (2): 275–94.

KAWAGUCHI, Takayuki 川口隆行 (2011): *Zōhoban: Genbaku bungaku to iu puroburematīku* 増補版：原爆文学という問題領域. Fukuoka: Sōgensha.

KAWAMURA, Minato 川村湊 (2013): *Shinsai, genpatsu bungakuron* 震災・原発文学論. Tōkyō: Inpakuto shuppankai.

KÖHN, Stephan (2022): "3 / 11 and the crisis of representing disaster—a rather polemic reflection on the responsibility of modern writers and scholars". In: MLADENOVA, D. / JAWINSKI, F. / GENGENBACH, K. (eds.): *Die Aufgabe der Japanalogie. Beiträge zur kritischen Japanforschung* (Leipziger Ostasien-Studien 21). Leipzip: Leipziger Universitätsverlag, 127–39.

KUROKO, Kazuo 黒古一夫 (2007): *Hayashi Kyōko ron: Nagasaki, Shanghai, Amerika* 林京子論：ナガサキ・上海・アメリカ. Tōkyō: Nihon tosho sentā.

KUZNICK, Peter ピーター・カズニック (2011): "Aizenhawā no kaku-seisaku アイゼンハワーの核政策". In: TANAKA, Toshiyuki 田中利幸 / KUZNICK, Peter ピーター・カズニック: *Genpatsu to Hiroshima* 原発とヒロシマ (Iwanami bukkuretto 岩波ブックレット 819). Tōkyō: Iwanami shoten, 9–22.

NAKANO, Kōji 中野孝次 / NAGAOKA, Hiroyoshi 長岡弘芳 (1985): "Taidan: Genbaku bungaku o megutte 対談：原爆文学をめぐって". In: *Kokubungaku kaishaku to kanshō* 国文学解釈と鑑賞 50 (9): 10–23.

SHIJŌ, Chie 四条知恵 (2015): *Urakami no genbaku no katari: Nagai Takashi kara Rōma kyōkō e* 浦上の原爆の語り：永井隆からローマ教皇へ. Tōkyō: Miraisha.

TREAT, John Whittier (1995): *Writing Ground Zero. Japanese Literature and the Atomic Bomb.* Chicago / London: University of Chicago Press.

WATANABE, Sumiko 渡邊澄子 (2005): *Hayashi Kyōko: Hito to bungaku* 林京子：人と文学. Nagasaki: Nagasaki shinbunsha.

WATANABE, Sumiko 渡邊澄子 / SURIANO, Manuela スリアーノ・マヌエラ (2009): *Hayashi Kyōko* 林京子 (Nihon no sakka 100 nin 日本の作家100人). Tōkyō: Bensei shuppan.

XIONG, Fang 熊芳 (2018): *Hayashi Kyōko bungaku: Sensō to kaku no jidai o ikiru* 林京子文学：戦争と核の時代を生きる. Tōkyō: Inpakuto shuppan.

YAMADA, Kan 山田かん (2001): *Nagasaki genbaku, ronshū* 長崎原爆・論集. Takaokachō (Miyazaki): Honda kikaku.

YAMAMOTO, Akihiro 山本昭宏 (2021): *Genshiryoku no seishinshi: <Kaku> to Nihon no genzaichi* 原子力の精神史:<核>と日本の現在地 (Shūeisha shinsho 集英社新書 1057). Tōkyō: Shūeisha.

Contributors

CONSTANCE, LAUREN
Education: 2015–19 Bachelor of Arts in Japanese and Spanish, Cardiff University. 2020–present PhD candidate in Japanese Studies, Cardiff University; *Research in Japan*: June–August 2022 Japan Society for the Promotion of Science Summer Program Fellowship; *Research Interests*: Eyewitness testimony in Japanese memorial museums, a-bomb legacy successor programmes, the ethics of museum displays.

DINITTO, RACHEL
Education: 1984–88 Bachelor of Arts in Oriental Studies, University of Pennsylvania; 1990–96 Master of Arts in Modern Japanese Literature, University of Washington; 1996–2000 PhD in Modern Japanese Literature, University of Washington; *Academic Positions*: 2000–06 Assistant Professor of Japanese Studies, The College of William & Mary; 2006–15 Associate Professor of Japanese Studies, The College of William & Mary; 2015–19 Associate Professor of Japanese Literature, University of Oregon; since 2019 Professor of Japanese Literature, University of Oregon; *Research Interests*: Post-disaster contemporary Japanese literature and film, environmental humanities, nuclear literature.

MARIE-CHRISTINE DRESSEN
Education: 2015–21 Bachelor of Arts in Japanese Studies, University of Cologne; since 2021 graduate student in Japanese Studies (Master of Arts), University of Cologne; *Research Interests*: Japanese cinema, film theory, female Japanese directors / authors / artists, Japan's war and postwar period, Japanese architecture.

HERTRICH, ANDRÉ
Education: 1994–2003 M. A. (Magister Artium) in Modern History and Japanese Studies, Ludwig-Maximilians University Munich; 2005–09 M. A. (Master of Arts) in Peace and Conflict Studies, Philipps-University Marburg; 2016–20 PhD in Japanese Studies, University of Hamburg; *Academic Positions*: 2019–

24 Post-Doc Researcher in the ERC-Project "Globalized Memorial Museums", Austrian Academy of Sciences Vienna; *Research Interests*: Memorialization and Musealization of War and Atrocities, War Crime Trials in Asia, Democratic Control of Armed Forces.

HÜLSMANN, KATHARINA
Education: 2006–10 Bachelor of Arts in Modern Japanese Studies, Heinrich-Heine-University; 2010–14 Master of Arts in Modern Japanese Studies, Heinrich-Heine-University; 2017 PhD Fellow at the German Institute for Japanese Studies, Tokyo; 2021 PhD in Modern Japanese Studies, Heinrich-Heine-University; *Academic positions*: 2015–18: Lecturer and Research Associate at Heinrich-Heine-University; 2021–22 Research Fellow in the DFG-funded project "Discursive Constitution between the Atomic Bombs (*genbaku*) and Nuclear Power Plants (*genpatsu*)", University of Cologne; 2022–23 Research Assistant at the Institute for Japanese Studies, University of Cologne; since 2023 Lecturer and Research Associate in Modern Japanese Studies, Heinrich-Heine-University; *Research Interests*: Japanese popular culture (with a focus on manga), transcultural phenomena in media and fan cultures, gender representations and autobiographical narratives in comics.

HOOD, CHRISTOPHER P.
Education: 1989–93 Bachelor in Business Studies and Japanese Studies, University of Sheffield; 1994–98 PhD in Japanese Studies, University of Sheffield; *Academic positions*: since 2000 Director of the Cardiff Japanese Studies Centre and Senior Lecturer; since 2007 Reader in Japanese Studies, Cardiff University; *Research Interests*: Disasters in Japan, transportation in Japan, symbolism in relation to Japan.

KÖHN, STEPHAN
Education: 1989–96 Bachelor and Master of Arts in Japanese Studies, University of Frankfurt; 1996–99 PhD in Japanese Studies, University of Frankfurt; 2004 Habilitation in Japanese Studies, University of Würzburg; *Academic positions*: 1997–2000, 2002 Research Assistant, University of Frankfurt; 2003–06 Assistant Professor, University of Würzburg; 2007–09 and 2011 Associate Professor, University of Erlangen; 2009–10 and 2011–12 Visiting Professor, Universities of Tübingen and Düsseldorf; 2012–13 Associate Professor, University of Leipzig; since 2013 Professor of Japanese Studies at the University of Cologne; *Research*

Interests: Popular and media culture, atomic bomb literature, premodern book publishing, social precarity.

KUZNICK, PETER
Education: 1970 Bachelor of Arts in History, Rutgers University; 1975 Master of Arts in History, Rutgers University; 1984 PhD, Rutgers University; *Academic Positions*: 1984–85 Joint Postdoctoral Fellow, Smithsonian Institution/Visiting Assistant Professor of American Studies, George Mason University; 1985–86 Visiting Assistant Professor of American Studies, University of Maryland, Baltimore County; since 1986 Assistant, Associate and Full Professor of History, American University, Washington DC; since 1995 Director of the American University Nuclear Studies Institute; *Research Interests*: History of U.S. culture, antiwar, antinuclear and civil rights movements, Hiroshima and Nagasaki, American radicalism, scientists and the Vietnam War, Cold War, history of the American Empire.

MASON, MICHELE M.
Education: 1984–89 Bachelor of Arts in Linguistics and Japanese, University of Oregon; 1992–95 Master of Arts in Modern Japanese Literature, University of California, Los Angeles; 1996–97 and 2000–02 Graduate Research Student, University of Tokyo; 2005 PhD in Modern Japanese Literature, University of California, Irvine; 2005–07 Postdoctoral Fellow, Stanford University; *Academic positions*: since 2007 Associate Professor, University of Maryland; *Research Interests*: Colonial and postcolonial studies, gender and feminist theory, masculinity studies, environmental humanities, contents tourism, Hiroshima and Nagasaki in literature and history, global *hibakusha* (atomic survivor) movements, and nuclear power.

OKUDA, HIROKO
Education: 1993–95 Master of Arts in Language and Information Sciences, University of Tokyo; 2001 PhD in Communication Studies, Northwestern University; *Academic Position*: 2001–04 Lecturer of Communication Studies, Akita University; 2004–13 Associate Professor of Communication Studies, Nanzan University; 2013–18 Associate Professor of Communication Studies, Kantō Gakuin University; since 2018 Professor of Communication Studies, Kantō Gakuin University; *Research Interests*: Japan's war memories, foreign policy discourse, peace studies.

SHIGESAWA, ATSUKO
Education: 1987–92 Bachelor of Arts in Foreign Studies at the Department of English Studies, Kobe City University of Foreign Studies (KCUFS); 2008–09 Master of Arts from the Faculty of International Studies, Hiroshima City University; 2019 Doctor of Philosophy from the Faculty of International Studies, Hiroshima City University; *Academic positions*: 2014–15 Visiting Scholar at the Department of History, American University in Washington DC (Fulbright Dissertation Program); since 2016 Associate Professor of American Studies / Journalism at the Department of English Studies, KCUFS; *Research Interests*: 20th century American history, especially discourses on the atomic bomb, media representations of nuclear issues, and race relations during the Cold War period.

WEBER, CHANTAL
Education: 1997–2003 Bachelor and Master of Arts in Japanese Studies, Archeology and Art History, University of Cologne; 2011 PhD in Japanese Studies, University of Cologne; *Academic positions*: 2003–05 Computer Center, University of Freiburg, Germany; 2006–08 International Office, University of Freiburg; since 2008 Research Assistant and now Assistant Professor for Japanese Studies, University of Cologne; 2023 Visiting Professor, Leipzig University; *Research Interests*: Culture-historical network analysis, history of *sadō* (Way of Tea) and *kōdō* (Way of Incense).

WEISS, TOBIAS
Education: 2004–12 Bachelor and Master of Arts in Japanese Studies, University of Hamburg; 2018 PhD in Japanese Studies, University of Zurich; *Academic positions*: 2017 Predoctoral Fellow, Waseda University; 2018–19 JSPS Postdoctoral Fellow, University of Zurich; 2019–20 JSPS Postdoctoral Fellow, Waseda University; 2020–22 Assistant Professor, Heidelberg University; since 2022 Associate Professor, Sophia University; *Research Interests*: Journalism and politics in Japan, civil society and social movements, political sociology.

WIEMANN, ANNA
Education: 2003–08 Bachelor of Arts in Japanese Linguistics and Roman Philology (French), Ruhr University Bochum; 2008–11 Master of Arts in Peace and Conflict Studies, Philipps-University Marburg; 2018 PhD in Japanese Studies, Hamburg University; *Academic Positions*: 2016–18 Lecturer for Japanese Studies, Hamburg University; 2018–19 Research Associate, Heinrich-Heine-

University Düsseldorf; since 2019 Assistant Professor at Ludwig-Maximilians University Munich; *Research Interests*: Social movements, civil society, social networks, collective memory.

Index

20th Century Media Information Database 56
Abe, Shinzō 288, 306, 318
activism 7, 227, 255, 263, 265
"Adventurous Dankichi" (*Bōken Dankichi*) 81, 85–87, 96, 97, 100
Advertising Council Japan (AC Japan) 273
"A Flash in Summer" (*Senkō no natsu*) 353
Agamben, Giorgio 15
akai kyōkasho → school textbook
Akai yuki → "Red Snow"
Akuma no mitsurin → "Demon Jungle"
Allied Powers 3, 11, 62, 88, 128, 133, 180, 195, 354
"Along With Those Who Can Hear My Mother's Voice" (*Haha no koe ga kikoeru hitobito to tomo ni*) 227
"American Shadow" (*Amerika no kage*) 7
Anders, Günther 5, 6, 200
Ando, Ryoko 182
"An Unfathomable Deep Light" (*Kaitei no yō na hikari*) 1, 195
anzen shinwa → safety myth
Arihara, Seiji 120
Ariyama, Teruo 7
Ariyoshi, Sawako 219
Aru wakusei no higeki → "Tragedy of a Planet"
Asada, Jirō 216
Asaoka, Kōji 100
ashes of death (*shi no hai*) 175
Assmann, Aleida 331
"A Summer's Afterimage: Nagasaki-August 9" (*Natsu no zanzō: Nagasaki no hachigatsu kokonoka*) 255
atomic bomb 1, 3, 4, 9, 36, 39, 41, 42, 45, 55, 57, 60, 62, 64, 65, 71–74, 81, 83, 87, 88, 91, 92, 99, 113, 115, 127, 132, 134, 151, 154, 161, 162, 168, 178, 195, 197, 199, 201–03, 205, 206, 208, 215–17, 220–22, 235–37, 240, 253, 272, 285, 353, 356–58

atomic bombing 2, 4–7, 13, 32, 33, 42, 55, 57, 64, 67, 69, 70, 72, 106–08, 117, 118, 120, 123, 137, 165, 169, 175, 176, 179, 180, 182, 188, 189, 216, 223, 225, 226, 235, 236, 240, 241, 243, 246, 254, 256, 271, 278, 353, 355
atomic bomb test 234, 241
atomic bomb tourism 68, 69
atomic weapon 105, 121, 188, 216, 235, 246, 253, 254, 263, 267, 271, 275
atomic bomb survivor (*hibakusha*) 4, 15, 32, 44, 105, 112, 157, 160, 175, 177, 179, 180–82, 184, 188, 189, 198, 205, 206, 217, 218, 221, 225, 226, 236, 246, 256, 262, 338, 353–59, 361
"Atomic Bomb Manga Collection" (*Gensuibaku manga korekushon*) 100
Atomic Energy Basic Act (*Genshiryoku kihon hō*) 9, 34
"Atomic Genkichi" (*Genshi no Genkichi*) 82, 96, 97, 99
"Atomic Story" (*Genshi monogatari*) 92
Atoms for Peace 8, 9, 81, 216, 254, 260, 263, 275, 278, 285
"August 6, 8:15" (*Hachigatsu muika, hachiji jūgofun*) 1, 2, 5

Baldwin, Hanson 237
"Barefoot Gen" (*Hadashi no Gen*) (Manga) 100
"Barefoot Gen" (*Hadashi no Gen*) (Anime) 114, 117, 118, 120, 121
"Barefoot Gen 2" (*Hadashi no Gen* 2) (Anime) 114
Barthes, Roland 106
Batto-kun 96
Bennett, Jane 257
Bharne, Vinayak 139
"Bikini Ashes of Death" (*Bikini shi no hai*) 100

Bikini Atoll 8, 42, 187, 207, 217, 219, 223, 260
  Bikini Incident 8, 9, 175, 176, 179–81, 184, 188, 190
Bikini shi no hai → "Bikini Ashes of Death"
Blackett, P. M. S. 241
black rain 32, 108, 223
"Black Rain" (*Kuroi ame*) 106, 120, 355
Blume, Lesley M. M. 34
Bockscar 109
Boissou, Jean-Marie 83
Bōken Dankichi → "Adventurous Dankichi"
Bordwell, David 107, 122
"Boundless Energy: Concrete" (*Mugen no enerugī: Konkurīto*) 135
Bourdaghs, Michael 222
Boyer, Paul 239
Braw, Monica 3, 60, 87–89
Broderick, Mick 105
Brown, Blain 107, 108
Buell, Lawrence 218
Bührmann, Andrea D. 15
Bungei shuto → magazine
Burchett, Wilfred 5

Cabinet Intelligence Bureau (Naikaku jōhōkyoku) 4
Carson, Rachel 219, 220
Caruth, Cathy 200
"Celebrating Half-Life" (*Hangenki o iwatte*) 215, 225
censorship 2–5, 7, 33, 34, 40, 43, 60, 61, 82, 87–90, 180, 195, 196, 203
  Civil Censorship Detachment (CCD) 3, 5, 6, 56, 59, 61, 87, 88, 90, 100, 180, 196
  Civil Information and Educational Section (CIE) 4, 6
  Civil Intelligence Section (CIS) 4
  press code 3, 61, 62, 88
  self-censorship / self-restraint 2, 3, 44, 74, 181, 195, 196, 203
Charter of Athens 129, 131, 134, 135, 138
Cheng Chua, Karl Ian Uy 85, 86
Chernobyl 31–33, 45, 46, 218, 219, 224, 260, 262, 278, 337, 352, 360, 361

"Children of the Atomic Bomb" (*Genbaku no ko*) 119
Chūgaku shakai → school textbook
Chilton, Paul 275
"City and People in the Evening Calm" (*Yūnagi no machi to hito to*) 197, 200, 201, 206
"City of Corpses" (*Shikabane no machi*) 2, 195
Cold War 6, 83, 101, 121, 167, 241, 244, 253, 275
commemoration 10, 13, 14, 57, 64, 67, 68, 70, 72, 154, 357
communism 5–7, 12, 202
Conde, David W. 6, 7
Congrès Internationaux d'Architecture Moderne (CIAM) 127, 131, 134, 138, 139
conversion (*tenkō*) 11, 204
Countdown 1945: The Extraordinary Story of the Atomic Bomb and the 116 Days that Changed the World 243
Cramer, Maria 243
criticality accident (*rinkai jiko*) 351, 358

Daigo Fukuryū Maru → Lucky Dragon Number 5
Daigo Fukuryū Maru (Film) 175
Daigo Fukuryū Maru tenjikan → Lucky Dragon Number 5
Daitōa kyōeiken → Greater East Asia Co-prosperity Sphere
"Dance of the Atoms" (*Genshi no odori*) 92
Day of the Western Sunrise 175, 180
Debiru fisshu (tako) → "Devil Fish (Octopus)"
"Deep Sea Fish" (*Shinkaigyo*) 254
Democratic Party of Japan (Nihon minshu tō) 202, 274
"Demon Jungle" (*Akuma no mitsurin*) 93
Dengen sanpō → Three Electric Power Laws
Der Mann auf der Brücke → "The Man on the Bridge"
"Devil Fish (Octopus)" (*Debiru fisshu (tako)*) 254
Diehl, Chad 63, 64
"Disappearing Girl" (*Kieyuku shōjo*) 100

disaster 14, 255, 272, 274, 277, 286, 288, 289, 318, 330, 331, 333, 335, 339, 344, 351
    disaster research 331
    earthquake 13, 140, 188, 253, 265, 277, 283, 284, 288, 289
    Great East Japan Earthquake (Higashi Nihon daishinsai) 14, 277, 285, 288, 329, 332
    Great Hanshin-Awaji Earthquake (Hanshin Awaji daishinsai) 14, 289
    Great Kantō Earthquake (Kantō daishinsai) 14, 130
    triple disaster 13, 14, 215–19, 221, 222, 224, 225, 227, 253, 265, 271, 277, 358
    tsunami 13, 253, 263, 265, 272, 277, 284, 298, 304, 320, 329, 331, 332, 359
Dower, John 357
Duffy, Terence 187

Ecology Without Culture 220
economy 272, 286
    economic growth 271, 286, 289
Edano, Yukio 312, 314
education system 201, 202
    Basic Act on Education (Kyōiku kihon hō) 11, 12, 202
    Double Education Act (Kyōiku nihō) 12
    Guidelines for Teaching and Learning (Gakushū shido yōryō) 12, 202, 203
Einstein, Albert 91
Eisenhower, Dwight D. 8, 81, 275
emonogatari → picture story
emperor 6, 11, 19, 60, 72, 73, 203, 354
energy 286, 358
    clean energy 272, 359
    energy policy 271
    nuclear energy 272, 288, 360
Enola Gay 109, 110, 115, 117, 160
Enola Gay, The Men, The Mission, and The Atomic Bomb 113
Erll, Astrid 330
Etō, Jun 4, 196
Evans, Medford 241
"Exhibition on the Peaceful Use of Nuclear Energy" (Genshiryoku heiwa riyō hakurankai) 8, 207

experience 178, 179, 182, 195, 196, 205, 208, 352, 354–56, 358, 360
eyewitness 8, 177, 178, 180, 183, 188, 195, 197, 200, 204, 205, 353, 354, 359

Fallout 34
fallout 108, 175, 253, 266
Farrell, Thomas F. 5
Fear, War, and the Bomb 241
"Festival Ground" (*Matsuri no ba*) 353, 354
Fifteen Years War 3, 7, 11, 201, 203, 204
Foucault, Michel 15
"From 'Dream Songs'" (*'Yume no uta' kara*) 215, 218, 219, 221, 222, 224
"From Trinity to Trinity" (*Toriniti kara Toriniti e*) 216, 356, 358, 359
fukkō → reconstruction
Fukuma, Yoshiaki 10
Fukushima 13, 33, 46, 81, 178, 181, 183, 189, 200, 226, 227, 255, 260–62, 268, 272, 277, 286–89, 316, 332–34, 336, 338–40, 344, 352, 358–61
    Fukushima Daiichi Nuclear Power Plant 13, 31, 178, 215, 255, 266, 272, 273, 275, 277, 280, 281, 283–85, 288, 289, 313, 329, 330, 359, 360
    Great East Japan Earthquake and Nuclear Memorial Museum in Fukushima (Higashi Nihon daishinsai, genshiryoku saigai denshōkan) 178, 183
Fushi no shima → "Island of Eternal Life"
Fussell, Paul 237
Futari no bohyō → "Two Grave Markers"
Futatabi Rui e → "To Rui, Once Again"

Gakushū shido yōryō → education system
Genshi bōru → "Nuclear Ball"
Genshi monogatari → "Atomic Story"
Genshi no Genkichi → "Atomic Genkichi"
Genshiryoku kihon hō → Atomic Energy Basic Act
Gallup, George 238, 241, 244
genbaku bungaku → literature
genbaku bungaku ronsō → literature
Genbaku dōmu → Hiroshima
Genbaku no ko → "Children of the Atomic Bomb"

Gen-chan no bōken → "The Adventure of Gen-chan"
Genpatsu wa naze kowai ka → "Why Nuclear Power is Scary"
Genshi bakudan → "The Atomic Bomb"
Genshi no odori → "Dance of the Atoms"
Genshiryoku anzen hoan'in → Nuclear and Industrial Safety Agency
Genshiryoku heiwa riyō hakurankai → "Exhibition on the Peaceful Use of Nuclear Energy"
Genshiryoku iinkai → Nuclear Energy Commission
Genshiryoku mondai to bungaku → "Literature and the Problem of Nuclear Energy"
Genshiryoku no kenkyū → "Nuclear Energy Research"
Genshiryoku saigai taisaku honbu → Nuclear Emergency Response Headquarters
genshiryokumura → nuclear power
Gensuibaku manga korekushon → "Atomic Bomb Manga Collection"
Genzai Nihon ni oite kindai kenchiku o ika ni rikai suru ka: Dentō no sōzō no tame ni → "How to Understand Modern Architecture in Japan Today: For the Creation of Tradition"
Gerster, Julia 332
gisei no shisutemu → sacrificial system
Gleiter, Jörg H. 128
Gluck, Carol 73
"Golden Dream Song" (*Ōgon no yume no uta*) 222
"Goodbye, Atomic Dragon: The Story of Atomic Weapons and Nuclear Power" (*Sayonara atomikku doragon: Kaku to genpatsu no ohanashi*) 255, 256, 258–63, 265, 266
Gordon W. Prange Collection 5, 56, 59, 61, 65, 74, 82, 89, 91, 92
Gotō, Shinpei 130
Greater East Asia Co-prosperity Sphere (Dai-tōa kyōeiken) 129
Groves, Leslie R. 32–36, 38–40, 42, 43, 45, 240
Guo, Qinghua 130, 131

gyaku kōsu → occupation period
Hachigatsu muika, hachiji jūgofun → "August 6, 8:15"
Hachigatsu kokonoka no Santakurōsu: Nagasaki genbaku to hibakusha → "The Santa Claus of August 9th: Nagasaki's A-Bomb and the Survivors"
Hadashi no Gen → "Barefoot Gen"
Haha no koe ga kikoeru hitobito to tomo ni → "Along With Those Who Can Hear My Mother's Voice"
Halbwachs, Maurice 331
"Half Human" (*Han ningen*) 10, 197
"Half Nomad" (*Han hōrō*) 207
Hamai, Shinzō 71
Hanada, Kiyoteru 205
Hanano, Kaoru 100
Hangenki o iwatte → "Celebrating Half-Life"
Hanshin Awaji daishinsai → disaster
Han ningen → "Half Human"
Han hōrō → "Half Nomad"
Hara, Kazushi 96
Hara, Tamiki 196, 201
harmful rumors (*fūhyō higai*) 14, 221, 227, 288, 360
Harō, jīpu → "Hello Jeep"
Harootunian, H. D. 131
"Harvest" (*Shūkaku*) 351–53
Hatoyama, Ichirō 3
Hayashi, Kyōko 4, 215–17, 219, 221, 226, 227, 351–56, 358, 359, 361
Hein, Carola 131, 132, 135
heiwa → peace
heiwa sengen → peace
Heiwa fukkōsai → peace
"Hello Jeep" (*Harō, jīpu*) 89, 90
Hersey, John 32–34, 36, 39, 40, 43, 44, 240
Hibaku Maria no inori: Manga de yomu sannin no hibaku shōgen → "The Prayer of the A-Bombed Maria: Manga Testimonies of Three Survivors"
hibakusha → atomic bomb
Higashi Nihon daishinsai → disaster
Higashi Nihon daishinsai fukkō kaigi → National Resilience Council

Higashi Nihon daishinsai, genshiryoku saigai denshōkan → Fukushima
hikaku sangensoku → Three Non-Nuclear Principles
Hirano, Kyōko 6
Hirata, Toshio 114
Hiroshima 1, 4, 5, 9, 10, 13, 32, 36, 38, 41, 45, 55, 56, 59, 62, 65, 66, 68, 73, 88, 92, 105, 107, 108, 110, 113, 114, 117, 118, 120–22, 127, 128, 132, 134, 135, 137, 139, 151, 154, 160, 165, 168, 175, 178, 180, 188, 195, 196, 199–203, 206, 208, 216, 225, 226, 234–37, 241–43, 246, 254, 262, 271, 272, 278, 289, 353, 355, 361
   Genbaku Dome (Genbaku dōmu) 71, 117, 118, 136, 223
   Hiroshima Children's Library (Hiroshima-shi jidō toshokan) 135, 207, 208
   Hiroshima Peace Memorial City Construction Law (Hiroshima heiwa kinen toshi kensetsu hō) 55, 68, 73, 206
   Hiroshima Peace Memorial Museum (Hiroshima heiwa kinen shiryōkan) 8, 10, 58, 64, 68, 127, 136, 138, 151, 153, 159, 164, 168, 169, 176, 178–80, 183, 206, 207, 235, 236, 242
   Hiroshima Peace Media Center 59, 64
   Hiroshima Peace Memorial Park (Hiroshima heiwa kinen kōen) 57, 70, 128, 135–38, 140, 178, 206
Hiroshima (Book) 33–35, 39, 40, 42, 43, 45, 240
Hiroshima (Film) 114, 116, 119–21
Hiroshima mon amour 55
Hiroshima Out of the Ashes 113
Hiroshima Traces: Time, Space, and the Dialectics of Memory 58
Holocaust 161, 162, 165, 167, 205, 206
   Holocaust literature 196, 205
   Holocaust memorial museum 152, 155, 156, 160, 161, 167, 169
Hood, Christopher P. 108
Hook, Glenn 60, 61
"Hope" (*Kibō*) 358
Hosono, Gōshi 306, 309

"How to Understand Modern Architecture in Japan Today: For the Creation of Tradition" (*Genzai Nihon ni oite kindai kenchiku o ika ni rikai suru ka: Dentō no sōzō no tame ni*) 138
Hurbis-Cherrier, Mick 107
hydrogen bomb 8, 175, 207, 241

Ibuse, Masuji 355
Ienaga, Saburō 202
Igarashi, Yoshikuni 203
Iida, Yukisato 91, 92
Ikeda, Hayato 12
Ima koso watashi wa genpatsu ni hantai shimasu → "Now More Than Ever, I Oppose Nuclear Power"
Imamura, Shōhei 106, 120
Inoue, Hisashi 4
Inoue, Kazuo 96
International Atomic Energy Agency (IAEA) 272, 275, 278, 280, 312
"I saw it" (*Ore wa mita*) 100
Ishida, Takeshi 12
Ishida, Yorifusa 133
Ishikawa, Hideaki 134
Ishimaru, Norioki 134
"Island of Eternal Life" (*Fushi no shima*) 225
Isozaki, Arata 131, 132, 138, 140
Itō, Chūta 131
Itō, Toyō 140
Itsuka, Nanohana hatake de ~ Higashi Nihon daishinsai o wasurenai~ → "Someday in a Rapeseed Field: We Will Not Forget Japan's Great East Disaster"

"Jacka Dofuni: A Tale of Oceanic Memory" (*Jacka Dofuni: Umi no kioku no monogatari*) 215, 224
Jacobson, Harold 39–41
Japanese Nuclear Safety Commission (JNSC) 297
Japanese Teachers' Union (Nihon kyōshokuin kumiai) 202
Jieitai → Self-Defense Forces
Jiyū minshu tō → Liberal Democratic Party

Kadota, Ryūshō 316, 317
Kaieda, Banri 306, 311, 313, 319
Kairiki genshi bōru → "Superhuman Power Nuclear Ball"
Kaisoku rokettodan → "Superfast Rocket Missile"
Kaitei no yō na hikari → "An Unfathomable Deep Light"
Kaizō → magazine
Kajiya, Kenji 138
Kamei, Fumio 6
Kan, Naoto 298–302, 304–06, 308–10, 316, 318, 352
Kantō daishinsai → disaster
kataribe → storyteller
Katō, Norihiro 7, 8, 204
Katsumata, Susumu 254, 255
Katsumata, Tsunehisa 303, 318
Kawaguchi, Takayuki 200, 361
Kawaguchi, Yūko 58, 60–62, 72
Kawamura, Minato 352
Kawara → "Riverbank"
Kawazoe, Noboru 137
Keizai sangyō shō → Ministry of Economy, Trade and Industry
Keller, Reiner 59, 73, 74
Kibō → "Hope"
Kieyuku shōjo → "Disappearing Girl"
Kim, Jong Un 234
Kimura, Hideaki 320
Kimura, Tsuneyuki 92
Kindai bungaku → magazine
kindai no chōkoku → overcoming modernity
kiroku bungaku → literature
Kitamura, Seibō 70, 75
Köhn, Stephan 83
Koizumi, Kashirō 133
Kondō, Hidezō 84
Konishi, Shirō 202
Kōno, Fumio 256
Koolhaas, Rem 140
Kōra, Tomi 134
Koyama, Buntarō 201
Kuan, Seng 127
Kuboyama, Aikichi 175, 181, 184
Kurihara, Sadako 196
Kurihara, Tadaichi 204
Kuroi ame → "Black Rain"

Kuroko, Kazuo 208, 352
Kusaka, Tatsuo 100
Kuznick, Peter 9, 359
Kyōiku kihon hō → education system
Kyōiku nihō → education system

LaCapra, Dominick 206
Lang, Daniel 34, 35, 43
Lang, Jessica 205
Lawrence, William L. 5
Le Corbusier 127, 131, 137
Liberal Democratic Party (Jiyū minshu tō) 274
Lin, Zhongjie 136
Lippit, Yukio 127
literature 11, 64, 204, 205, 353, 356
    atomic bomb literature (*genbaku bungaku*) 32, 64, 196, 201, 205, 207, 353, 354, 356, 361
    disputes on atomic bomb literature (*genbaku bungaku ronsō*) 196
    documentary literature (*kiroku bungaku*) 204
    I-novel (*shishōsetsu*) 197, 204
    Japanese Association for the Promotion of Literature (Nihon bungaku hōkoku kai) 204
    postwar literature 204, 205
"Literature and the Problem of Nuclear Energy" (*Genshiryoku mondai to bungaku*) 196
Los Alamos 40, 256, 356–58, 361
Lucky Dragon Number 5 (Daigo Fukuryū Maru) 5, 8, 175–77, 179, 182, 184, 186, 216, 218
    Daigo Fukuryū Maru Exhibition Hall (Daigo Fukuryū Maru tenjikan) 176, 177, 179, 180, 182, 183, 188, 189

MacArthur, Douglas 6, 7, 11, 60, 71, 180
Macdonald, Dwight 238
Macdonald, Sharon 177
Madarame, Haruki 301, 305, 306, 308, 315, 319, 320
Maekawa, Kunio 127, 131, 132, 140
magazine 3, 83, 84, 87, 88, 96, 205, 355, 356
    "Literary Capital" (*Bungei shuto*) 217, 353

Comix Box 255
Fortune Magazine 238, 241
GARO 254
Gunzō 225, 351, 353, 355, 356, 359
Harper's 240
Japan Punch 83
Manga shōnen 82, 96, 97
"Modern Literature" (*Kindai bungaku*) 11, 204
National Review 241
"New Literature of Japan" (*Shin Nihon bungaku*) 11, 204
"Novel" (*Shōsetsu*) 1, 2
"Reconstruction" (*Kaizō*) 1, 2
Shōjo kurabu 83
Shōnen kurabu 81, 83, 85–88, 96
The New Yorker 33–35, 37, 39, 43, 45, 240
"Women's Literature" (*Nyonin geijutsu*) 195
Malloy, Sean 45
Maly, Elizabeth 332
Manchuria 128–33, 139
manga 81–83, 88, 90, 93, 97, 99–101, 215, 253–55, 258, 259, 262, 264, 265, 267
Manhattan Project 5, 32, 243
Marchi, Leonardo Zuccaro 138, 139
Marran, Christine 220–22
Marxism-Leninism 202
Masaki, Mori 114
Matsumoto, Hiroshi 13, 206
Matsuri no ba → "Festival Ground"
Matsuura, Sōzō 7
Mazza, Michael 235, 236
McCarthy, Mary 240
McKnight, Anne 222
media 3–5, 13, 271, 275, 277, 287, 301, 304, 305, 307, 318, 319, 337, 340, 342, 344
   media discourse 271, 273, 275, 277, 285, 287, 289
   news media 271, 278
meltdown 189, 219, 253, 255, 266, 272, 273, 278–81, 283, 288, 298, 352, 359
Memorial Museums: The Global Rush to Commemorate Atrocities 167
memory 4, 10, 200, 271, 330, 331, 358
   collective memory 13, 202, 330
   cultural memory 331, 332

Metabolism (*metaborizumu*) 128, 140
Ministry of Economy, Trade and Industry (Keizai sangyō shō) 273, 297–300, 306, 311, 319
Minobe, Ryōkichi 201
Minshu shugi to akarui seikatsu → school textbook
Misukoso 265, 266
Miyazaki, Tomomitsu 64, 105
Mohan: Chūgaku shakai → school textbook
Mugen no enerugī: Konkurīto → "Boundless Energy: Concrete"
musekinin no taikei → system of irresponsibility
mushroom cloud 105, 106, 108, 109, 112, 113, 115–18, 121, 122, 135, 155, 208, 217, 278

Nagai, Jirō 203
Nagai, Takashi 64, 67, 70, 72, 74, 75, 112, 178, 353, 354
Nagaoka, Hiroyoshi 207
Nagaoka, Shōgo 183
Nagasaki 1, 4, 5, 10, 13, 38, 41, 45, 55, 56, 59, 62, 66, 72, 81, 88, 92, 105, 107, 108, 110, 116–18, 120, 122, 132, 151, 154, 155, 159, 165, 168, 175, 178, 180, 188, 200–02, 216, 225, 226, 236, 237, 241–43, 246, 254–56, 262, 267, 271, 272, 278, 289, 353, 354, 356, 357, 359, 360, 361
   Nagasaki Atomic Bomb Museum (Nagasaki genbaku shiryōkan) 10, 56, 151, 155, 156, 168, 169, 176, 178, 183
   Nagasaki International Culture City Construction Law (Nagasaki kokusai bunka toshi kensetsu hō) 56, 69, 73
   Nagasaki Peace Park (Nagasaki heiwa kōen) 56, 70, 155
   Nagasaki's Testimony Committee (Nagasaki no shōgen no kai) 255
Nagasaki 1945—Anzerasu no kane → "The Bells of Angelus"
Nagasaki no kane → "The Bells of Nagasaki"
Naikaku jōhōkyoku → Cabinet Intelligence Bureau
Nakagami, Kenji 355, 356

Nakasone, Yasuhiro 9
Nakazawa, Keiji 100
Nan'yō no o-shōgatsu → "New Year in the South Sea"
National Resilience Council (Higashi Nihon daishinsai fukkō kaigi) 272
Natsu no hana → "Summer Flowers"
Natsu no zanzō: Nagasaki no hachigatsu kokonoka → "A Summer's Afterimage: Nagasaki-August 9"
New Architects' Union of Japan (Shin Nihon kenchikuka shūdan) 133
New Manga group (Shin mangaha shūdan) 84
newspaper 3, 4, 56, 58, 62–65, 68, 74, 83, 84, 275, 280, 300, 316
   Asahi shinbun 1, 3, 300, 302–04, 306–08, 310–12, 314, 317, 318
   Chūgoku shinbun 56, 58–60, 63–65, 68, 70, 72, 73
   Detroit Free Press 244
   Mainichi shinbun 300, 317
   Nagasaki min'yū 67, 69, 72
   Nagasaki shinbun 56, 64
   New York Herald Tribune 239
   Sankei shinbun 300, 301, 306, 316, 318, 319
   St. Louis Post Dispatch 239
   The Daily Express 5
   The New York Times 5, 237, 240–43, 247
   The Wall Street Journal 235, 243, 305
   Yomiuri shinbun 300–03, 306–08, 317–19
"New Treasure Island" (*Shin Takarajima*) 86
"New Year in the South Sea" (*Nan'yō no o-shōgatsu*) 85
Nietzsche, Friedrich 138
Nihon bungaku hōkoku kai → literature
Nihon kyōshokuin kumiai → Japanese Teachers' Union
Nihon minshu tō → Democratic Party of Japan
Nippon no higeki → "The Tragedy of Japan"
Nishimura, Sey 60, 61, 74
Nishioka, Yuka 255, 256, 258, 259, 261–63, 265–67
Nishiyama, Uzō 133

Noguchi, Isamu 133, 137
Norakuro 86, 87, 100
Nora, Pierre 331
Nornes, Markus 105
"Now More Than Ever, I Oppose Nuclear Power" (*Ima koso watashi wa genpatsu ni hantai shimasu*) 215
Nuclear and Industrial Safety Agency (Genshiryoku anzen hoan'in) 273
"Nuclear Ball" (*Genshi bōru*) 90
nuclear dispositive 14, 15, 218, 220, 226
Nuclear Emergency Response Headquarters (Genshiryoku saigai taisaku honbu) 333
Nuclear Energy Commission (Genshiryoku iinkai) 9
"Nuclear Energy Research" (*Genshiryoku no kenkyū*) 92
Nuclear Industry Safety Agency (NISA) 297, 306, 308, 311, 312, 319, 320
nuclear power 1, 8, 9, 13, 14, 45, 83, 91, 93, 95, 100, 101, 181, 215–17, 253–55, 259, 262, 263, 267, 271–75, 278, 285, 288, 290, 352
   nuclear age 8, 15, 241
   nuclear colonialism 219, 225
   nuclear industry 216, 218, 219, 297, 300, 319
   nuclear power plant 8–10, 14, 216, 218–20, 222, 254, 262, 278–83, 285, 319, 351, 353, 358
   nuclear testing 34, 116, 223, 253, 356
   nuclear umbrella 235
   nuclear village (*genshiryokumura*) 297, 300, 301, 303, 305, 318, 319
   nuclear war 32, 220, 234, 244
   nuclear waste 225
Nyonin geijutsu → magazine

Obama, Barack 235, 236, 242, 243, 245, 246
Obrist, Hans Ulrich 140
occupation period 6–8, 11, 12, 56, 60, 61, 63, 73, 74, 81, 82, 89, 90, 100, 101, 181, 201
   Allied Occupation 3, 4, 5, 60
   reverse course (*gyaku kōsu*) 7, 11

Odagiri, Hideo 196
Ōe, Kenzaburō 216, 217

Ōgon no yume no uta → "Golden Dream Song"
Ōishi, Matashichi 175, 177, 179, 181, 183, 184, 188
Okada, Yuzuru 201, 202
Okuda, Hiroko 66, 70, 73
Oppenheimer (Film) 123
Oppenheimer, Robert J. 40
Ore wa mita → "I saw it"
O'Reilly, Bill 242, 245
Osada, Arata 201
Ōsuga, Akira 203
Ōta, Yōko 1, 2, 4, 5, 10, 195–99, 201, 203–08, 353, 354
Otsuki, Tomoe 70
overcoming modernity (*kindai no chōkoku*) 131, 132, 140

peace (*heiwa*) 12, 57, 58, 65, 70, 73
    Peace Constitution 202
    peace declaration (*heiwa sengen*) 72
    Peace Reconstruction Festival (Heiwa fukkōsai) 70, 71, 206
Pearl Harbor 169, 236, 237
picture story (*emonogatari*) 97
pika/pikadon 117, 119, 120, 122
postwar period 7, 9–12, 15, 72, 81, 132, 196, 224, 239, 271, 272, 274, 275, 277, 285, 288, 353, 354
    postwar generation 202
    postwar history 197
    postwar society 11, 196, 200
publisher 195, 196, 203, 207

Rabiger, Michael 107
radioactivity 5, 13, 14, 40, 261, 272, 333, 360
    radiation 31, 32, 37–41, 43, 44, 109, 175, 178, 181, 220, 221, 223, 225, 255, 263, 272, 280, 285, 302, 333, 335, 337, 338, 340, 342, 345, 352, 353, 358
    radioactive 5, 13, 31, 219, 220, 223, 225, 280, 282, 284, 288, 310
    radioactive contamination 219, 222, 224–26, 254, 255, 260, 262, 266, 332, 333, 340, 351

radioactive exposure 14, 41, 217, 337, 351, 360
Raymond, Antonin 133, 137
Reagan, Ronald 244
reconstruction (*fukkō*) 40, 68, 69, 73, 128, 130, 132, 134, 199, 200, 206, 272, 286
"Red Snow" (*Akai yuki*) 254
Reimink, Keith 175, 180
Resnais, Alain 55
Reynolds, Jonathan M. 133, 138
Rice, Susan 236, 242, 243, 245
Riki-san 84
rinkai jiko → criticality accident
"Riverbank" (*Kawara*) 1, 2
Robertson, Walter S. 12
Roland Barthes by Roland Barthes 107
Roosevelt, Franklin D. 84
Roper, Elmer 238, 241, 244
Ross, Harald 34, 41, 43
Rostand, Jean 220
Runit Dome 222–24

sacrificial system (*gisei no shisutemu*) 203
safety myth (*anzen shinwa*) 13, 216, 221, 286, 353
Sakakura, Junzō 131, 132, 140
Sano, Toshikata 130
Sasaki, Sadako 178
Sayonara atomikku doragon: Kaku to genpatsu no ohanashi → "Goodbye, Atomic Dragon: The Story of Atomic Weapons and Nuclear Power"
SCAP (Supreme Commander of the Allied Powers) 60, 61, 63, 64, 66, 74
Schäfer, Stefanie 64, 68, 70, 72
Scherer, Anke 130
Schneider, Werner 15
school textbook 201–03
    "Democracy and Cheerful Life" (*Minshu shugi to akarui seikatsu*) 201
    "Model for Social Studies in Junior High Schools" (*Mohan: Chūgaku shakai*) 201
    red textbook (*akai kyōkasho*) 202, 203
    "Social Studies in Junior High Schools" (*Chūgaku shakai*) 201, 202
Seaton, Philip 332

Seibo no iru tasogare → "The Blessed Virgin in the Twilight"
Self-Defense Forces (Jieitai) 279, 280, 282, 283
Senba, Nozomu 206
Senkō no natsu → "A Flash in Summer"
Sensai fukkōin → War Reconstruction Agency
Seo, Gijae 87–90
Sharp, Patrick B. 33
Shawn, William 34, 35, 42
Sheftall, M. G. 169
Sherif, Ann 62, 63, 65, 71, 74
shi no hai → ashes of death
Shigesawa, Atsuko 4
Shiina, Etsusaburō 129
Shijō, Chie 353
Shikabane no machi → "City of Corpses"
Shimada, Keizō 81, 82, 85, 93, 96, 97, 100, 101
Shimizu, Isao 84
Shimizu, Masataka 299, 302, 309, 311, 312, 314
Shindō, Kaneto 119, 175
Shinkaigyo → "Deep Sea Fish"
Shinkyō (Changchun) 128, 130, 132, 139
Shin Nihon bungaku → magazine
Shin Nihon kenchikuka shūdan → New Architects' Union of Japan
Shin mangaha shūdan → New Manga group
Shin Takarajima → "New Treasure Island"
Shinohara, Ken'ichi 92
Shirato, Sanpei 100
shishōsetsu → literature
Shōriki, Matsutarō 9
Shōsetsu → magazine
Shūkaku → "Harvest"
Silent Spring 220
Slater, David 330, 333, 334
Smith, Jeff 107, 122
"Someday in a Rapeseed Field: We Will Not Forget Japan's Great East Disaster" (*Itsuka, Nanohana hatake de ~ Higashi Nihon daishinsai o wasurenai~*) 265
sōteigai → unforeseeable
State Secrecy Law (Tokutei himitsu no hogo ni kan suru hōritsu) 267

Stewart, David B. 128, 133
Stewart, John R. 130
Stimson, Henry 240, 241
Stone, Oliver 108
storyteller (*kataribe*) 177, 178
Sugiura, Yukio 84
"Summer Flowers" (*Natsu no hana*) 196
"Superfast Rocket Missile" (*Kaisoku rokettodan*) 93
"Superhuman Power Nuclear Ball" (*Kairiki genshi bōru*) 82, 93–95, 97
Suriano, Manuela 358
survivor 2, 10, 56, 64, 197, 200, 205, 206, 240, 353, 355
system of irresponsibility (*musekinin no taikei*) 204

Tagawa, Suihō 85, 86, 96
Takahashi, Hiroko 267
Takahashi, Tetsuya 14, 203
Takakuwa, Kōkichi 4
Takayama, Eika 133
Takeda, Shinpei 82, 90, 93, 95, 97, 100, 101
Takeguro, Ichirō 299, 307–09, 318
Takemine, Seiichirō 267
Tanaka, Yuki 9
Tange, Kenzō 127, 128, 131, 133–40, 208
Tange Kenzō no Nihonteki seikaku → "The Japanese Character of Tange Kenzō"
Tatsumi, Yukako 65
Tawada, Yōko 225
TEPCO → Tokyo Electric Power Company
testimony 162, 163, 165, 176, 178, 180, 182, 185, 196, 255, 256, 355
Tezuka, Osamu 81, 86
Thank God for the Atom Bomb 237
"The Adventure of Gen-chan" (*Gen-chan no bōken*) 95
"The Atomic Bomb" (*Genshi bakudan*) 91, 92
The Atomic Bomb Suppressed 60, 87
The Atomic Plague 5
The Beginning or The End 113, 114
"The Bells of Angelus" (*Nagasaki 1945—Anzerasu no kane*) 120, 121
"The Bells of Nagasaki" (*Nagasaki no kane*) 64, 68, 178

The Birth of Tragedy 138
"The Blessed Virgin in the Twilight" (*Seibo no iru tasogare*) 195
The Day the Sun Rose in the West: Bikini, the Lucky Dragon, and I 176
The End of Man: A Feminist Counterapocalypse 258
"The Japanese Character of Tange Kenzō" (*Tange Kenzō no Nihonteki seikaku*) 136
"The Man on the Bridge" (*Der Mann auf der Brücke*) 200
The Politics of Display: Museums, Science, Culture 177
"The Prayer of the A-Bombed Maria: Manga Testimonies of Three Survivors" (*Hibaku Maria no inori: Manga de yomu sannin no hibaku shōgen*) 256
"The Problem with Alarming Textbooks" (*Ureu beki kyōkasho mondai*) 202
"The Santa Claus of August 9th: Nagasaki's A-Bomb and the Survivors" (*Hachigatsu kokonoka no Santakurōsu: Nagasaki genbaku to hibakusha*) 256
"The Tragedy of Japan" (*Nippon no higeki*) 6
The Untold History of the United States 247
Thompson, Kristin 107, 122
Three Electric Power Laws (Dengen sanpō) 218
Three Mile Island 218, 219, 254, 287
Three Non-Nuclear Principles (*hikaku sangensoku*) 218
Tōhoku 13, 140, 226, 283, 287, 334, 340
Tōhoku kara no koe → "Voices from Tohoku"
Tōkaimura 9, 216, 218, 279, 284, 351–53, 358, 360, 361
Tokutei himitsu no hogo ni kan suru hōritsu → State Secrecy Law
Tokyo Electric Power Company (TEPCO) 31, 262, 275, 277, 279–82, 284, 287, 297, 298, 300–02, 304–07, 309, 310, 312, 314, 317, 318
Tokyo War Crimes Trial 7, 11, 203, 205
"To Rui, Once Again" (*Futatabi Rui e*) 216, 220, 221, 359, 360, 361
Toyokawa, Saikaku 136

"Tragedy of a Planet" (*Aru wakusei no higeki*) 100
trauma 167, 205, 206, 253, 331, 332
Treat, John Whittier 353, 355
Trinity test site 113, 216, 246, 356, 357, 359, 361
Toriniti kara Toriniti e → "From Trinity to Trinity"
Truman, Harry S. 4, 36, 43, 233, 236, 237, 242, 246
Trump, Donald 233–35, 245
Tsurumi, Shunsuke 11
Tsushima, Kai 227
Tsushima, Yūko 215, 217–26
Tucker, David 129
"Two Grave Markers" (*Futari no bohyō*) 355

Uchida, Yoshikazu 129
UNESCO 57, 68
unforeseeable (*sōteigai*) 284, 352, 358, 359, 360
United States 5, 7–9, 63, 275, 277, 278, 280, 289, 355, 357
United States Holocaust Memorial Museum 152, 161, 165–67
Ureu beki kyōkasho mondai → "The Problem with Alarming Textbooks"

Van Sijll, Jennifer 107, 108
Vibrant Matter: A Political Ecology of Things 257
Voices from Tohoku" (*Tōhoku kara no koe*) 333

Wallace, Chris 243, 245
Warburg, Aby 331
War Reconstruction Agency (Sensai fukkōin) 132, 134
war responsibility 11, 204, 205
war guilt 6, 7, 11
Watanabe, Sumiko 358
Wendelken, Cherie 131
"Why Nuclear Power is Scary" (*Genpatsu wa naze kowai ka*) 254
"Wildcat Dome" (*Yamaneko dōmu*) 215, 222–24
Williams, Paul 167, 168

Winther, Bert 137
Wirgman, Charles 83
world peace 55, 58, 60, 67, 69, 74, 75, 199, 206, 207
World War II 81, 85, 91, 233, 236, 241, 242, 332

Xiong, Fang 358
Xu, Bin 332

Yad Vashem 152, 161, 165–67
Yamaneko dōmu → "Wildcat Dome"
Yamamoto, Akihiro 360
Yavenditti, Michael J. 33–35

Yokoi, Fukujirō 84
Yokoyama, Ryūichi 84
Yoneyama, Lisa 58, 182
Yoshida, Masao 299, 302, 304, 305, 307, 308, 310, 312, 315
Yoshimi, Shun'ya 216
'Yume no uta' kara → "From 'Dream Songs'"
Yūnagi no machi to hito to → "City and People in the Evening Calm"

Zakaria, Fareed 243
Zborowski, James 108
Zwigenberg, Ran 179
Zylinka, Joanna 258, 263, 265